Optimization in Signal and Image Processing

Optimization in Signal and Image Processing

Edited by
Patrick Siarry

ISTE

WILEY

First published in France in 2007 by Hermes Science/Lavoisier entitled *Optimisation en traitement du signal et de l'image* © LAVOISIER, 2007
First published in Great Britain and the United States in 2009 by ISTE Ltd and John Wiley & Sons, Inc.

ISTE Ltd
27-37 St George's Road
London SW 19 4EU
UK

www.iste.co.uk

John Wiley & Sons, Inc.
111 River Street
Hoboken, NJ 07030
USA

www.wiley.com

Library of Congress Cataloging-in-Publication Data

Optimisation en traitement du signal et de l'image. English.
Optimization in signal and image processing / edited by Patrick Siarry.
p. cm.
Includes bibliographical references and index.
ISBN 978-1-84821-044-8
1. Signal processing. 2. Image processing. I. Siarry, Patrick. II. Title.
TK5102.9.O6813 2009
621.382'2--dc22

2009017635

British Library Cataloguing-in-Publication Data
A CIP record for this book is available from the British Library
ISBN: 978-1-84821-044-8

Printed and bound in Great Britain by CPI Antony Rowe, Chippenham and Eastbourne.

Table of Contents

Introduction

Engineers constantly encounter technological problems which are becoming increasingly complex. These problems may be encountered in different domains such as transport, telecommunications, genomics, technology for the healthcare sector and electronics. The given problem can often be expressed as one which could be solved by optimization. Within this process of optimization, one or several "objective functions" are defined. The aim of this process is to minimize the "objective function" in relation to all parameters concerned. Apart from problems of optimization, i.e. the problem's objective function which is part of this topic (e.g. improving the shape of a ship, reducing polluting emissions, obtaining a maximum profit), a large number of other situations of indirect optimization can be encountered (e.g. identification of a model or the learning process of a new cognitive system). When looking at this issue from the angle of available methods used to resolve a given problem, a large variety of methods can be considered. On the one hand, there are "classic methods" that rely purely on mathematics, but impose strict application conditions. On the other hand, digital methods that could be referred to as "heuristic" do not try to find an ideal solution but try to obtain a solution in a given time available for the calculation. Part of the latter group of methods is "metaheuristics", which emerged in the 1980s. Metaheuristics has many similarities with physics, biology or even ethology. "Metaheuristics" can be applied to a large variety of problems. Success can, however, not be guaranteed. The domain of optimization is also very interesting when it comes to its functions within the field of application. In the domain of optimization, the processing of signals and images is especially varied, which is due to its large number of different applications as well as the fact that it gave rise to specific theoretical approaches such as the Markov fields, to name just one example.

These ideas have influenced the title of this book *Optimization in Signal and Image Processing*. This book has been written for researchers, university lecturers and engineers working at research laboratories, universities or in the private sector. This book is also destined to be used in the education and training of PhD students as well as postgraduate and undergraduate students studying signal processing, applied mathematics and computer science. It studies some theoretical tools that are used in this field: artificial evolution and the Parisian approach, wavelets and fractals, information criteria, learning and quadratic programming, Bayesian formalism, probabilistic modeling, the Markovian approach, hidden Markov models and metaheuristics (genetic algorithms, ant colony algorithms, cross-entropy, particle swarm optimization, estimation of distribution algorithms (EDA) and artificial immune systems). Theoretical approaches are illustrated by varied applications that are relevant to signals or images. Some examples include: analysis of 3D scenarios in robotics, detection of different aggregates in mammographic images, processing of hand-written numbers, tuning of sensors used in surveillance or exploration, underwater acoustic imagery, face recognition systems, detection of traffic signs, image registration of retinal angiography, estimation of physiological signals and tuning cochlear implants.

Because of the wide variety of different subjects, as well as their interdependence, it is impossible to structure this book – which contains 13 chapters – into distinct divisions, which might, for example, separate traditional methods and metaheuristics, or create a distinction between methods dealing with signals or with imagery. However, it is possible to split these chapters into three main groups:

– the first group (Chapters 1 to 5) illustrates several general optimization tools related to signals and images;

– the second group (Chapters 6 to 10) consists of probabilistic, Markovian or Bayesian approaches;

– the third group (Chapters 11 to 13) describes applications that are relevant for engineering in the healthcare sector, which are dealt with here through the use of metaheuristics.

Chapter 1 deals with the benefits of modelization and optimization in the analysis of images. After the introduction of modelization techniques for complex scenes, the analysis of images has become much more accurate. In particular, traditional means of image analysis, such as the segmentation of an image, need to be revised. Jean Louchet creates a link between two domains that have been developing independently. These are the synthesis and the analysis of images. The synthesis of images relies on a wide range of different modelization techniques which are based on geometrics, depiction and movement. The author shows that some of these techniques can also be used for the analysis of images, which would broaden the possible applications of these techniques. Jean Louchet also shows how artificial evolution can lead to a better exploitation of models, create new methods of analysis and push back the limits of Hough transform using a stochastical exploration of the model's parameter space.

In Chapter 2 Pierre Collet and Jean Louchet present the so-called "Parisian" approach of evolutionary algorithms and how these algorithms are used in applications when processing signals and images. Evolutionary algorithms are reputed to take a long time to perform calculations. The authors, however, show that it is possible to improve the performance of these algorithms by – if possible – splitting the problem into smaller sub-problems. When using the "Parisian" approach to analyze a scene, the objects which have been modified by genetic operators are not the vectors of the parameters that determine a complete model of an image. These objects are elementary entities which only make sense when merged together as a representative model of the scene that will be studied. In other words, a problem cannot be represented by a single individual but by several individuals, or even the entire population. The "Parisian" approach is successfully used in the field of robotics when analyzing 3D scenes via stereovision. The so-called "Fly algorithm" allows for the detection of obstacles in real time and much more quickly than when using traditional approaches. Other visual applications based on models can be processed by evolutionary methods. Here, the authors discuss the identification of models of mechanical systems based on sequences of images.

Chapter 3 deals with the use of wavelets and fractals when analyzing signals or images. The application of these techniques is becoming increasingly frequent in natural science as well as in the study and research carried out in the scientific fields of engineering and economics. Abdeljalil Ouahabi and Djedjiga Ait Aouit show that multifractal analysis and the exploitation of techniques of multiresolution based on the concept of wavelets lead to a local as well as global description of the signal's singularities. On a local level, the criterion of punctual regularity (rugosity) based on Hölder's inequality can be characterized by the decrease of the wavelets' coefficients of the analyzed signal. On a global level, the distribution of a signal's singularities can be estimated by global measures when using the auto-similarity of multifractals. In other words, the spectrum of singularities is obtained when localizating the maxima of the module of the wavelet transform of a signal. The authors give two examples of the aims and applications of this formalism. One example in the healthcare sector is a multifractal analysis which allows for the detection of different aggregates in mammographic images. The second example is fracture mechanics. In this field the formalism described above is used to study the resistance of materials.

Chapter 4 deals with the information criteria and their applications when processing signals and images. Here, the model of a random signal should be optimized. An information criterion is a description or formulation of an objective function that should be minimized. The information criteria are an improvement on the traditional technique of the maximum likelihood. This improvement is due to the focus being shifted towards simultaneous research on the optimal number of free parameters in the model as well as the ideal values for these parameters. Christian Olivier and Olivier Alata first give a general overview of the main information criteria as well as the relevant literature. The majority of the criteria were introduced for research using 1D auto-regressive (1D AR) models. In Chapter 4, this case is illustrated by an application that involves the segmentation of natural images. The information criteria were then transferred to the 2D AR model. Two applications resulted from this. These are the modelization of the image's texture and the unsupervised segmentation of textured images. The authors then look at the extension of the information criteria to other models based on parameters. These are a mix of Gauss's laws n-D, which are here applied to unsupervised classification as well as Markov's modes. Last but not least, this chapter deals with the application of information criteria in the case of non-parametrical problems, such as the estimation of distribution via histograms or the search for antiderivatives that carry a maximum amount of information depending on the form of the information. The information criteria finally offer a means to justify the choice of parameters which are linked to a large number of problems when processing signals or images. The information criterion deals with a high number of observations. This is why the time required to carry out the calculation might be high (particularly in an unsupervised context). Dynamic algorithms, however, are able to reduce the number of operations that need to be carried out.

Chapter 5, written by Gaëlle Loosli and Stéphane Canu, deals with an aspect of optimization that can currently be encountered within signals and images, for example in shape recognition, i.e. learning processes. More precisely, the chapter focuses on the formulation of learning as a problem in convex quadratic programmation on a large scale (several million variables and constraints). This formulation was obtained by the "nucleus methods", which emerged about a decade after neural networks. Its main aim is linked to the fact that the solution in question is often "parsimonious", i.e more than 99% of all unknown variables are zero. Using this specific feature enables learning algorithms to solve large scale quadratic programming problems within a reasonable amount of time. The best-performing methods, known as "active constraints", work "off-line". In other words, they are able to determine the exact solution of a given problem if they are provided with all the data that is used in the learning process. To carry out an "online" learning process, a method of iterative stochastic optimization is used, which allows us to obtain an approximate solution. This chapter describes one of the methods which is part of the "support vector machine" (SVM) type. The efficiency of this technique is illustrated by results of experiments which focused on the problem of recognizing handwritten numbers.

Chapter 6 deals with the problem of planning within time and space the use of sensors with the aim of optimizing the exploration and surveillance of a specific zone; given the rather low number of available sensors as well as their capacity, this zone is large. Due to the problem being rather extensive, exact methods cannot be used. An approximate solution can, however, be obtained with the help of metaheuristics. In this case, Frédéric Dambreville, Francis Celeste and Cécile Simonin, the authors of this chapter, recommend the use of "cross-entropy". This method was initially created to evaluate the probability of rare events and has been adapted to "difficult optimization" problems (many local minima need to be considered). The solution is obtained with the help of a probability law that continually approaches the global optimum. This method is applied to the problem of planning sensors via *a priori* modeling mainly under the form of different groups of probability laws, of possible planning policies. In this chapter, three examples are explained in detail. The first example looks at how to ideally array search units in the context of military operations. The aim is to maximize the probability of locating the target which does not move but is hidden. In the second example, cross-entropy is used for an exploratory mission. The movement of the vehicle needs to be planned based on maps that show the environment. The third example is the problem of optimal control in an environment where only certain parts of the environment can be observed. Cross-entropy is particularly useful when dealing with data that are very difficult to formalize. Optimization via cross-entropy therefore means to "learn" an optimal strategy.

The topic of Chapter 7 is linked to that of the previous chapter. Chapter 7 deals with a surveillance system such as a maritime patrol aircraft that needs to locate a moving target. In order to do this, all resources, i.e. passive as well as active sensors (e.g. a radar), need to be used. Passive measures do not involve any cost. However, they only determine the direction of the target. Active measures provide much more information since they can evaluate the distance to the target. These measures, however, need to be used sparsely because of their cost (emitting a wave) and with discretion. The author of this chapter, Jean-Pierre Le Cadre, gives a general outline of the problem of optimal and temporal repartition when using active measures. He futhermore describes the general mathematical tools (e.g. multilinear algebra) that allow for the analysis of this problem. The study focuses on the explicit calculation of objective functions while expressing the quality of the estimation (or tracking) of the trajectory's location by using non-linear observations of state. First of all, this chapter examines the case of targets that contain a determined trajectory. Their movement is rectilinear and uniform, or in other words the target is "maneuvering". When dealing with certain types of approximations, the problem of convex optimization comes into play. This problem can easily be resolved. The author also looks at the stochastic evaluation of this case. He shows that it is possible to directly calculate the objective function of a target of Markovian trajectory without having to use simulations.

Chapter 8 deals with segmentation methods of images which exploit both the Markovian modeling of images and the Bayesian formalism. For every image under observation there is an infinite number of combinations of objects that can be associated with it. These combinations of objects represent, or in other words create ,the image. To reduce the number of possible solutions that should be integrated in the stage of segmentation, prior local or global knowledge is required. The aim of Markovian modeling lies precisely in its capacity to locally describe global properties. Due to the equivalence between Markov's field and Gibbs's distribution, the optimal segmentation can be obtained by the minimization of a function linked to energy. Christophe Collet, the author of this chapter, applies this formalism to the context of underwater acoustic imagery. To detect small objects on the seabed, the author exploits images that have been taken by a lateral multibeam sonar. The images that were obtained were distorted by noise. A segmentation of good quality therefore requires the nature of noise to be taken into consideration during the process of image modeling. This chapter shows different examples of application. These are the segmentation of sonar images into two different groups (shadow, reflections of the seabed) or segmentation into three different groups (shadow, seabed and echo). Due to the third group, echo, physics, which forms the basis of the creation of sonar images, is also taken into consideration. Two other examples are the differentiation between manufactured and natural objects, as well as the subdivision of the seabed into different regions (sand, mid-ocean ridges, dunes, stones and rocks). All tasks linked to detection and classification are first of all united in the fact that the function of energy, which integrates the prior knowledge required to obtain a solution, needs to be minimized. The technique used for this optimization is a deterministic method or a genetic algorithm, depending on whether an initial good quality solution is available or not.

Chapter 9 was written by Sébastien Aupetit, Nicolas Monmarché and Mohamed Slimane and describes the use of hidden Markov models (HMM) for the recognition of images. Hidden Markov models are statistical tools which allow for the modelization of stochastic phenomena. This type of phenomenon may, for example, consist of several sequences of images. Images of the same sequence are taken from different angles but show the same scene, e.g. a person's face. After a learning phase, HMM is prepared for the process of recognition. During this learning phase several sequences of images, let us say four sequences of four photographs each showing the faces of four different people are processed. When confronted with a new photograph of a face, HMM is able to distinguish which person is shown in the picture from the four previous pictures. At the same time, the risk of HMM making a mistake is minimized. More precisely, a discrete HMM corresponds to the modeling of two stochastic processes. The first process is hidden and perfectly modeled by a discrete Markov chain while the second observed process is dependent on the state of the first process, i.e. the hidden process. This chapter focuses on learning processes, a crucial aspect of HMM. It provides an overview of the main criteria of existing learning processes and the possible solutions for HMM learning processes. Furthermore, the principles of three metaheuristics inspired by biology and population-based are also addressed by the authors and analyzed in light of HMM learning processes. These three metaheuristics are a genetic algorithm, ant colony algorithm and particle swarm optimization (PSO). Several versions of these types of metaheuristics (which are different to one another because of the mechanisms which are implemented, or simply due to the settings of the respective methods) are examined and tested in great detail. These tests are carried out on a set of test images as well as samples of literature. The chapter emphasizes the fact that results can be improved if metaheuristics used for learning processes are combined with a method dedicated to local optimization.

In Chapter 10 Guillaume Dutilleux and Pierre Charbonnier use different metaheuristics inspired by biology for the automatic detection of traffic signs. The aim is to make an inventory of road signs currently used in the French secondary road network. The data used are images that have been collected by vehicles inspecting the roads that are part of the respective network. The application does not face any real time constraint. However, the application needs to be robust when faced with changes in the conditions under which the images are collected. Problems might occur due to differences in light, backlighting, worn out or partially hidden traffic signs. The method that has been proven to be successful includes the technique of "deformable models". This technique consists of a mathematical model, a prototype of which the object research is carried out upon. This model's shape can be manipulated and changed to such an extent that it is adapted to the respective image that should be analyzed. The quality of this adjustment and to what extent manipulation can be accepted are, in the case of Bayesian formalism, respectively measured by a likelihood and an *a priori*. The problem of localizing an object therefore comes down to the problem of optimization in the sense of a maximum *a posteriori*. The residual value of a minimized objective function gives an indication of the effective presence of the object in the scene which is to be analyzed. In practice, the presence of numerous local minima justifies the use of metaheuristics. The authors have carried out experiments with three different techniques in the field of metaheuristics. These are an evolutionary strategy, PSO and a method of clone selection (the latter is relevant to a more general field of "artificial immune systems"). The performance of automatic detection is compared to a number of different algorithms when dealing with a sequence of traffic signs. (For these test images the real data had already been obtained manually.)

The majority of metaheuristics were initially created for the processing of problems that arise when dealing with discrete optimization. Chapter 11, written by Johann Dréo, Jean-Claude Nunes and Patrick Siarry, looks at their adaptation to applications with continuous variables, which are encountered frequently, especially in the field of signals and images. The techniques suggested in the literature for this adaptation are linked to each specific form of metaheuristics. These techniques cannot be generalized, i.e. it is not possible to apply these techniques to another application. Furthermore, no metaheuristic, whether it is continuous or discrete, is the ideal technique, i.e. most efficient, for all possible sorts of problems. This is why hybrid methods, which combine different forms of metaheuristics or metaheuristics with downhill simplex techniques, often need to be used. This chapter describes two "continuous metaheuristics". These are an ant colony algorithm and EDA. Furthermore, a local technique, which is frequently used in continuous cases to refine the search within a "promising valley" of solutions, is Nelder and Mead's downhill simplex method. These methods are used for image registration in the field of retinal angiography. Before a doctor can actually interpret a sequence of images, the problem of inevitable eye movement during the procedure needs to be dealt with. In the example given in this chapter, image registration is carried out by using only translatory motions between different images. Metaheuristics were found to be particularly appropriate for image registration in angiography with a high resolution. The time required for calculations only increases a little when increasing the resolution of images.

Chapter 12, written by Amine Naït-Ali and Patrick Siarry, describes the introduction of a genetic algorithm used for the estimation of physiological signals, the Brainstem Auditory Evoked Potentials (BAEP). BAEP is an electric signal which is generated by the auditory system as a response to acoustic stimulation. Studying this signal allows for the detection of pathologies such as acoustic neuroma. Measuring BAEP is, however, a problem as this signal is of a very low energy and covered by electric noise that stems from spontaneous electric activity of the cerebral cortex (these signals can be measured using electroencephalograms (EEG)). To identify a patient's effective BAEP, several hundred signals need to be exploited. These signals are obtained as a result of acoustic stimulation. They also have to be synchronized before being simply added to one another in order to eliminate the noise. The synchronization process is expressed in the form of an optimization problem in which unknown variables are the random delays of different signals. Here, the problem is solved with the help of a genetic algorithm. The authors show that a significant acceleration of this technique can be obtained when creating a model for the variation law of these delays. This can, for example, be performed using a set of sinusoids.

Chapter 13, written by Pierre Collet, Pierrick Legrand, Claire Bourgeois-République, Vincent Péan and Bruno Frachet, presents an evolutionary algorithm that allows for the adjustment of parameters for a cochlear implant. This adjustment is carried out in interaction with the patient using the device. This type of implant enables deaf people, whose cochlear plate is still intact, to hear. The device works as follows: a group of electrodes is implanted into the patient's cochlear plate. These electrodes stimulate the auditory nerve. The electrodes are connected to a digital signal processor (DSP) that receives the sound as signals through a microphone situated next to the patient's ear. The parameters of DSP need to be adjusted in a way that reconstructs the patient's auditory ability to a point that he/she might even be able to understand spoken language. Adjusting these parameters is usually undertaken by a human and becomes increasingly complicated as technology progresses. A current implant consists of 20 electrodes and several hundred parameters. The effort for adjusting these parameters is dependent on the patient's ability to understand spoken language. This is why this study looks at the performance of an interactive evolutionary algorithm which should take over the task of adjusting the parameters of a cochlear implant. There are a large number of difficulties that lie within this application. These are the subjective evaluation of every single patient, the quality of every single solution produced by the algorithm, the necessity of a rapid convergence of the algorithm in order to strictly limit the amount of solutions to be evaluated by the patient (as every evaluation takes a few minutes) as well as the fact that the search space is very broad. This chapter presents experiments undertaken by the authors with the help of a small number of patients following a methodical protocol. The first results are promising. They show the disadvantages of manual adjustment in cochlear implants which is increasing because the number of available electrodes is currently increasing.

Chapter 1

Modeling and Optimization in Image Analysis

1.1. Modeling at the source of image analysis and synthesis

From its first days, image analysis has been facing the problem of modeling. Pioneering works on contour detection led their authors to refer to explicit models of edges and noise [PET 91], which they used as a conceptual basis in order to build their algorithms. With an opposite approach to these phenomenological models, a physical model of light diffusion on surfaces has been used as the basis for Horn's works [HOR 75] on shape from shading. More generally, a phenomenological model aims at describing a directly computable property of the geometric configurations of gray levels on an image; the physical model then tries to use the knowledge corpus of physics, or even sometimes to create an *ad-hoc* conceptual system, as we will see later. Between these two extremes, there is a large number of approaches to modeling. Here, we shall try to illustrate them using some examples.

It is important to first show the links between image analysis and synthesis. For a number of years, these two domains have been undergoing largely independent development processes. In spite of their conceptual similarity, they have been dealt with by two separate scientific communities, with different origins and centers of interest, which did not address the same applications. Robotics is one of the few fields of application that has played an important role in moving them closer to one another. Image synthesis addresses another large panel of approaches to modeling, in particular in the fields of geometry, rendering and motion modeling – to such an extent that there are important international communities, journals and conferences that specialize in each of these approaches to modeling. One of the benefits of

---

Chapter written by Jean LOUCHET.

connecting image analysis and image synthesis comes from the fact these two domains often use common modeling techniques which they use as bridges to their constructive interaction.

1.2. From image synthesis to analysis

One of the difficult points in image analysis and machine vision is algorithm validation methods. In most cases it is not possible to access the "ground truth" that corresponds to the image or the image sequence on which we want to evaluate the quality of the analysis process. A frequent compromise consists of evaluating an algorithm by referring to another algorithm, which is seldom acceptable. The ideal, straightforward solution would consists of creating a synthesis tool – and therefore a model – able to build dependable test data from any ground truth.

Once the effort of building such a model has been carried out, we naturally arrive at the idea of incorporating into the image analysis algorithm, the knowledge of the physical world and its rules, following an "artificial intelligence" approach. This can be done implicitly (using the same knowledge corpus and coding it in a way suitable to the analysis algorithm) or alternatively by explicitly incorporating the model into the analysis process. Of course, this raises the delicate ethical question of mutually validating two algorithms running the same model, and therefore susceptible to containing the same errors or clumsy simplifications. In any case, this is, when pushed to its extreme, the basis of the so-called "analysis by synthesis" approaches where rather than the model (whether photometric, geometric, kinematic or physical), it is the whole image synthesis process that is embedded into the analysis algorithm. Between merely using general physical knowledge in the analysis process, and at the other end, embedding a complete synthesis process into the analysis algorithm, it appears that the image analysis techniques explicitly exploiting a model are undergoing an important development, particularly thanks to modern optimization techniques, as we will see in several examples.

First, we will examine the classical approaches to image segmentation and show how they have built their organization, often implicitly, after the way scene models used in image synthesis are naturally organized.

In the next section, we will revisit the Hough transform [HOU 62], which is probably the best known example of image analysis and model inversion, through deterministic, exhaustive search in a parameter space; the model used here is phenomenological (visual alignment). We show the Hough transform and its generalizations may be rewritten into an evolutionary optimization version; as a stochastic exploration of a parameter space, here each point represents a particular instance of a model. This considerably widens the field of potential applications of the original Hough transform.

The following part will quickly examine the contribution of physical models to image analysis. This is a promising yet little known topic we will discover through two examples using photometric and dynamic models.

In some applications, the model underlying the analysis technique may be taken apart into elementary objects whose collective behavior actually represents the object to be modeled. A specific evolutionary optimization method, called "Parisian Evolution", can then be implemented. This is a change in the semantics of the evolved population but a classical evolutionary process is still applied to the elementary objects. This will be the subject of Chapter 2.

1.3. Scene geometric modeling and image synthesis

As discussed earlier, image contour segmentation took its foundations from hypotheses about image signals, resulting into a wide use of differential operators as main analysis tools. Region segmentation, which was developed later, probably because of its greater need for computational power, brought more evidence of the strong link between the structures of a 3D scene and the image entities directly accessible to calculation. It is therefore tempting to revisit the notion of image segmentation [COC 95, GON 92] through its possible interpretations in terms of scene models.

Seen from this point of view, segmentation into regions could be defined as any image partitioning technique such that each region entity it extracts is a good candidate image projection of a 3D or space-time varying physical object in the scene.

Similarly, it is possible to give a new definition of contour segmentation: it describes any image line extraction technique such that each line extracted is likely to be the image projection of an edge of a physical object present in the scene.

With each level of primitives in the polyhedral model (vertex, edge, facet, etc.) it is possible to associate a *probable* local property of the image, such as the contrast along a line, the homogenity of a region, etc., and a corresponding calculation technique. Classical segmentation techniques are often a decisive step in the process of instantiating the model as efficient model exploration heuristics; contours usually are the projection of the subset of the scene where the probability of finding an edge is highest, thus the knowledge of contours contributes to the efficiency of the exploration of the space of parameters which describe the possible positions of edges. Similarly, interest points give useful hints on where to look for polyhedron vertices, and so on.

One of the consequences is that the pertinence of a segmentation technique on a class of images essentially depends on whether it actually corresponds to an observable characteristic of the model underlying the class of scenes and how the images have been captured, An illustration of this is given by fluid flow imaging, where polyhedral models are irrelevant and classical contour or region segmentation techniques are just as irrelevant. Segmentation techniques are the translation of the scene-specific description language.

The primary role of image analysis is to instantiate or identify the parameters of a general scene model. If we consider scenes made from opaque objects, which is not too bad in most familiar scenes, the most widely used modeling language in image analysis, as in synthesis, is based on polyhedral objects. The ultimate goal of image segmentation should ideally be to provide a description of the scene using the same primitives and language as in image synthesis: a geometrical description (polyhedra, facets, edges, vertices), completed (if useful to the application) with a photometric description (light sources, radiance factors, diffusion coefficients, etc.). In the case of time-dependent sequences, it will be necessary to include object motion and deformations, and the analysis may even include the building of a description of the scene in terms of agents, individual behaviors and physical interaction [LUC 91].

In all these cases, "informing the model" means optimizing the likeness between real data and data synthesized from the scene model, and therefore will generally involve the optimization of a cost or resemblance function.

1.4. Direct model inversion and the Hough transform

1.4.1. *The deterministic Hough transform*

One of the main motivations of the development of image segmentation techniques is the difficulty of directly resolving the problem of optimizing a scene model using classical methods. However, a well known exception to the rule is given by the Hough transform [HOU 62, BAL 82], which may be described as a direct parameter space exploration technique. It consists of filling in the space of model parameters (also known as the *Hough accumulator*) using a vote technique, where each locally detected primitive in the image results in incrementing a sub-variety of the parameter space. The parameter vector that will be eventually elected is the one which has received the greatest number of votes, and therefore is likely to represent the best possible model, i.e. the most satisfying *a priori* global explanation of all the primitives that have been previously extracted from the image.

In spite of a reasonable success story, the Hough transform and its priority-to-image philosophy imply that for each image primitive considered to be relevant, the accumulation process will go into the n-dimensional parameter space to modify the values of all the points belonging to a variety with dimension $n - 1$ (within the limits of the search space). The heuristics that have been found in order to improve the algorithm's speed have a limited effect and the generalized Hough transform becomes unusable when the dimension of the search space becomes greater than 3 or 4, mainly due to available memory space, memory access time and the complexity of the dual space incrementation task.

1.4.2. *Stochastic exploration of parameters: evolutionary Hough*

If we consider the Hough technique, which involves calculating voting scores throughout the parameter space then exhaustively exploring this space in order to find the best optima, it becomes an attempt to directly explore the search space [ROT 92, LUT 94]. Evolutionary programming provides us with very welcome exploration techniques which, in our case, allow us only to calculate the values in the dual space where individuals of the evolving population actually are, rather than on all of the dual space.

According to the general principles of artificial evolution (also known as evolutionary programming), a function to be optimized is given, though not explicitly. We then consider an arbitrary set of individuals ("population") which belong to the space where this function is defined. This population is then evolved, in a way reminiscent of biological evolution, using genetic operators such as mutation crossover and selection. The selection criterion is the individual's performance ("cost" or "fitness") as given by the function to be optimized. The expression "evolutionary strategy" [REC 94, BÄC 95] refers to the artificial evolution algorithm, where gene coding is performed using real numbers as variables, unlike genetic algorithms which, strictly speaking, use Boolean variables.

Thus it is possible to define an evolutionary version of the Hough transform:

– the population is a finite subset of the parameter space;

– for each individual in the population, the fitness function gives a measurement of the pertinence of the corresponding image pattern;

– classical selection operator (tournament);

– barycentric crossover;

– mutation: Gaussian noise.

The fitness function used in the evolutionary version of the Hough transform is calculated according to the same criterion as in the classical versions of Hough; for example, if the criterion for a point to participate into the incrementation of the accumulator is that its contrast is greater than a given threshold, then in the evolutionary version the fitness of an individual (representing e.g. a straight line) will be the number of points on this line with a contrast higher than the same threshold.

In the case of the classical Hough transform, there is no very clear advantage to either approach. Indeed, the classical Hough method in which each image point (x, y) votes for the set of points (θ, ρ) in the dual space, such that $\rho = x \cos \theta + y \sin \theta$, is relatively fast and needs a reasonable memory allocation (Figures 1.1 and 1.2); on

Figure 1.1. *Result of the classical Hough transform (image 288×352)*

Figure 1.2. *The (θ, ρ) Hough accumulator corresponding to the same image (image 628×300)*

the other hand, the evolutionary process which, thanks to the sharing operator, is able to find several different solutions (Figure 1.3), suffers from not having a canonically defined ending: it is therefore difficult to compare processing times. In practice, the number of generations needed to ensure convergence results in similar calculation times.

Figure 1.3. *Result of the evolutionary version of Hough*

However, when the dimension of the parameter space becomes higher, dual space storage and exploration soon become prohibitive, while the evolutionary version is much less greedy in terms of memory and computing time. The evolutionary version of the Hough transform really becomes interesting when considering more complex parametric optimization problems as shown in the following examples, where a classical Hough approach would fail.

1.4.3. *Examples of generalization*

The following example[1] consists of detecting circles with unknown diameters (a moving ball) in an image sequence. The individuals are triples (a, b, r) which define circles with equation $(x-a)^2 + (y-b)^2 = r^2$. The fitness of a particular individual is defined as the average gradient norm taken on 40 points randomly distributed on the circle. The algorithm parameters are:

population size	100
selection	2-tournament
mutation rate (%)	15
r mutation amplitude	10
a, b mutation amplitude	40
barycentric crossover rate (%)	5
number of generations per frame	init. 800 then 240

1. These results were produced by A. Ekman, a PhD student at KTH Stockholm in October 2004.

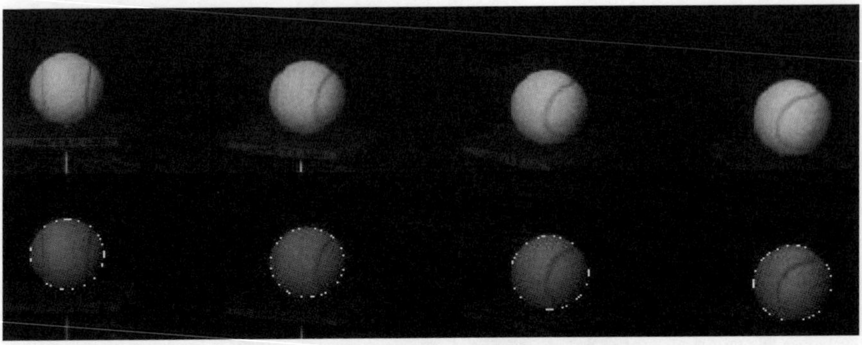

Figure 1.4. *Four original images from the "tennis ball" sequence (top) and results of the "evolutionary Hough" detection of circles with unknown radius (bottom)*

It is worth noting an interesting property of this approach: if motion is small enough between two consecutive frames, the evolutionary algorithm is able to track the object's motion, unlike its deterministic counterpart which has to resume its calculations from the very beginning, even with a tiny image change, whatever the degree of redundancy in their information contents. In other words, in spite of the fact artificial evolution is often regarded as slow, the evolutionary version of the Hough transform possesses true real time properties that its deterministic version does not have.

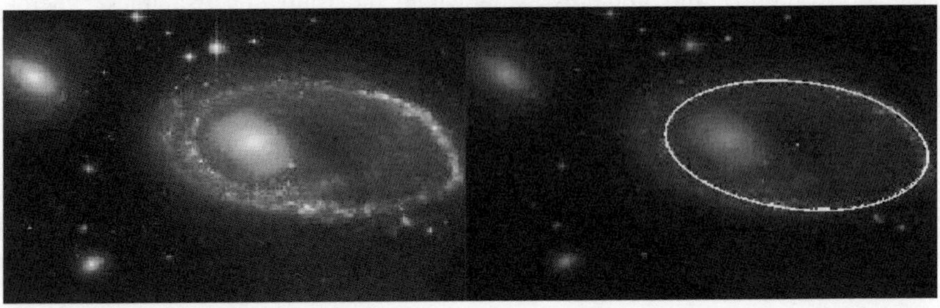

Figure 1.5. *Image of the galaxy AM 0644-741 taken by the Hubble telescope (left) and the result of the evolutionary Hough ellipse detection algorithm (right)*

Another example (Figure 1.5) consists of detecting a conical section (ellipse) in an image, using the same method and the same adjustment of genetic parameters as with the circle detection, but with a 5-parameter genome corresponding to the ellipse parametric equation:

$$x = a + r_x \cos\alpha \cos\theta + r_y \sin\alpha \sin\theta$$

$$y = b - r_x \cos\alpha \sin\theta + r_y \sin\alpha \cos\theta$$

It is not easy to compare the theoretical performance of classical (filling up parameter spaces) Hough methods to their evolutionary versions, as strictly speaking the number of generations required for convergence depends on analytical properties of the fitness function, which is image-dependent. However, let us re-examine the practical examples given above. With a square $N \times N$ image and an accumulator (parameter space) with dimension n, where each parameter can take P different values, the classical Hough transform needs a memory with size P^n, and roughly $N^2 \times P^{n-1}$ calculations to fill the accumulator. The evolutionary version does not require large memory resources, and it needs $G \times E \times N$ calculations, where G is the number of generations and E the population size. In the case of ellipse detection $(n = 5)$, the images (213×161 pixels) correspond to $N \approx 200$ and the precision of quantization to $P = 200$; the classical Hough method would use a 300 GB accumulator and about 6×10^{12} calculations, against about 5×10^6 calculations with the evolutionary method, which gives a gain factor about 10^6. In the simpler case of circles (3 parameters) the computing time ratio falls to about 100, with a Hough accumulator size about 10 MB. From these examples it is safe to say that with three parameters, the evolutionary method is more efficient than the classical Hough method, but with 4 or more parameters, the only realistic method is the evolutionary version. This gives quite an important extension to the field of potential applications of the Hough transform.

1.5. Optimization and physical modeling

In the discussion above, we considered image analysis as the instantiation of a scene model. This leads us to wonder whether all the aspects of modeling used in image synthesis and computer graphics are still relevant in image analysis. While it still looks premature to give an exhaustive answer, we will examine two aspects of this question: the photometric models and the motion models.

1.5.1. *Photometric modeling*

Photometric modeling is essential to image synthesis. Rather surprisingly, photometry is not yet in wide use in image analysis. The main explicit application of photometry to image analysis is shape from shading, which consists of building the shape of a surface from the variations of the luminance. In his pioneering works, Horn [HOR 75] showed that it is possible, through the resolution of differential systems initialized on local luminance extrema, to recover the 3D shape of a surface, under relatively strong hypotheses: the surfaces are assumed to have Lambertian scattering properties, the light sources to be at infinity, etc. Practical applications have been limited by these constraints and the ill-conditioning of the problem. We may however note that recent research by Prados [PRA 04] showed that taking into account the distance-dependent reduction of lighting (light source at a finite distance) allows us to obtain a well-conditioned problem and become free of some of the usage restrictions.

Another example has been given by G. Maurin [MAU 94] with the 3D location of a light source. In this preliminary study, the author exploits one of the images from a stereo pair of an interior scene which has previously been analyzed in three dimensions; thanks to homogenity hypotheses concerning the regions detected, he calculates the position of the (single) light source in a room.

Another original example of how to exploit a simple photometric model in image analysis was given by J.B. Hayet [HAY 98] with the Robocop project (obstacle location through observation and calculation of shadows). In this application, rather than using a second camera, one or several computer-controlled light sources are set on the robot: detecting the shadow edges in each lighting configuration enables a cheap and fast 3D analysis of the scene and an elementary obstacle detection.

1.5.2. *Motion modeling*

1.5.2.1. *Kinematic modeling*

Motion modeling is the heart of image animation. Motion can be modeled at several levels, the kinematic level being the simplest and the most widely used. It consists in analyzing, in a purely geometric way, the motion of an object in the scene. It may be analyzed as a planar motion (e.g. for a pure translation, through the exploitation of the apparent motion constraints equation [HOR 81]); concerning more general planar movements, it is possible to exploit an equation which delivers the instantaneous planar rotation centers [LOU 96a]; 3D movements can also be analyzed directly [HUA 83].

However, this is not all about motion analysis and modeling. To each possible level of motion modeling, there is a corresponding image sequence analysis technique. Classical motion analysis (resulting in optical flow data) corresponds to a purely kinematic modeling of motion; similarly, physical or behavioral modeling approaches may be matched to a corresponding semantic level of image sequence analysis.

1.5.2.2. *Physical modeling*

Here, we will examine how physical modeling of motion may be used in image sequence analysis, using the mass-link modeling paradigm as it was developed by the Acroe team in Grenoble, France [LUC 91]. The CORDIS-ANIMA model and system were created in the 1970s with the main application objective of multi-modal man-machine interfacing, incorporating acoustical visual and gestural modalities. In addition to specialized peripherals such as their retroactive gestural transducer, the conceptual heart of the system is a physical modeling language based on two primitives: the mass (fitted at each instant, with a position and a velocity) and the link (between two visco-elastic and generally non-linear masses). An important conceptual, theoretical and experimental construction has allowed us to demonstrate the power of this approach through its applications in particular to the interactive synthesis of sounds and images. One of the problems to have been addressed around this project is the inverse problem which consists of building a physical model able to reproduce as accurately as possible the given motion, deformation and interaction of an object. This problem has been partially resolved through a decomposition of the mass-link structures and a multi-objective evolutionary strategy to optimize the model [LOU 94, STA 02] using individual cost functions associated with each of the masses.

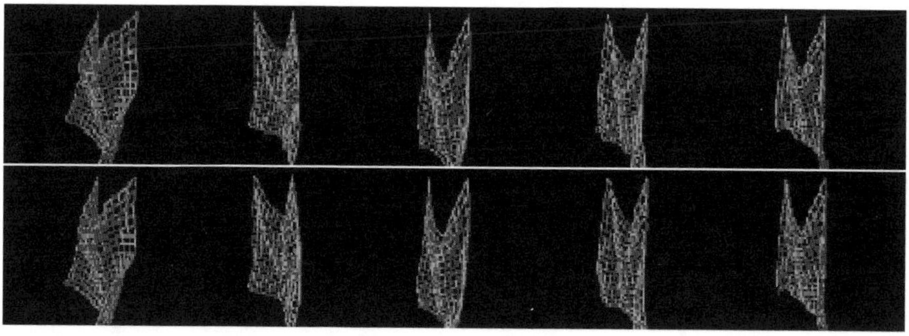

Figure 1.6. *Reconstruction of an image sequence of a cloth: original synthetic sequence (first line) and reconstructed sequence (second line) using cloth physical parameters identified from the original image data (modeling and images: X. Provost)*

One of the applications consisted of identifying the internal mechanical parameters of a cloth from an image sequence of a hanging cloth sample, then from these parameters re-synthesizing images of cloth from the same fabric (Figure 1.6) in more complex configurations [LOU 95]. In a similar way, from an image sequence representing a compressible viscous fluid flow, Jiang Li showed [LOU 96b] it is possible to identify the Cordis-Anima parameters that characterize the fluid's viscosity and compressibility and re-synthesize other flows from the same fluid (Figure 1.7).

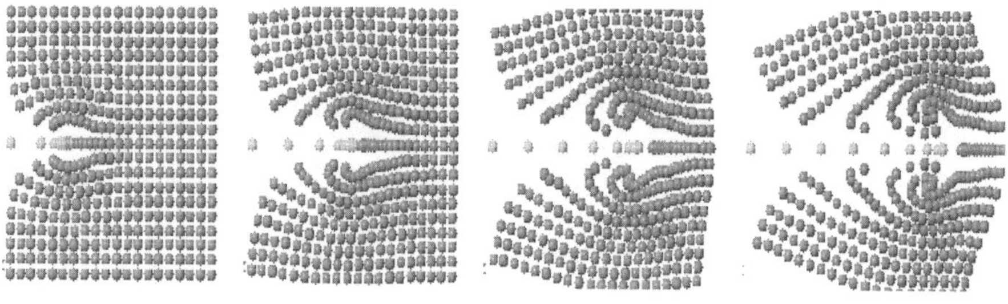

Figure 1.7. *Four successive synthesized images of the turbulent meeting of two compressible fluids (images: L. Jiang)*

Another application in the same spirit, consisted of detecting heart stroke scar zones from X-ray scanner image sequences. The heart is modeled as a mass-link system where the internal parameters are then identified using image sequence data. Anomalous values of the internal parameters of viscous-elastic links indicate a high probability of having a scar zone in the corresponding region (Figure 1.8).

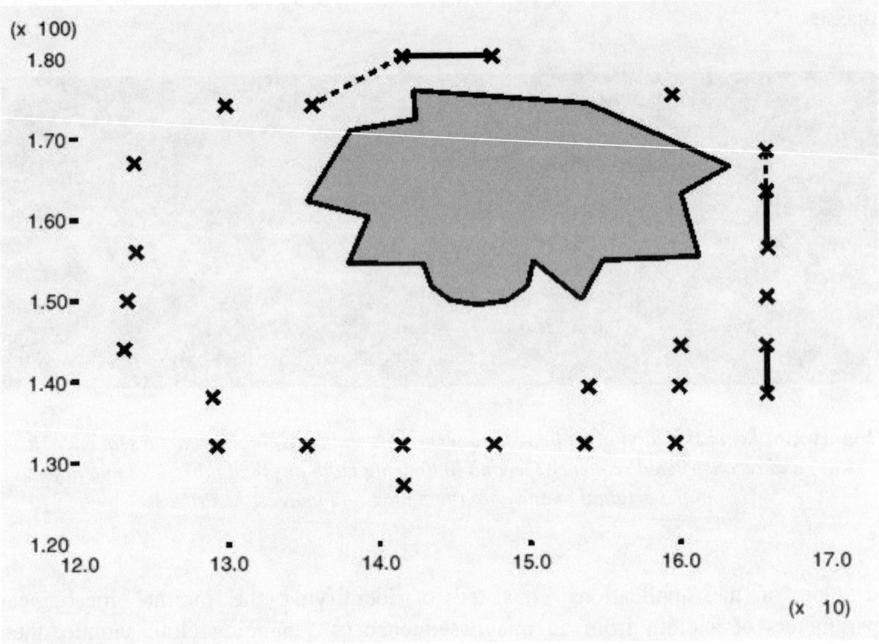

Figure 1.8. *Heart X-ray image sequence analysis: the gray zone represents the necrosed area, the crosses are calibration points. Dotted lines show where the algorithm found high stiffness using a planar model based on a single slice of the 3D data. Continuous lines represent links with high stiffness in a 3D model based on all the slices*

1.6. Conclusion

In this chapter, we revisited some classical approaches to image processing in the light of implicit or explicit scene modeling. This allowed us to outline a rarely described logical organization of the existing methods in image processing, setting some light into certain research directions still little exploited and sometimes promising. It looks like many modeling techniques developed by researchers in other communities, in particular by the image synthesis community, have potential applications to the domains of image or image sequence analysis. In a way, the challenge of modeling is not so much adding a specific technique to the existing panoply of image processing, but rather reconsidering the extension of the semantic field of what is commonly called "image processing" and "machine vision".

1.7. Acknowledgements

Thank you to Pierre Collet, Anders Ekman, Jean-Loup Florens, André Gagalowicz, Jiang Li, Annie Luciani, Evelyne Lutton, Xavier Provot, Georges Stamon, Bogdan Stanciulescu and all the colleagues who contributed the ideas, algorithms and results described or quoted in this chapter.

1.8. Bibliography

[BÄC 95] BÄCK T. and SCHWEFEL H.-P., "Evolution Strategies I: variants and their computational implementation", *Genetic Algorithms in Engineering and Computer Science*, John Wiley & Sons, 1995.

[BAL 82] BALLARD D.H. and BROWN C.M., *Computer Vision*, Prentice Hall, 1982.

[CAD 94] CADOZ C., *Les réalités virtuelles*, Dominos/Flammarion, 1994.

[COC 95] COCQUEREZ J.-P., PHILIPP S. *et al.*, *Analyse d'images: filtrage et segmentation*, Masson, 1995.

[COL 00] COLLET P., LUTTON E., RAYNAL F. and SCHOENAUER M., "Polar IFS + Parisian Genetic Programming = Efficient IFS inverse problem solving", *Genetic Programming and Evolvable Machines*, vol. 1, pp. 339–361, 2000.

[DEL 93] DELNONDEDIEU Y., LUCIANI A. and CADOZ C., "Physical elementary component for modelling the sensory-motricity: the primary muscle", *4th Eurographics Workshop on Animation and Simulation*, pp. 193–207, Barcelona, September 1993.

[GON 92] GONZALEZ R.C. and WOODS R.E., *Digital Image Processing*, John Wiley & Sons, 1992.

[HAR 80] HARALICK R.M., "Using perspective transformations in scene analysis", *Computer Graphics and Image Processing*, vol. 13, pp. 191–221, 1980.

[HAY 98] HAYET J.-B. and TADJADIT M., Projet ROBOCOP: repérage d'obstacles par observation et calcul des ombres portées, rapport de stage ENSTA, PPL 99/11, 1998.

[HOR 75] HORN B.K.P., "Obtaining shape from shading information", *The Psychology of Computer Vision*, McGraw-Hill, 1975.

[HOR 81] HORN B. and SCHUNCK B., "Determining optical flow", *Artif. Intell.*, vol. 17, pp. 185–203, 1981.

[HOU 62] HOUGH P.V.C., Method and Means of Recognising Complex Patterns, U.S. Patent no. 3 069 654, 18 December 1962.

[HOU 92] HOUSE D.H., BREEN D.E., GETTO P.H., "On the dynamic simulation of physically-based particle-system models", *Proceedings of EuroGraphics'92 Workshop on Animation and Simulation*, Cambridge England, 5-6 September 1992.

[HUA 83] HUANG T.S., *Image Sequence Processing and Dynamic Scene Analysis*, Springer Verlag, Berlin, 1983.

[LOU 94] LOUCHET J., "An evolutionary algorithm for physical motion analysis", *British Machine Vision Conference*, York, BMVA Press, pp. 701–710, September 1994.

[LOU 95] LOUCHET J., PROVOT X. and CROCHEMORE D., "Evolutionary identification of cloth animation models", *Computer Animation and Simulation '95, Proc of the Eurographics Workshop*, Maastricht, Springer, pp. 44–54, September 1995.

[LOU 96a] LOUCHET J. and BOCCARA M., CROCHEMORE D. and PROVOT X., "Building new tools for synthetic image animation using evolutionary techniques", *Evolution Artificielle/Artificial Evolution 95*, Brest, September 1995, Springer Verlag, 1996.

[LOU 96b] LOUCHET J. and JIANG L., "An identification tool to build physical models for virtual reality", *IWSIP Manchester*, UK, November 1996.

[LUC 91] LUCIANI A., JIMENEZ S., CADOZ C., FLORENS J.-L. and RAOULT O., "Computational physics: a modeler-simulator for animated physical objects", *EuroGraphics '91 Conference*, Vienna, Elsevier Science Ed., 1991.

[LUT 94] LUTTON E. and MARTINEZ P., "A genetic algorithm for the detection of 3D geometric primitives in images", *12th ICPR*, Jerusalem, Israel, October 9-13, 1994/INRIA technical report # 2210.

[MAU 94] MAURIN G. and GAGALOWICZ A., Localisation 3-D de sources de lumière par utilisation des variations lentes d'illumination dans les images, rapport de stage INRIA/ENSTA, EPR, 1994.

[PRA 04] PRADOS E. and FAUGERAS O., "Unifying approaches and removing unrealistic assumptions in shape from shading: mathematics can help", *Proc. ECCV04*, 2004.

[PET 91] PETROU M. and KITTLER J., "Optimal edge detector for ramp edges", *IEEE Pattern Analysis and Machine Intelligence*, vol. 13, no. 5, pp. 1483–1491, 1991.

[REC 94] RECHENBERG I., "Evolution strategy", in ZURADA J.M., MARKS R.J. and ROBINSON C.J. (Eds.), *Computational Intelligence Imitating Life*, IEEE Press, pp. 147–159, 1994.

[ROU 81] O'ROURKE J., "Motion detection using Hough technique", *IEEE Conference on Pattern Recognition and Image Processing*, Dallas, pp. 82–87, 1981.

[REY 87] REYNOLDS C., "Flocks, herds and schools: a distributed behavioural model", *Computer Graphics (Siggraph)*, vol. 21, no. 4, pp. 25–34, 1987.

[REE 83] REEVES W.T., "Particle systems – a technique for modelling a class of fuzzy objects", *Computer Graphics (Siggraph)*, vol. 17 no. 3, pp. 359–376, 1983.

[ROT 92] ROTH G. and LEVINE M.D., "Geometric primitive extraction using a genetic algorithm", *IEEE CVPR Conference*, pp. 640–644, 1992.

[SER 99] SER P.K., CLIFFORD S., CHOY T. and SIU W.C., "Genetic algorithm for the extraction of nonanalytic objects from multiple dimensional parameter space", *Computer Vision and Image Understanding*, vol. 73, no. 1, pp. 1–13, 1999.

[STA 02] STANCIULESCU B., Modèles particulaires actifs pour la synthèse d'images animées, PhD Thesis, Joseph Fourier University, Grenoble, 2002.

Chapter 2

Artificial Evolution and the Parisian Approach. Applications in the Processing of Signals and Images

2.1. Introduction

This chapter aims to present the so-called Parisian approach and one of its applications in the field of signal and image processing. Modeling, particularly in the fields of geometry and physics, has already been covered in Chapter 1. However, this may also come into play in image processing applications. Optimization through artificial evolution also plays a role in efficiently instantiating these models. In certain cases, it might be better to decompose the model into its elementary entities as they are easier to manipulate. Objects that have been manipulated by genetic operators are no longer vectors of parameters that describe a complete model of a scene. The elementary entities, however, take on a similar role. They only come together as a whole when the representative model of the scene is being studied.

2.2. The Parisian approach for evolutionary algorithms

In traditional evolutionary algorithms, the aim is to find the best possible solution for a given problem. However, certain problems (NP-complete, or not) can turn out to be extremely complex, especially if the search space is very large. To conceptualize the problem the evolutionary algorithm is confronted with, it makes sense to look at some examples. Let us assume that, for instance, we want to segment (isolate)

Chapter written by Pierre COLLET and Jean LOUCHET.

the different objects of an interior scene using a standard evolutionary algorithm. Assuming a contour extraction algorithm is available, and that the genome of an individual contains a representation of the set of objects to be segmented, the first problem to solve is to determine how many genes should be present in the genome (i.e. how many objects the individual should find in the scene). Assuming the number of genes is fixed at 10 and that all objects of the scene are simple mathematical figures (straight line, ellipsis, triangle, trapezoid), a simple algorithm could randomly take 10 different shapes among these and choose their orientations, sizes, positions at random, in order to create an individual.

The task of the evolutionary algorithm is then to evolve a population of such individuals in order to maximize the intersection between the shapes encoded by an individual and the outlines that have been provided by the contour extraction algorithm. The search space is very large as for every gene of an individual (who possesses 10 genes in total) four different shapes can be chosen. These shapes can be centered on any of the 786,432 pixels of a 1024×768, and orientated on 360 degrees (if orientation is discretized into degrees), with a homothetic factor that we may choose to be an integer between 1 and 1,000. Even with such extreme simplifications and discretizations, there are already 11×10^{12} possibilities, which do not even take into account the basic dimensions of all considered shapes!

With this algorithm, the chances of finding an individual whose genes represent the outlines of the scene in reasonable time are therefore extremely slim.

We might therefore wonder whether the problem has been expressed correctly. Indeed, the search space is very large if every individual is supposed to find a solution for the entire scene. Would it therefore not be possible to introduce the old concept of divide and conquer?

Conventional evolutionary algorithms follow a Pittsburgh approach similar to what is used in classifier systems [SMI 80] (in which every individual encodes all rules). In the Michigan approach, however [HOL 78], an individual only encodes one single rule. The Parisian approach of evolutionary algorithms suggests doing the same, by encoding the solution with the help of a group of individuals.[1] Similar to the Michigan approach, the Parisian approach needs a complex retribution. If the problem fits this approach, benefits can be significant because principles apply which can be found in data-level parallelism, emergence and in optimization paradigms such as ant colony optimization (ACO) or particle swarm optimization (PSO).

1. This is a type of co-evolution. The term co-evolution, however, was marked by strong connotations which made it impossible for the term to be used. The approach has therefore been named after its "place of origin" similar to the approaches from Pittsburgh and Michigan.

We should note that in a recent article [WAL 06], Julian Miller and his team make a connection between the Parisian approach and polyploid representations of individuals where several alleles are used to encode a gene. In this article the results and the acceleration that have been obtained due to a polyploid presentation are comparable to those obtained by the Parisian approach. A possible conclusion might be that the two approaches are in fact very similar.

For a detailed explanation of the Parisian approach concept, the example of the original paper will be briefly recalled, before showing how the algorithm finds other developments in the field of image and signal processing.

2.3. Applying the Parisian approach to inverse IFS problems

Iterated functions systems (IFS) [BAR 85, HAR 85, HUT 81] allow us to define a fractal attractor (see Figure 2.1) if the starting point is situated within the attractor defined by the set of functions and if all functions of the set are *contractant* (i.e. the distance between the image of point x of the attractor by any of the functions and the central point of the attractor – known as the fixed point – needs to be lower than the distance between x and the fixed point).

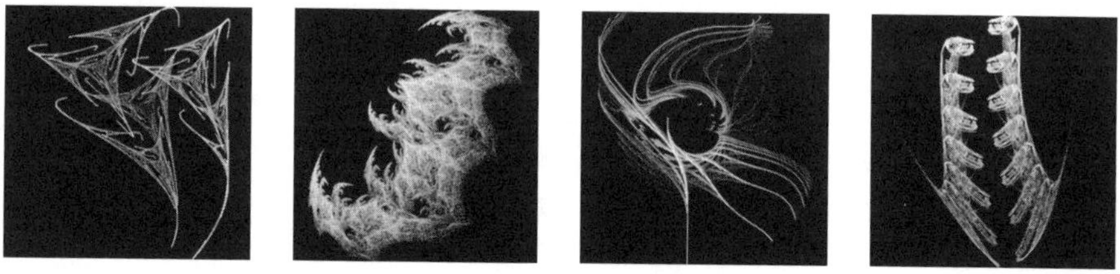

Figure 2.1. *Examples of IFS attractors*

Let us take a system of three functions f_1, f_2, f_3 that take a point on the plan as a parameter and a starting point $P_0(x_0, y_0)$ situated within in the attractor of the system (f_1, f_2, f_3), the construction of the attractor can then be carried out by the so-called toss-coin algorithm:

```
P=P0
While (true) {
  Choose f randomly among f1, f2, f3 // here is the toss-coin operation
  P=f(P)
  display P
}
```

The inverse problems for IFS consists of finding which system of functions has a pre-determined target image as an attractor. This inverse problem has many important applications in the field of fractal compression.

The standard evolutionary approach uses genetic programming, where individual genomes consist of n trees (one per function). Now, the attractor is in fact a collage of several attractors represented by each function [VRS 91]. It might now be interesting to find out if this complex inverse problem could be decomposed into a series of subproblems that would individually be much easier to solve. If the problem is subdivided into 10 subproblems that are each 1,000 times easier to solve than the original problem, we obtain a $100x$ speedup.

In the case of IFS, rather than evolving individuals that contain several functions that are supposed to encode the complete problem, we have tried in [COL 00] to evolve a population of individuals that would each contain a single function, even if this meant that several individuals (a sub-population) are needed for the evaluation of a complete solution.

Several problems therefore need to be solved:

– which individuals should be chosen when evaluating the complete IFS?

– how can each individual be rewarded correctly depending on whether it has participated in the complete IFS or not?

2.3.1. *Choosing individuals for the evaluation process*

Using the Parisian approach requires a sharing strategy; a way to preserve diversity. In the case of IFS, the final attractor being a collage of different attractors that are defined by all the functions that make up the system, the attractors need to be distributed as efficiently as possible in the target. The dynamic niche sharing technique [MIL 96] allows for the distribution of the population into different ecological niches, based on a distance between the individuals and a maximum number of niches. The distance can be calculated on the genotype or phenotype. In the case of polar representation, IFS used in [COL 00] each function (i.e. individual) has a fixed point in the center of the attractor that it determines. The coordinates of the fixed point can be used to determine the minimal distance between the individuals (radius around the fixed point of individuals). The maximum number of niches established by dynamic niche sharing was fixed at $n/2$ (with n being the number of individuals in a population) and the fittest individual is selected by the best found niche.

2.3.2. *Retribution of individuals*

In the Parisian approach, a number of n individuals cooperate to obtain a potential solution to be evaluated. To the standard individual fitness retribution, we must add (for the individuals that have been chosen to form a complete IFS) a retribution on the quality of the obtained attractor. IFS are remarkably well adapted to the evaluation of every individual before producing a complete IFS. The IFS's contracting properties ensure that the image of a point in an attractor refers to another point of the attractor and is convergent towards the center of the attractor. Therefore, if the IFS exactly represents the target, the image of any point within the target must lie inside the target. This property is present in each of the functions that are part of the IFS. It is therefore possible to evaluate the benefit of each (individual) function by checking if the target's image of points inside the target remains inside the target or not. The relationship between these two forms of distribution (global and individual) is of course highly empirical.

2.3.2.1. *Individual retribution*

In the standard retribution in evolutionary algorithms, where every individual is evaluated separately, the individual retribution [COL 00] is calculated by function:

$$\mathcal{R}_{\text{loc}}(w_i) = \mathcal{F}_1(w_i) + \mathcal{F}_2(w_i) + (1 - s_i)$$

where w_i is the function/individual of an IFS. $\mathcal{F}_1(w_i)$ is a ratio between the number of pixels (#) inside target A and the number of pixels outside the target:

$$\mathcal{F}_1(w_i) = \frac{\#[w_i(A) \bigcap A]}{\#[w_i(A) \bigcap A] + \#[w_i(A) \backslash A]}$$

with a maximum of 1.

$\mathcal{F}_2(w_i)$ rewards the w_i individual increasingly with the size of surface $w_i(A) \bigcap A$ with a maximum of 1:

$$\mathcal{F}_2(w_i) = \frac{\#[w_i(A) \bigcap A]}{\#[A]}$$

Finally, s_i is the estimated contraction factor of w_i. $(1 - s_i)$ is antagonist of $\mathcal{F}_2(w_i)$, in order to avoid obtaining an identity function (which would maximize $\mathcal{F}_2(w_i)$ without being of any interest).

Each individual now being rewarded by its contribution to the target, we must now reward the team as a whole for the obtained attractor.

2.3.2.2. *Global retribution*

Once all individuals have been evaluated, an IFS Ω will be constructed made of the N fittest individuals of all niches. The attractor A_Ω of this IFS is evaluated with the help of the following two values:

$$\text{Inside}_\Omega = \frac{\#[A_\Omega \bigcap A]}{\#[A]} \quad \text{proportions of points of } A_\Omega \text{ inside the target } A$$

$$\text{Outside}_\Omega = \#[A_\Omega \backslash A] \quad \text{number of points of } A_\Omega \text{ outside } A$$

Again the recipe of retribution is highly empirical. Rather than giving as a global reward an absolute value determined by the quality of the attractor, the evaluation is a comparison between the created attractor and that of the previous generation:

$$R_{\text{glob}}(n) = \big[\text{Inside}_{\Omega(n)} - \text{Inside}_{\Omega(n-1)}\big] - \big[\text{Outside}_{\Omega(n)} - \text{Outside}_{\Omega(n-1)}\big] + \alpha\big[\text{Nb\_functions}\,(\Omega(n)) - \text{Nb\_functions}\,(\Omega(n-1))\big]$$

In the next step, the global retribution for every individual w_i is calculated as:

$$\text{Fitness}\,(w_i) = R_{\text{loc}}(w_i) + R_{\text{glob}}(n)$$

– If w_i is already part of $\Omega(n-1)$, then:

$$\text{Fitness}\,(w_i) = R_{\text{loc}}(w_i) + \frac{R_{\text{glob}}(n) + R_{\text{glob}}(n-1)}{2} \times \frac{1}{[\text{age}\,(w_i)]^2}$$

where $\text{age}(w_i)$ represents the number of generations for which w_i was part of the IFS.

– If w_i has just been removed from the IFS (i.e. has participated to the IFS until $\Omega(n-1)$ but does not participate in $\Omega(n)$):

$$\text{Fitness}\,(w_i) = R_{\text{loc}}(w_i) - R_{\text{glob}}(n)$$

– If w_i does not belong to the IFS:

$$\text{Fitness}\,(w_i) = R_{\text{loc}}(w_i) + \frac{R_{\text{glob}}(n-1)}{2}$$

(the global reward added to each individual decreases throughout the generations).

In other words, the quality of the created IFS is redistributed to the entire population. This process of redistribution considers the past of the individuals as well as the number of generations in which this individual has been selected.

2.4. Results obtained on the inverse problems of IFS

Until [COL 00], the best solutions obtained by a standard approach were obtained by [LUT 95] (see Figure 2.2). In order to compare the computational effort with the Parisian approach (where every individual represents a function) it makes more sense to count the number of evaluated functions. The standard GP algorithm uses a generational replacement with a population of 30 individuals (30 children are created per generation). Since 1,500 generations were required to obtain the best individual of Figure 2.2, and since each individual had to encode five functions, this individual necessitated the evaluation of 30 individuals $\times 5$ functions $\times 1,500$ generations = 225,000 in order to be found.

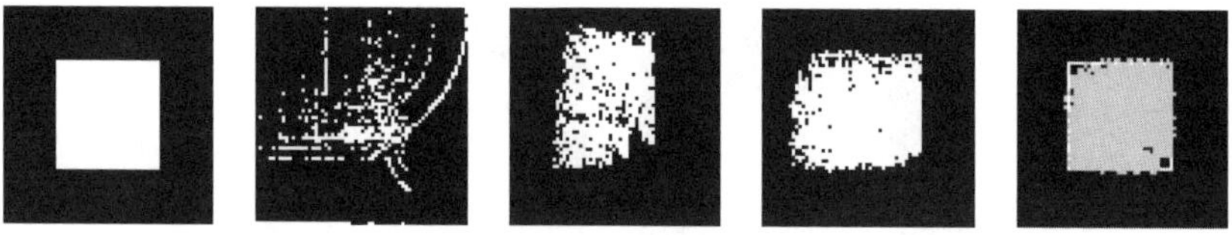

Figure 2.2. *Optimization of a 64×64 square with mixed IFS using standard Genetic Programming. From left to right, we can see the target (full 64×64 square) and then, the best individuals from generations 10, 100, 300 and 1,500. Size of the population: 30 individuals containing 5 functions each. The best individual on the right fills the target by 91%. 4% of the pixels are out of the target*

On the same problem and to obtain a similar result (Figure 2.3 left) the Parisian approach of [COL 00] required only 49 generations and 60 individuals. In terms of the number of evaluations, since a $(60+30$, i.e. 60 parents and 30 children per generation) replacement strategy (borrowed from Evolutionary Strategies [BÄC 97]) was used, the necessary number of function (individual) evaluations was $60 + 49 \times 30 = 1,530$ to which 49 evaluations of complete IFS must be added.

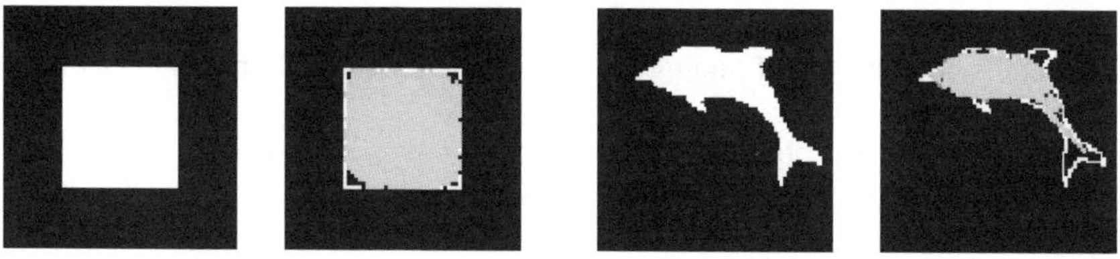

Figure 2.3. *Left: optimization of the same 64×64 square with polar IFS, by genetic programming using the Parisian approach. The best IFS contains 14 individuals and is obtained after 49 generations of only 60 individuals. It fills in 85% of the target and 0 pixels are outside the target. Right: the dolphin (a more complex form) needed 400 generations of 60 individuals and is composed of 13 individuals. The dolphin is filled in up to 60% and 0.36% of the pixels are outside the target*

Given that on average 50% of the individuals participated in the IFS (which is overestimated, as the best IFS of the 49th generation only required 14 individuals) these 49 IFS evaluations amounted to another $49 \times 30 = 1470$ evaluations (i.e. nearly as many evaluations as for the personal evaluation of each individual).

Overall the Parisian approach required around 3,000 evaluations, compared to the 225,000 evaluations of the standard approach, which is a difference of nearly two orders of magnitude.

This allows us to envisage an approximation of more complex shapes with a reasonable level of precision (e.g. dolphin in Figure 2.3 right) that would not be obtainable with a traditional algorithm.

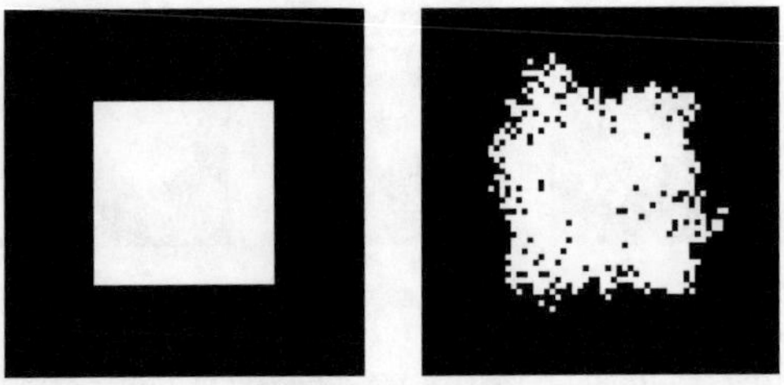

Figure 2.4. *Optimization of a 64×64 square with a genetic algorithm that was used to optimize 24 real parameters for 4 functions. This result was obtained after evaluating 800,000 functions (20 individuals encoding 4 functions for 10,000 generations). The square is filled by 88.4% and 17% of the pixels are located outside the target*

2.5. Conclusion on the usage of the Parisian approach for inverse IFS problems

Two main obstacles arise if a standard method is to be used on the inverse IFS problem:

– the search space is very large, as the individuals must find a complete solution to the problem;

– there is no easy way of finding out how many functions an individual will need.

For the square, a simple solution exists with four individuals. Even though this solution is known, it is very difficult to obtain with a genetic algorithm [GOE 94] (see Figure 2.4). Five functions were empirically chosen for the standard GP approach of Figure 2.2. As the problem is well adapted to the Parisian approach, this approach allows us to decompose the problem into subproblems that can be solved much more easily than the original problem. Furthermore, the number of functions used is obtained dynamically and automatically.

However, the price to be paid is an even higher complexity of the algorithm with:

– two evaluation functions (a local and a global function) that must be skillfully intertwined;

– a strategy that maintains the diversity and *sharing* of ecological niches;

– a $(\mu + \lambda)$ replacement, borrowed from evolutionary strategies, that preserves a large number of parents in the following generation. These parents could also be kept due to a strong elitist strategy (see the fly algorithm later on in this chapter).

Despite these disadvantages, the results obtained with the Parisian approach are worthwhile.

According to Julian Miller *et al.* [WAL 06] this approach is very similar to the polyploid approach (every gene is represented by several alleles that are or are not all considered in the production of the phenotype). Many living beings (humans included) have diploid chromosomes. This phenomenon is very common amongst plants. Many of them are even triploid (bananas, apples), tetraploid (coffee, cotton), hexaploid (wheat), etc.

From an evolutionary point of view, the real advantage that polyploidy has over haploidy is unknown (otherwise it would be used even more often in artificial evolution). In a recent PhD on this topic [TOU 04], it is observed that in two comparable haploid and diploid contexts for a form of yeast, the mutation rate was 18 times higher in the haploid context. This may show that the diploid phase would allow for fewer modifications even though it allows for more variety in the nature of these modifications. The two last sentences of this thesis are the following

> This thesis has shown among other issues that the diploid phase allows for both a large genomic flexibility that would favor the creativity that is essential for the evolutionary process, while ensuring enough stability to avoid a too large mutation rate that would be disruptive for the species. Evolution is therefore the result of a fair equilibrium between creativity and stability.

The Parisian approach was developed in order to divide a complex problem into independent subproblems that would be easier to solve. If this approach proves to be very similar to a polyploid approach, this would shed new light on the advantage of using polyploidy.

2.6. Collective representation: the Parisian approach and the Fly algorithm

In the introduction, we saw an example of how the principles of evolutionary computation may give new life to older image processing techniques based on model optimization methods. In the following sections, we will get a look at how the Parisian approach may also be applied to image processing, through the example given by the Fly algorithm. While in many cases the image or image sequence to be processed can be viewed as a simple addition of separate entities, each one corresponding to a particular point in the model's parameter space, there are many cases where this is untrue. It may then be more efficient to exploit the model's rather than the image's separability, and describe the image as the result of the combination of a *subset* of suitable parameter space. This is the basic idea of the Parisian approach in evolutionary computation: using evolutionary methods the normal way with the usual tools, but with a major change in their semantics. Each individual in the parameter space now only represents a small part of the problem to be solved: rather than looking after one particular point in the parameter space, we are now interested in the resulting population as a whole.

2.6.1. *The principles*

Stereovision is defined as the construction of a 3D model of a scene, using the images delivered by two cameras with well known geometric characteristics. Most classical stereovision methods are based on image segmentation, primitive extraction and matching. They may provide detailed accurate scene descriptions but at a high computational cost. Voting methods described in Chapter 1 are generally not relevant due to the complexity of the image primitives involved. Moreover, the accuracy they are capable of is seldom a requirement in robotics applications – in particular in the field of mobile robotics where obstacle avoidance and trajectory planning rarely need such a high precision in scene primitive location.

The main idea here is to represent the scene using a population of 3D points and evolve them in such a way that they are likely to concentrate on the visible surfaces of the objects present in the scene. Here, using the Parisian approach means that the algorithm's result be expressed as a large set of simple primitives: points in the 3D space, rather than a smaller set of more complex primitives. This principle of simplicity[2] is followed even into the fitness function – which has a major contribution to the total computation load.

Basically, the method consists in creating a population of points (the "flies") randomly spread over the common field of vision of the cameras, then evolve this population so that the points tend to locate themselves on the surfaces of the objects we are trying to locate spatially. To this end, all the cameras' geometric parameters being known, the algorithm will calculate for each fly in the population, a fitness value which estimates the degree of similarity between the fly's calculated projections into each camera image.

A fly is defined as a point in space, with coordinates (x, y, z). As we are using (at least) two cameras, each fly will project onto the left camera image as a point with coordinates (x_L, y_L) and similarly as (x_R, y_R) into the right camera. Using the cameras' calibration parameters, it is an easy task to calculate x_L, y_L, x_R, y_R according to x, y, z using the projective geometry formulas:

$$\begin{pmatrix} x_L \\ y_L \\ 1 \end{pmatrix} \equiv F_L \begin{pmatrix} x \\ y \\ z \\ 1 \end{pmatrix}, \quad \begin{pmatrix} x_R \\ y_R \\ 1 \end{pmatrix} \equiv F_R \begin{pmatrix} x \\ y \\ z \\ 1 \end{pmatrix}$$

where F_L and F_R are the $(3, 4)$ projective matrices of the left and right cameras.

If a fly is located at the surface of an object, then the image pixels where it projects into each camera image will usually have similar attributes: gray level, color, local

2. Eberhart's particle swarms [EBE 95] are another optimization tool inspired by the "artificial life" techniques, which also exploits the separability of the optimization problem in a similar way as the Fly algorithm.

texture, etc.[3] Conversely, if the fly is not on a visible object surface, how much its projections will look like one another will randomly depend on the textural properties of the first objects aligned with the fly and each camera's focal point. The fundamental idea of the algorithm is to translate this principle into a fitness function able to control the fly population's evolution process successfully.

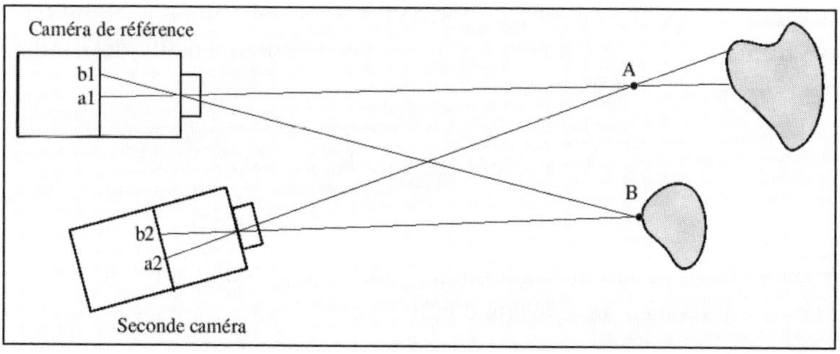

Figure 2.5. *Pixels b1 and b2, projections of fly B, have identical gray levels. Pixels a1 and a2, projections of fly A, generally have different gray levels, because they correspond to two different points of the visible surface of the object*

The fitness function should be chosen in order to efficiently recognize the degree of similarity of the flies' projections and if possible use some neighborhood-related properties (contrast, texture) not to be fooled by apparently similar pixel gray values. For example, a fitness function only based on the identity of the projections' gray levels, would give undesirable high fitness values to flies located in front of a uniform object, even if far from its surface. To get around this, the fitness function includes a normalization term at its numerator:

$$\text{fitness}(indiv) = \frac{G}{\sum_{(i,j)\in N}\left(L(x_L+i, y_L+j) - R(x_R+i, y_R+j)\right)^2}$$

where

– $L(x_L+i, y_L+j)$ is the gray level of the left image at pixel (x_L+i, y_L+j);

– N is a small neighborhood used to calculate the fly's projections' likeness;

– usual techniques are used to avoid division by zero.

3. Specular reflections and, more generally, any important discrepancy between the real photometric properties of the object's surface and the default Lambertian reflectance model, which is the basis of the assumption that a given point in the scene will appear with the same gray level independently of the point of view, may perturb this criterion and alter the algorithm's performance; however this is equally true with most surface-based stereovision algorithms.

The denominator evaluates the square deviation between the pixel configurations around the two projections of the fly. Thus, a fly whose projections are similar will give rise to a small denominator value and a high fitness value.

The normalization term G gives a measurement of the contrast around the current pixel of the reference image; it has to achieve a good balance between giving too high a fitness value to flies in front of a uniform object, or only giving a high fitness value to flies in front of a contrasted contour. It has been experimentally determined that a good trade-off is obtained with:

$$G = \sqrt{\sum_{(i,j)\in N} \left(L(x_L + i, y_L + j) - L(x_L, y_L)\right)^2}$$

Moreover, the evaluation function includes a correction term to eliminate the constant by subtracting a local average gray level.

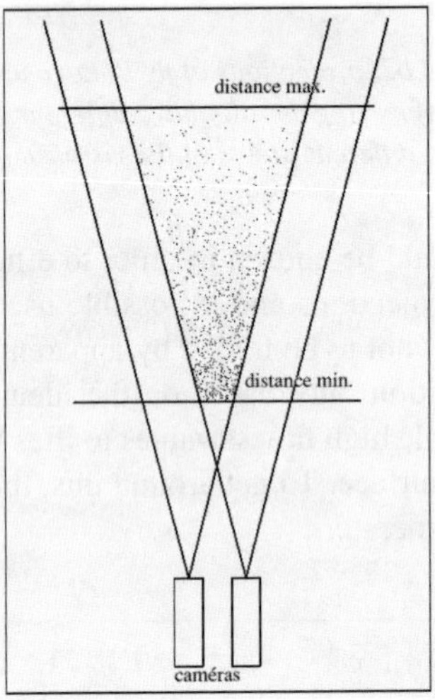

Figure 2.6. *The fly population is initialized inside the truncated intersection of the camera vision cones*

Let us now examine the evolution operators for the (x, y, z) genome.

The initial population is created inside the intersection of the vision cones of the two cameras, with optional maximum and minimum clipping distances (see Figure 2.6). An individual's chromosome is the triplet (x, y, z) of its coordinates, Oz corresponding to the left camera's (used as the reference camera) axis. The flies are given a uniform distribution in z^{-1} beyond the minimal clipping distance, which gives an initial density which decreases with depth.

As the calibration parameters of the cameras are known, the coordinates of each fly's projection pixels are calculated and the fitness is evaluated. The main genetic operators are the standard ones found in evolution strategies: Gaussian mutation and barycentric crossover. In the following examples, the selection process is a 2-tournament. As significant populations are normally used, somewhat to compensate for the simplicity of the individuals, an interesting speed improvement is obtained by calculating a percentile on a random sample of the population and using it as a threshold, without any measurable loss of performance.

A 2D sharing function is use to reduce the fitness values of flies that project into crowded areas in the images [BOU 01]. A 2D fly projection density is calculated and used into the calculation of a sharing penalty.

2.6.2. *Results on real images*

2.6.2.1. *Classical stereovision*

The 760 × 560 stereo pair used in the following examples was obtained using a monochrome camera with a sideways motion. The genetic parameters are similar to the preceding ones: 5,000 individuals, 100 generations 40% mutation probability, 10% crossover, sharing radius 2, sharing coefficient 0.3.

On the results image, it is possible to see the two sides of the chest, a part of the wall on the right and the front half of the round stool.

Figure 2.7. *Left and right images*

Genetic algorithms and evolution strategies are often seen as slow processes, unable to cope with real-time applications. In reality, things are much more subtle:

– "real time" may not be reduced to a question of execution speed. It would be more realistic to see it as the ability of an algorithm to exploit the flow of input data and adapt itself to the response time needed by the final user. Rather interestingly, evolutionary strategies are capable of adaptation, and may work on a problem described by a fitness function and accept the fitness function being modified and updated during execution [SAL 97]; this is far from being a common property among optimization methods.

Figure 2.8. *Result, seen from above (384×288 image)*

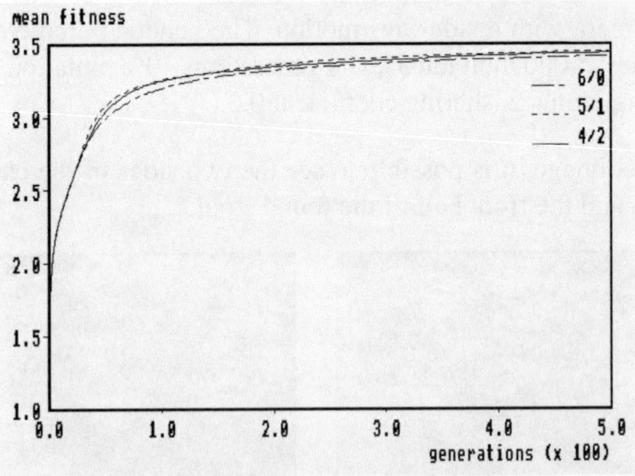

Figure 2.9. *Average fitness of a population of 5,000 individuals depending on the number of generations, for three combinations of growth and mutation rates (60%/0%; 50%/10% and 40%/20%)*

– the execution speed of an evolution strategy depends to a large extent on the computational complexity of the fitness function – which is quite simple in this case.

This is why we worked at extending the algorithm to the processing of stereo image sequences, with a special interest in sequences taken by moving cameras, as it is the case in mobile robotics. The results show that if motion is slow enough, convergence is significantly speeded up if at each new image pair, the population of flies is initialized using the result – the old positions of flies – obtained at the previous step rather than using purely random initial positions.

In order to cope with slightly faster motion, we experimented with an extension of the flies' chromosomes, introducing three velocity parameters into each fly's genome. This allows each fly, now with 6 parameters, to keep a memory of its relative speed in the robot's coordinate system, in a similar way to a Markov process. This technique, called the "dynamic flies" technique, does not overload the fitness calculation process, as the three extra genes are not expressed in the fitness, but are only used to update each fly's position between successive frames more efficiently. On the other hand, the genetic operators become more complex, with more variables to manipulate, which results in a non-negligible "administrative overhead" in the algorithm. The interest of dynamic flies is very much dependent on the particular problem to be solved, in particular the robot's velocity and the computer's speed, as the benefits of convergence in a smaller number of generations may be worse than annihilated by the extra cost of running the operators – which suggests a sad metaphor in management sciences. These methods are described in detail in [LOU 02].

In the next section, we will show some results of the fly algorithm running in real time on a basic laptop computer aboard a car equipped with two cameras[4]. This is an application of the flies to a project aimed at developing driving assistance systems, in a framework of cooperation between the IMARA and COMPLEX teams in INRIA, France. With each fly is associated an "alarm value", which is higher when the fly's fitness is high. The fly is close to the collision trajectory and the fly's distance is low. The alarm function depends on multiple adjustment parameters, e.g. such as the car's width. The sum of alert values is used to trigger an emergency braking command.

Figure 2.10. *Real-time evolutionary image processing on a highway (the image comes from a monochrome camera)*

4. Images obtained by O. Pauplin within the framework of an INRIA project.

Figure 2.11. *A pedestrian*

Figure 2.12. *Image of alarm values (pedestrian)*

2.6.3. *Application to robotics: fly-based robot planning*

Classical navigation and obstacle avoidance methods used in robotics use as input data, scene descriptions built by the stereo analysis process, usually 3D structures such as polygons or facets. Here, the stereovision algorithm produces flies. It has therefore been objected that the task of converting the flies into a more classical representation usable as input data to existing navigation systems would cost more time than the time saved thanks to the simplicity of the fly algorithm. This is why we decided not to reinvent the wheel, but to write new navigation algorithms with the same spirit as what had already been developed, in particular concerning the resolution of robot blockage situations, but re-think them so that they can directly use the flies as input data.[5] Boumaza integrated the complete algorithmic construction (fly-based stereovision and navigation) into an *ad hoc* robot simulator, aiming at the simulation of the complete robot's perception-action loop (Figures 2.13 and 2.14), containing:

– a double camera simulator (image synthesis);

– stereovision (the fly algorithm);

– fly-based planning algorithm;

– a simple robot platform kinematic simulator.

Figure 2.13. *A synthetic image of the scene, as seen by the robot*

As could be predicted, running the simulator showed that the robot frequently encounters blockage situations where the controlling force (which acts on the robot's steering) oscillates around zero when the robot becomes trapped in a local potential minimum.

To get around this difficulty, A. Boumaza adapted and implemented three classical heuristic methods [ZEL 98, KOR 91] allowing us to generate a new steering strategy which drives the robot out of the local minimum: the random walk method, the wall following method and the harmonic function method.

5. The results presented in this section result from the PhD Thesis work of A. Boumaza at INRIA/René Descartes University, Paris.

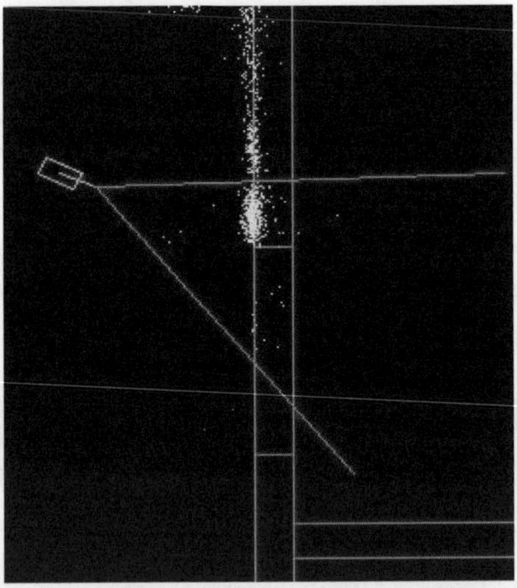

Figure 2.14. *The robot is facing a wall and a door. The bright points represent the flies found during the preceding steps*

In the random walk method, secondary targets are created at random places whenever a blockage situation is diagnosed. Then the robot uses what it knows about obstacles to choose the secondary target with the smallest number of obstacles, first between itself and the robot, and then between itself and the main target [BOU 01]. This secondary target plays for a short time the role of the primary target (see Figures 2.15 and 2.16).

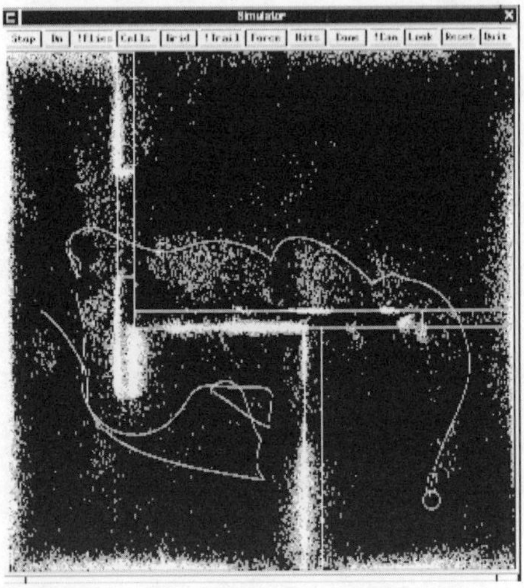

Figure 2.15. *Getting around a wall obstacle and through a door using secondary targets*

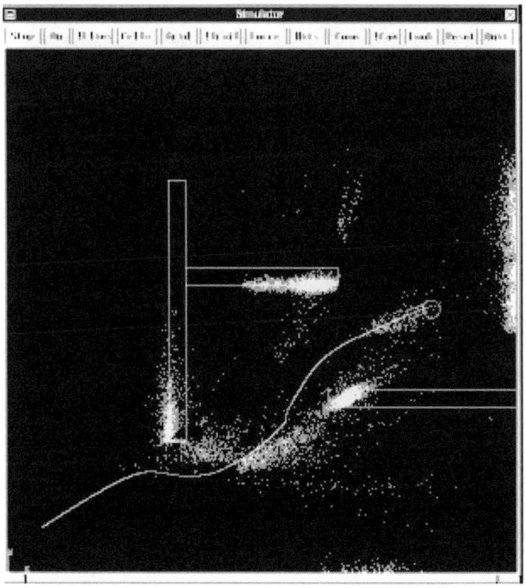

Figure 2.16. *Direct trajectory without secondary targets*

The wall following method only uses a change in the attraction vector, modified so that the new resulting force be roughly parallel to the direction of the obstacle.

The harmonic function method iteratively builds a harmonic attraction function with a Laplacian equal to zero except at the singularities given as initial conditions: the target (where the function value is -1) and the regions with a high density of flies with a high fitness (fitness valued at $+1$). The robot uses the gradient of this constantly updated function to control its direction along the steepest gradient lines (Figures 2.17 and 2.18).

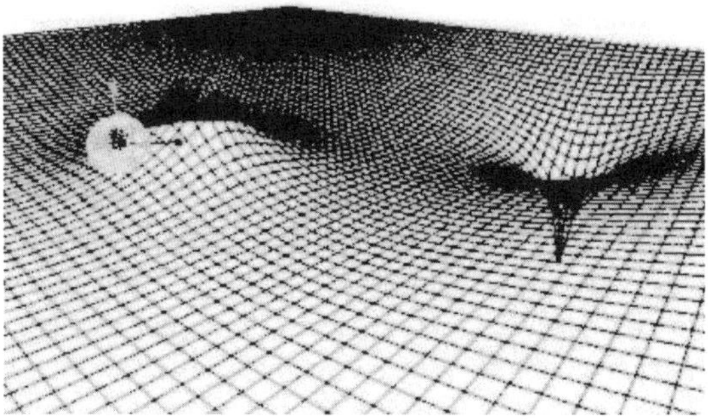

Figure 2.17. *Harmonic function used for obstacle avoidance*

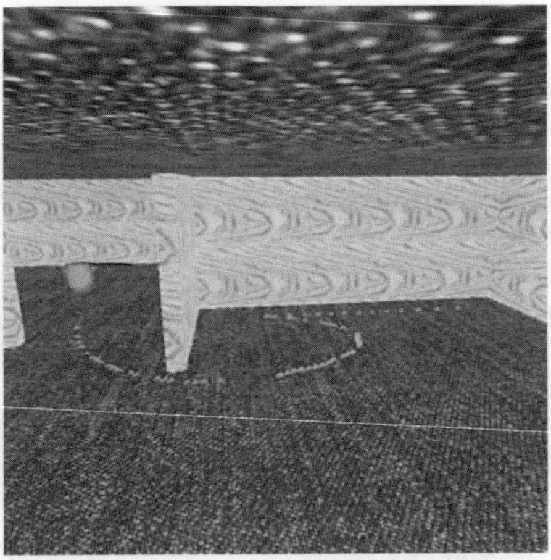

Figure 2.18. *Obstacle avoidance using the harmonic function*

2.6.4. *Sensor fusion*

One type of sensor is generally unable to provide a mobile robot with enough reliability in variable environments. Simultaneous use of different sensors, e.g. based on different physical principles (cameras sensitive to different wavelengths, acoustic sensors, odometers, etc.) requires us to be able to fuse different information sources. Sensor fusion is a key element in a robot's perception system. Many classical approaches use statistical sensor fusion based on Bayes' theorem; this technique cannot be used here, as the flies do not provide any probabilistic information. We have to follow another line.

Here, we will make the classical distinction between exteroceptive sensors (giving the robot information about the external world) and proprioceptive sensors (giving the robot information about its own state and position). We will show how it is possible to fuse exteroceptive and proprioceptive information by introducing them into the fly evolution strategy itself.

2.6.4.1. *Exteroceptive sensor fusion: a multisensor fitness*

We created an extended fitness function which now integrates the exteroceptive sensors. In the example taken in this simulation, the exteroceptive sensors are a set of six short-range ultrasonic sonars, with a field angle of 15 degrees. In the simulator, the sonar simulation simply uses the data from the Z-buffer taken out of the image synthesis program to calculate the distances to obstacles. The smallest value inside the detection angle is then disturbed by noise and returned as the simulated sonar's output. As discussed above, the lack of a probabilistic interpretation of the fly data prevents any strong mathematical justification of a fitness function which would integrate all the exteroceptive sensors. However, the idea is to increase the fitness of any fly whose position has been confirmed by one or several extra sensors. Conversely, if a fly has been attributed a good fitness by the vision process alone and has not been "seen" by the ultrasonic sensors, then it would be hazardous to reduce its fitness – it may come e.g. from an obstacle covered with an acoustically absorbent material. Therefore we define the multisensor fitness function the following way: if a fly has not been confirmed by any extra sensor, its fitness will be left unchanged as given by the main sensor (here the stereo cameras); if a fly lies within the field of vision of an extra sensor and its properties (depth) are consistent with the sensor's data, then its fitness will be increased by a given percentage B:

$$\text{newfitness} = \text{oldfitness}(1 + B)$$

Here, we take into account the poor angular resolution of acoustic sensors: it is safer not to give a high fitness value to a fly which did not obtain a good enough fitness value from the vision sensor in the first place.

In parallel to this inclusion of sonar sensor information into the fitness function, we also introduced an *immigration* operator, which creates new flies in space with a bias in favor of the sonars' suggestions.

This method allows us to integrate an arbitrary number of exteroceptive sensors, but is not suitable for proprioceptive sensors.

2.6.4.2. *Proprioceptive sensor fusion*

Proprioceptive sensors provide the robot with information about its own state, in particular its position. If this information is missing, when the robot moves slowly enough, the quasi-continuous fly optimization process will still be able to follow the scene's relative motion. If the robot moves faster, even an approximate knowledge of the robot's position enables us to continuously update the flies' positions and speed up the algorithm's convergence significantly.

The robot simulator thus integrates an odometric sensor simulation module. In the real world, the robot trajectory planner would control the wheel rotation angles, which are therefore perfectly known by the robot, but the actual robot's trajectory differs from the theoretical one due to external factors such as tire deformation, ground irregularity and gliding. Thus, to simulate the robot's trajectory we add a Gaussian noise to the trajectory command given by the planner, but the sensor fusion algorithm only receives the robot's internal odometric information as given by the trajectory controller to the robot motors. The fusion of odometric information is performed when updating the flies' 3D coordinates. Our experiments showed the convergence of flies, otherwise achieved after about 10 generations, now only requires 2 or 3 generations when the flies' position are updated, despite the poor precision of odometric information.

In the following example, the robot in open loop mode was turning 1 degree per frame, with only one algorithm generation being executed at each new frame. Without proprioceptive fusion (Figure 2.19) there is a delay in the detection of the close obstacle as the flies tend to stay on the remote wall already detected with the previous images. When introducing proprioceptive fusion (Figure 2.20), updating the flies' positions allows faster convergence and a better detection of the closest obstacle.

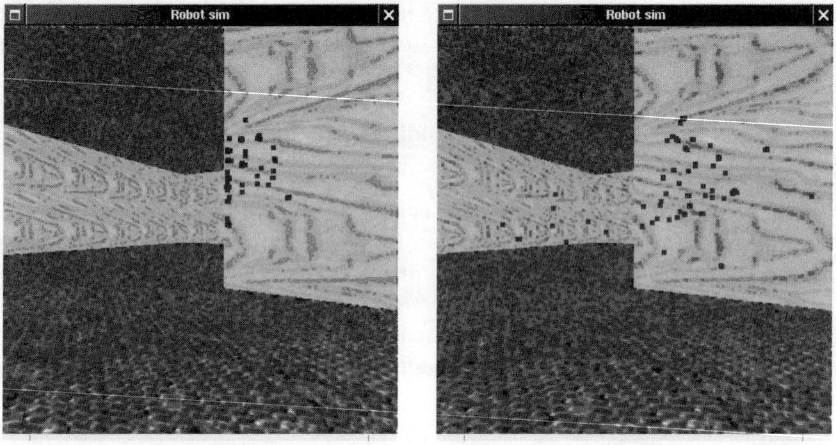

Figure 2.19. *Images N and N+5, without proprioceptive fusion*

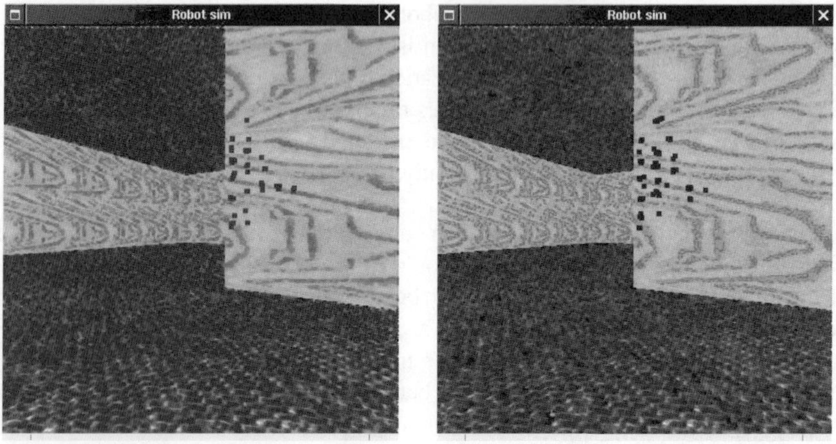

Figure 2.20. *Images N an N+5, with proprioceptive fusion.*
The flies become stabilized on close obstacles

2.6.5. *Artificial evolution and real time*

It is a common belief that artificial evolution, in spite of its robustness, suffers from poor speed and performance. It would be possible to analyze the reasons for this reputation, but let us re-examine this question in light of the fly algorithm.

The first important characteristic of the fly algorithm is that rather than reading pixels sequentially as in most conventional image processing algorithms, the pixel reading sequence is a random sequence of neighborhood readings. It is interesting to note that the CMOS cameras now commercially available, allow an asynchronous, random pixel reading from the photosensitive detectors[6] and seem to be well adapted to the execution of algorithms similar to the fly algorithm in embedded applications. In particular, it should be possible to get rid of the otherwise incompressible delay before the next synchronization signal, as image data are permanently available and updated by the incoming photon flux.

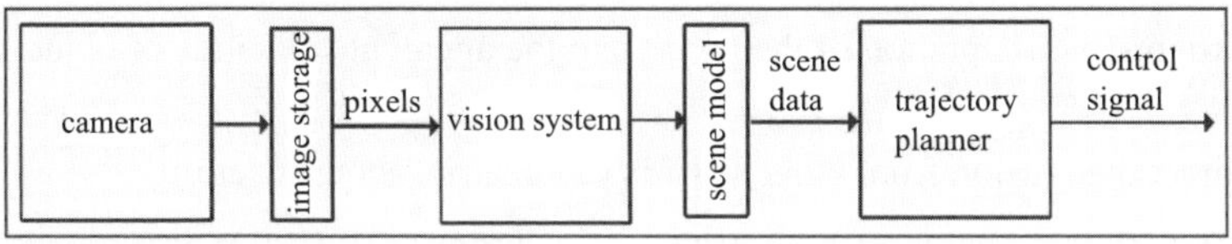

Figure 2.21. *Classical architecture of a vision-trajectory planning system. If T is the image renewal period, a classical vision algorithm synchronized to the camera will use input data with a delay up to 2T from the real world events, and (independently of its speed) the vision's output is a scene model which has to be completed before the planner is able to use it to calculate control signals. The resulting delay is therefore at least equal to 3T*

Second, unlike in the classical scene analysis approaches (see Figure 2.21), where a conventional vision algorithm would first end its work and then give the results to the trajectory planner (ideally, exactly at the new frame synchro which enables it to process the next frame[7]), here the fly algorithm permanently maintains its results file which the planner may read whenever it needs them, without having to wait until a new synchronization signal. This is a typical application of the concept of "anytime algorithms".

6. This is generally done by transferring blocks of image data corresponding to pixel neighborhoods, typically 8×8 pixel blocks. Sadly, most commercial camera-computer interfaces do not allow this random access mode, probably due to the absence of algorithms needing it, due to the absence of camera interfaces allowing it.

7. This delicate problem of synchronization may still be worsened by the conventional decomposition of the vision module itself into low-level (filtering and primary segmentation) and high-level processes (3D vision and scene analysis).

Moreover, the convergence of the fly algorithm is progressive, which means the results of the calculation are usable at any stage of their development. The calculation results are not lost when the scene undergoes a slight modification or upon the delivery of a new frame. This property is even enhanced by the use of proprioceptive data into the quasi-continuous updating of the flies' positions.

For all the reasons above, the Parisian approach in image analysis should enable us to be freed of the usual synchronization constraints, which otherwise result in delays at best equal to three periods of image acquisition. Here, image processing is done through requests to the camera and requires no digital filtering or image segmentation. Similarly, the navigation system associated with the fly algorithm uses asynchronous requests to the vision system and thus exploits the anytime properties of the algorithm. The benefits that can be anticipated are:

– fast execution without any delay resulting from a clocked vision system;

– programming flexibility in the framework of a given algorithmic architecture;

– progressive adaptation of the process to the apparent velocities of objects in the scene;

– direct data fusion from other sensors (odometric, acoustic, etc.);

– optimal exploitation of the intrinsic asynchronism of CMOS imagers, which is not achievable with classical "data flow" algorithms.

It seems therefore that artificial evolution algorithms may provide convenient and powerful solutions to some real-time applications:

– Evolutionary algorithms can accept modifications of the fitness function even while the algorithm is running. This is precisely what "real time" refers to: the ability of a system to exploit data as soon as they are available, and permanently keep fresh, updated results at the user's disposal, in a quasi-continuous way.

– Conventional image processing and stereovision algorithms, based on image segmentation, need a complete image scan before they are able to begin to sequentially process the image and eventually deliver a result. Here, segmentation-free image processing is compliant with continuous input data refreshing during its execution and always uses fresh data. This is particularly interesting with CMOS cameras which allow asynchronous access to pixels.

– The results of the fly algorithm are continuously updated (at the fast pace of generations), which allows the user (which can be e.g. a robot trajectory planner) to react faster to new events.

– Usually with evolutionary algorithms, the heaviest part of calculation is the calculation of the fitness function. Here, the Parisian approach allows us to split up the representation into many extremely simple primitives, which results in a simple fitness function and a globally fast algorithm.

The usual way to write an image processing algorithm is to custom design an arrangement of existing operators, in a specific way to the application. Here, the algorithm's structure is essentially independent from the application, most of the problem-specific knowledge is concentrated in its representation in the fitness function; apart from the fitness function, there is not much to modify if the algorithm code has to be transferred to another image processing application. Rather interestingly in our case, while general metaknowledge is classically built into the evolutionary resolution engine and on the other side, the image processing-specific knowledge is written into a fitness function, some of the intermediate, "robotics" knowledge about proprioceptive sensor fusion or *a priori* knowledge about the environment[8] is coded at a third place, into the genetic operators. This can be summarized in the following table.

	CCD + conventional algorithm	CMOS + Parisian approach
Image sensor	Delay between acquisition and restitution.	Asynchronous pixel access.
Image processing (input)	Segmentation requires a complete image and must wait until the end of the next cycle.	Reads pixels whenever it is needed.
Image processing (output)	Cannot deliver any result before end of cycle.	Results available at any time.
Planner	Must wait until end of the current image processing cycle.	Saves two acquisition cycles.

Table 2.1. *Comparison of the classical vision and the Parisian approach*

2.6.6. *Conclusion about the fly algorithm*

The principles of the Parisian approach described in the beginning of this chapter allowed us to build an evolutionary strategy able to provide a rough 3D description of a scene from stereo image pairs. Unlike conventional approaches to stereovision, no preliminary segmentation is required, and the precision of output results continuously improves, which is of interest to roboticians as it provides a welcome flexibility in the trade-off between speed and accuracy.

In Chapter 1, we showed how the Hough transform, where each pixel or image segment votes for a subspace of a parameter space, could be revisited in the light of evolutionary computation. Here, it is an evolving population that explores this

8. We showed that the optimal crossover rate essentially depends on the nature of the physical environment and is higher in the presence of manmade objects with planar surfaces.

parameter space, each individual testing a pixel-level predicate. There is no obvious, general rule to say which approach would be the most efficient in all cases, but in the example of flies where each individual is represented by a chromosome with only three parameters, a vote approach would obviously be way more costly. The main advantages over the vote approach are:

– fast processing[9] thanks to the non-exhaustive search;

– no preliminary segmentation required – this allows us to use the algorithm on poorly structured scenes;

– progressive accumulation of knowledge about the scene, allowing exploitation of results at any stage and an open trade-off between speed and precision, without any intervention on the algorithm;

– real-time compliance as the fitness function that contains all the useful information about the scene can be updated at any time while the algorithm is running.

2.7. Conclusion

Thanks to its ability to speed up optimization by several orders of magnitude, the Parisian approach opens the door to evolutionary computation into domains already considered as solved using more conventional methods, as we saw it in the example of obstacle detection.

Moreover, the Parisian approach may be a first step to echo the presence of polyploid chromosomes in natural life: these composite chromosomes might allow greater variety in the possible reconfigurations, but also allow us to decompose problems into subproblems that are easier to solve, in the way the Parisian approach does.

In the case of image processing, the radically innovative nature of the evolutionary approach allows us to anticipate several interesting research avenues: extending the traditional application field of traditional parametric approaches as we saw in Chapter 1 with the Hough transform and its extensions, or through the introduction of a frankly different programming style in image processing. The intrinsically asynchronous character of artificial evolution opens up possibilities of real-time applications which have been little explored as yet. Another interesting aspect of evolutionary image processing is the explicit reference to a model, in a way reminiscent of knowledge-based systems. This allows us to override some of the traditional image processing gear (segmentation operators) and open the way to new applications and new techniques using more general (and sometimes more efficient) tools.

9. Processing time depends on the population size rather than the image size.

In addition to this, we could address other applications in computer vision using evolutionary approaches. In particular, identifying mechanical (physically-based) animation models from image sequences, an operation reminiscent of reverse engineering, is a highly complex problem which could only be solved using multi-objective evolution strategies. These techniques allowed us to infer the mechanical internal parameters of animated objects from image data only: passive mechanical structures [LOU 95, LOU 96a], turbulent fluid flows [LOU 96b] and motion-controlled mechanical structures ("muscle models") [STA 02, STA 03]. Now, the asset is no so much to find some (marginal or not) improvement to the performance of computer vision algorithms, but actually to question the limits of the semantic domain of what is usually called "image analysis" or "computer vision".

2.8. Acknowledgements

We express our special thanks to Amine Boumaza, Baudoin Coppieters de Gibson, Anders Ekman, Jean-Loup Florens, Maud Guyon, Marie-Jeanne Lesot, Annie Luciani, Evelyne Lutton, Olivier Pauplin, Marc Schoenauer and Bogdan Stanciulescu, who contributed the ideas, algorithms and results described and quoted in this chapter.

2.9. Bibliography

[BÄC 97] BÄCK T., HAMMEL U. and SCHWEFEL H.-P., "Evolutionary computation: comments on the history and current state", *Transactions on Evolutionary Computation*, vol. 1, no. 1, pp. 3–17, 1997.

[BAR 85] BARNSLEY M. and DEMKO S., "Iterated function system and the global construction of fractals", *Proceedings of the Royal Society*, vol. A-399, pp. 243–245, 1985.

[BOU 01] BOUMAZA A. and LOUCHET J., "Using real-time evolution in robotics", *EVOIASP2001, Applications of Evolutionary Computing*, Como, Italy, vol. Springer LNCS 2037, pp. 288–297, 2001.

[COL 99] COLLET P., LUTTON E., RAYNAL F. and SCHOENAUER M., "Individual GP: an alternative viewpoint for the resolution of complex problems", in BANZHAF W., DAIDA J., EIBEN A.E., GARZON M.H., HONAVAR V., JAKIELA M. and SMITH R.E. (Eds.), *Proceedings of the Genetic and Evolutionary Computation Conference (GECCO-1999)*, Morgan Kaufmann, pp. 974–981, 1999.

[COL 00] COLLET P., LUTTON E., RAYNAL F. and SCHOENAUER M., "Polar IFS + Parisian Genetic Programming = Efficient IFS inverse problem solving", *Genetic Programming and Evolvable Machines*, vol. 1, pp. 339–361, 2000.

[EBE 95] EBERHART R.C. and KENNEDY J., "A new optimiser using particles swarm theory", *Proc. 6th Int. Symposium on Micro Machine and Human Science*, Nagoya, Japon, IEEE service Centre, Piscataway, NJ, pp. 39–43, 1995.

[GOE 94] GOERTZEL B., "Fractal image compression with genetic algorithms", *Complexity International*, vol. 1, 1994.

[HAR 85] HARDIN D.P., Hyperbolic iterated function systems and applications, PhD Thesis, 1985.

[HOL 78] HOLLAND J. and REITMAN J., "Cognitive systems based on adaptive algorithms", *Pattern-Directed Inference Systems*, Academic Press, 1978.

[HUT 81] HUTCHINSON J., "Fractals and self-similarity", *Indiana University Journal of Mathematics*, vol. 30, no. 5, pp. 713–747, 1981.

[KOR 91] KOREN Y. and BORENSTEIN J., "Potential field methods and their inherent limitations for mobile robot navigation", *Proc. of the IEEE Conf. on Robotics and Automation ICRA91*, pp. 1398–1404, April 1991.

[LOU 95] LOUCHET J., PROVOT X. and CROCHEMORE D., "Evolutionary identification of cloth animation models", *Computer Animation and Simulation '95, Proc. of the Eurographics Workshop*, Maastricht, Netherlands, Springer, pp. 44–54, September 1995.

[LOU 96a] LOUCHET J., BOCCARA M., CROCHEMORE D. and PROVOT X., "Building new tools for synthetic image animation using evolutionary techniques", *Evolution Artificielle 95*, Brest, France, Springer, September 1996.

[LOU 96b] LOUCHET J. and JIANG L., "An identification tool to build physical models for virtual reality", *Proc. IWISP96 International Workshop on Image and Signal Processing*, Manchester, UK, Elsevier, pp. 669–672, 1996.

[LOU 02] LOUCHET J., GUYON M., LESOT M.-J. and BOUMAZA A., "L'algorithme des mouches: apprendre une forme par évolution artificielle. Application en vision robotique", *Extraction des Connaissances et Apprentissage*, Hermes, January 2002.

[LUT 95] LUTTON E., VÉHEL J.L., CRETIN G., GLEVAREC P. and ROLL C., "Mixed IFS: resolution of the inverse problem using genetic programming", *Complex Systems*, vol. 9, pp. 375–398, 1995.

[MIL 96] MILLER B.L. and SHAW M.J., "Genetic algorithms with dynamic niche sharing for multimodal function optimization", *International Conference on Evolutionary Computation*, pp. 786–791, 1996.

[SAL 97] SALOMON R. and EGGENBERGER P., "Adaptation on the evolutionary time scale: a working hypothesis and basic experiments", *EA'97*, Nîmes, France, vol. Springer LNCS 1363, pp. 251–262, 1997.

[SMI 80] SMITH S.F., A learning system based on genetic adaptive algorithms, PhD Thesis, University of Pittsburgh, 1980.

[STA 02] STANCIULESCU B., Identification de modèles physiques et de contrôleurs en animation, PhD Thesis, INPG, June 2002.

[STA 03] STANCIULESCU B., FLORENS J.-L., LUCIANI A. and LOUCHET J., "Physical modeling framework for robotics applications", *IEEE Systems, Man and Cybernetics Conference*, Washington DC, October 2003.

[TOU 04] TOURRETTE Y., Sélection et analyse de remaniements chromosomiques chez Saccharomyces cerevisiae en contexte diploïde: origine des délétions et des translocations réciproques, PhD Thesis, Louis Pasteur University, 2004.

[VRS 91] VRSCAY E.R., "Moment and collage methods for the inverse problem of fractal construction with iterated function systems", in PEITGEN H.-O., HENRIQUES J.M. and PENEDO L.F. (Eds.), *Fractal 90 Conference, Fractals in the Fundamental and Applied Sciences*, North Holland, pp. 271–289, 1991.

[WAL 06] WALKER J., MILLER J. and CAVILL R., "A multi-chromosome approach to standard and embedded cartesian genetic programming", *Gecco'06*, Seattle, 2006.

[ZEL 98] ZELEK J.S., "Complete real-time path planning during sensor-based discovery", *IEEE RSJ Int. Conf. on Intelligent Robots and Systems*, 1998.

Chapter 3

Wavelets and Fractals for Signal and Image Analysis

3.1. Introduction

The determination of singularities and of scaling laws by multifractal analysis (which uses multiresolution techniques based on the concept of wavelets) can be found in more and more applications in the fields of natural science, engineering and economics.

In this chapter we highlight the different multifractal methods by analyzing 1D and 2D generic signals; the self-similarity of fractals using wavelet transform modulus maxima has allowed scientists to determine the distribution of the singularities of the very different complex signals which can be found in the domains of material physics, biology and medicine.

The different themes in the chapter have been chosen for their relevance in relation to the further improvements and advances in scientific know-how as well as for the improvement in the quality of results that have been obtained during scientific experiments. The aim of including these themes is to help the reader better understand and apply certain purely mathematical theories which are normally difficult to carry out.

Chapter written by Abdeldjalil OUAHABI and Djedjiga AIT AOUIT.

3.2. Some general points on fractals

They exist everywhere around us. These luminous, unusual, beautiful shapes are known as fractals.

3.2.1. *Fractals and paradox*

It is not easy to give a correct definition of fractals. Nevertheless, in terms of etymology the word fractal leads us to the idea of *fractus* meaning irregular or broken shape [MAN 82].

The analysis of unusual mathematical objects such as curves of infinite length, which possess a finite area, or non-derivable continuous functions, has considerably developed thanks to the work carried out by Mandelbrot, the founder of fractal geometry.

We can see fractal geometry around us in our everyday lives, in nature, in biology or in physics. Generally speaking, the term fractal is used for mathematical objects whose shape and complexity are governed by the inherent omnipresence of irregularities.

The scale invariance and the self-similarity of an object or signal must be taken into consideration in order to facilitate the exploration of fractals. An object is said to be scale-invariant when it is left unchanged following expansion; it is thus symmetric for this transformation process. Consequently, this property means that the change in the observation scale does not change the statistics which have been calculated from the signal. In other words, the general shape of the object remains the same regardless of the change in scale.

Unlike a Euclidean geometric figure, a fractal does not possess any scale or any characteristic of size. Each portion of the fractal reproduces the general shape of the signal regardless of the scale enlargement: this is the property known as self-similarity. This idea means that the information which comes from observation of the fractal is independent of the resolution at which the fractal's measurement is taken.

Fractal analysis was borne from the need to have a tool which could be adapted to the study of complex and irregular natural or artificial phenomena.

For the most part, the recent success of (multi)fractal analysis in the optimization of signal and image processing does not stem from the fact that the signals which are studied are fractal. In reality, and with some rare exceptions, the signals do not possess any self-similarity nor any attributes which are normally associated with fractal objects (except the idea of irregularity on all scales). The development of the methods used in fractal analysis enables scientists to describe the structure of the singularity of complex structures in fine detail. The development of these methods has also enabled scientists to study models which are the result of fractal analysis and which have led to significant progress in areas such as turbulence, growth models, finance, vibration phenomena, material rupture, biomedical signals, satellite images, earthquakes, etc.

3.2.2. *Fractal sets and self-similarity*

A set $A \subset R^n$ is said to be self-similar if the combination of unrelated sub-sets $A_1,\ldots A_k$ can be obtained from A by expansion, translation and rotation. This notion of self-similarity often implies an infinite multiplication of points, which leads to the creation of irregular structures. The Von Koch curve and Cantor's triadic set are simple examples of these irregular structures.

DEFINITION 3.1.– *Let f be a signal with compact support A. A signal f is self-similar if unrelated sub-sets $A_1,\ldots A_k$ exist, so that the restriction of the graph of f to each point A_i is an affine transformation of f. Thus a scale of $s_i > 1$, a translation u_i, a weight p_i and a constant c_i exist, so that*

$$\forall t \in A_i \quad , \quad f(t) = c_i + p_i f\left(s_i(t - u_i)\right)$$

It is assumed that *f* is constant when it is outside these sets.

Figures 3.2 and 3.4 illustrate that if f is self-similar then its wavelet transformation is also self-similar.

3.2.2.1. *The Von Koch snowflake*

The Von Koch curve was published in an article in 1904 which was entitled: "On a continuous curve without tangents, constructible from elementary geometry", [VON 04]. The Von Koch curve is a fractal set which is obtained by recursively dividing a segment of length l into four segments of length $l/3$ as can be seen in Figure 3.1. The length of each subdivision is then multiplied by 4/3. The limit of this process of subdivision thus results in a curve with infinite length.

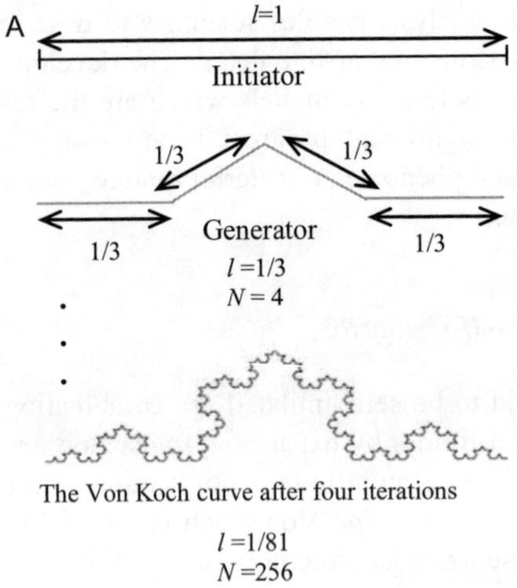

Figure 3.1. *Some iterations of a subdivision of Von Koch curve. The Von Koch curve is a fractal which is the result of an infinite number of subdivisions*

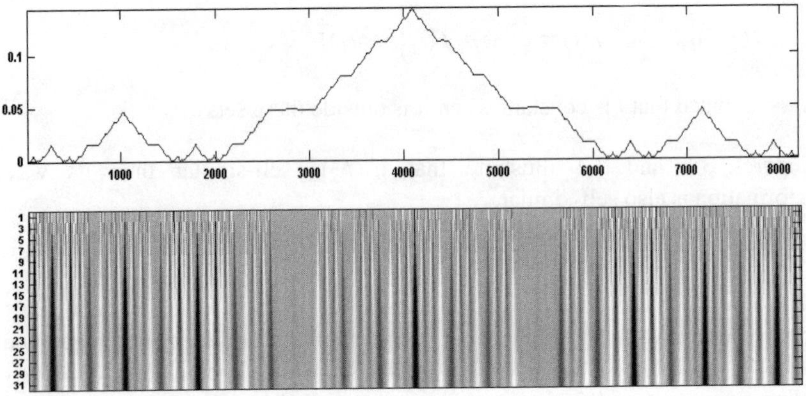

Figure 3.2. *The analysis of the Von Koch curve by wavelet transformation with* $\psi = -\theta'$ *where* θ *is a Gaussian function shows self-similarity*

Wavelet analysis detects the singularities of the signal and shows the initial pattern of the signal on a fine scale. This initial pattern is identically reproduced over and over again.

3.2.2.2. *Cantor's set*

The triadic set, introduced by Georg Cantor in 1884 [CAN 84], is created by recursively dividing intervals of size *l* into two intervals *l*/3 and a central hole, as can be seen in Figure 3.3. The triadic set which is the result of the limit of the subdivisions of the intervals is a set of points known as Cantor's set.

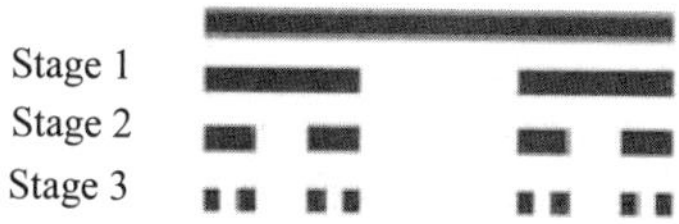

Figure 3.3. *Three iterations using Cantor's set*

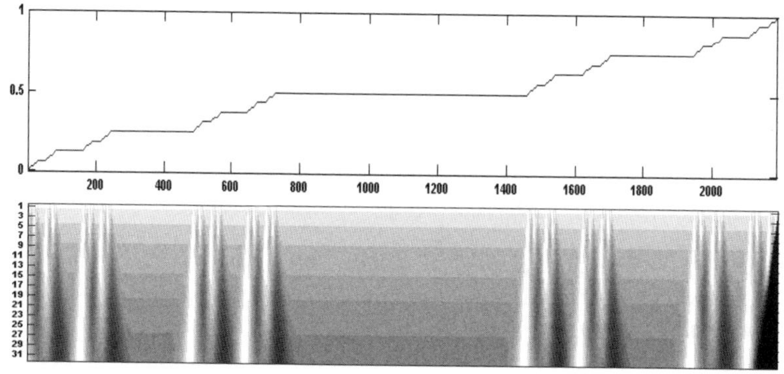

Figure 3.4. *The analysis of the Cantor set signal by wavelet transformation with* $\psi = -\theta'$ *where* θ *is a Gaussian function shows self-similarity*

3.2.3. *Fractal dimension*

Determining the dimension is to establish the relationship between the way in which an object uses its space, and its scale variation.

We tend to attribute whole dimensions to Euclidean geometric shapes: straight lines, circles, cubes, etc.

Fractals possess complex topological properties, for example the Von Koch snowflake, which is of infinite length but which resides in a finitely sized square. Faced with these difficulties, Hausdorff came up with a new definition of dimension, which was based on the variations of the size of the sets during the change in scale.

This idea became the basis of his fundamental work in 1919 [HAU 19], which was developed further in 1935 by Besicovitch [BES 35].

3.2.3.1. *Some definitions*

The fractal dimension (sometimes called the capacity dimension) can be defined as follows.

DEFINITION 3.2.– Let *A be a bounded part of* $\mathbb{R}^n$ *and N(s) the minimum number of balls of radius s which are necessary to cover A, then the fractal dimension can be defined as follows:* $d = \liminf_{s \to 0} \frac{-\log N(s)}{\log s}$.

The measurement of A is $M = \limsup_{s \to 0} N(s) s^d$. *The measurement of A can be either finite or infinite.*

With this formula it is possible to identify the idea of the box-counting dimension which is very useful in certain applications (see Figure 3.6).

Another definition which can be attributed to Hausdorff-Besicovitch gives a new perspective on the Hausdorff-Besicovitch dimension

DEFINITION 3.3.– *The Hausdorff-Besicovitch dimension* $d_{H\text{-}B}$ *is defined as the logarithmic quotient of the number of internal homothetic transformations N of an object, by the inverse proportion of this homothety (1/s):*

$$d_{H-B} = \frac{\log(N)}{\log(1/s)}$$

EXAMPLES 3.1

– For a point:

$$d_{H-B} = \frac{\log(1)}{\log(n)} = 0, \text{with } n \text{ natural} > 1$$

– For a segment, it is possible to establish two internal homothetic transformations of ratio 1:2:

$$d_{H-B} = \frac{\log(2)}{\log\left(\frac{1}{1/2}\right)} = \frac{\log(2)}{\log(2)} = 1$$

– For the Von Koch snowflake:

At each iteration and from each side, four new similar sides are generated in a homothetic ratio of 1:3.

$$d_{H-B} = \frac{\log(4)}{\log\left(\frac{1}{1/3}\right)} = \frac{\log(4)}{\log(3)} = 1.2619$$

– For Cantor's triadic set

At each iteration the current intervals are divided into two smaller intervals and a central hole in a homothetic ratio of 1:3.

$$d_{H-B} = \frac{\log(2)}{\log\left(\frac{1}{1/3}\right)} = \frac{\log(2)}{\log(3)} = 0.6309\,.$$

– For the following fractal curve:

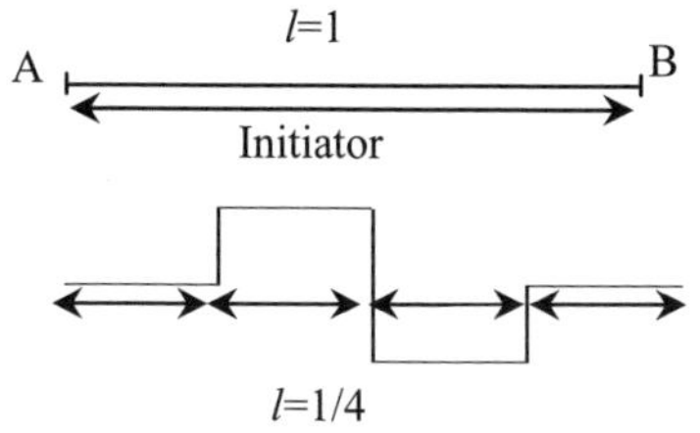

Generator:
- homothetic ratio: r = 1:4
- 8 affine transformations applied
to the new segment => N= 8 segments

$$d_{H-B} = \frac{\log(8)}{\log(\frac{1}{1/4})} = 1.5$$

Figure 3.5. *Calculation of the Hausdorff-Besicovitch dimension from any given fractal*

NOTE 3.1.– The fractal dimension of the Von Koch curve is not equal to its unit size as is the case for all traditional linear geometric shapes. In the case of Cantor's set, the limit of its subdivisions is a set of points in an interval [0,1]. Therefore, the Euclidean dimension of a point is zero whilst the fractal dimension of this set is 0.63.

As a consequence, the fractal dimension is greater than the topological or Euclidean dimension.

EXAMPLE 3.2.– The following application gives some insight into the principal use of the box-counting method by using linear regression to calculate the fractal dimension of a curve which represents a given physical phenomenon:

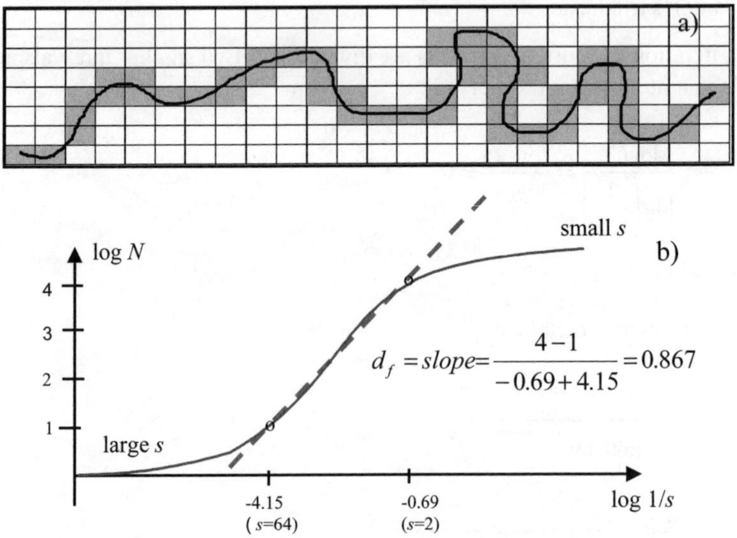

Figure 3.6. *Calculation of the fractal dimension by linear regression. (a) Paving followed by the covering of the function by N boxes of size s; (b) evolution of log (N) vs log (1/s)*

EXAMPLE 3.3.– Estimation of the fractal dimension from rupture profile using the Fraclab toolbox.

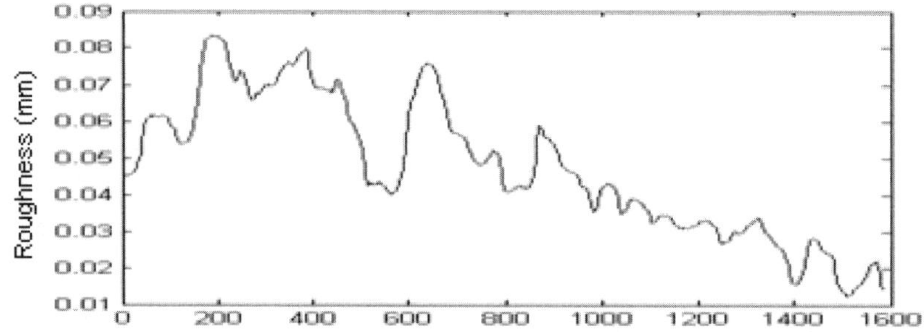

Figure 3.7. *Rupture profile*[1]

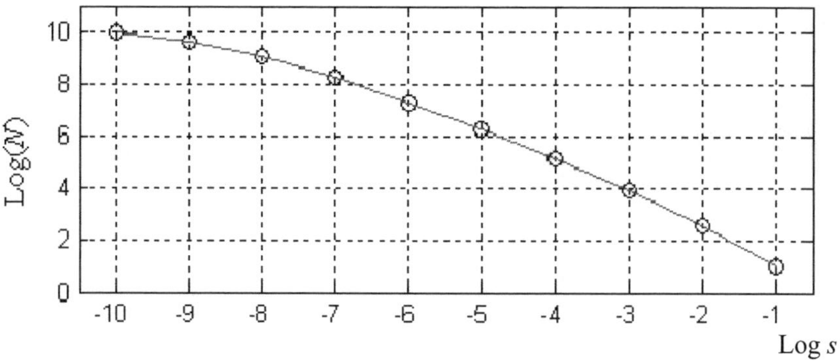

Figure 3.8. *Estimation of the fractal dimension using box-counting method* $d_f = 1.0082$

Some software for calculating the fractal dimension:

1. Matlab FRACLAB toolbox (free);
2. FDC (Fractal Dimension Calculator);
3. HarFA (Harmonic and Fractal image Analyzer);
4. BENOIT Fractal Analysis System;
5. FRACTALYSE.

1. After a number of stress cycles due to strains, a material ends up breaking and two rupture surfaces (or rupture faces) are recovered.

3.3. Multifractal analysis of signals

The objective of multifractal analysis is to describe and analyze phenomena whose regularity, which is measured by a particular indicator (the Hölder exponent), can vary from one point to another. Thus, multifractal analysis provides both a local and global description of a signal's singularity; a local description is obtained using the Hölder exponent, and the global description is obtained thanks to multifractal spectra. The multifractal spectra geometrically and statistically characterize the distribution of singularities which are present on the signal's support.

3.3.1. *Regularity*

EXAMPLE 3.4.– The Weierstrass function: self-similarity and singularities

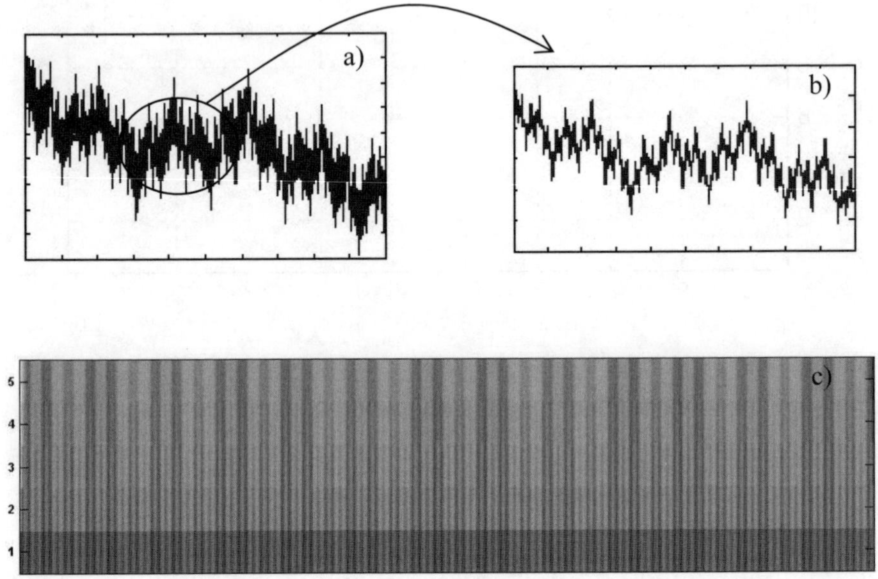

Figure 3.9. *A signal generated by the Weierstrass function where $\alpha = 0.9$ and $\beta = 0.3$: (a) the complete signal; (b) an extraction from a part of this signal; (c) its WT modulus*

The signal $f(t) = \sum_{k=1}^{10} \alpha^k \cos(2\pi\beta^k t)$, in which α, and β are real numbers and β is odd (under the condition $0 < \alpha < 1 < \alpha\beta$), is a continuous signal which is non-derivable. Its Hölder function is constant. Figure 3.9(b) shows the self-similarity of the signal; the shape of the signal remains the same regardless of the scale of the representation used. Figure 3.9(c) represents the wavelet transform modulus of *f(t)* which is calculated from a wavelet which has been derived from a Gaussian style wavelet: in other words confirming this notion of self-similarity on all scales and also confirming the detection of singularities (presence of impulses).

3.3.1.1. *The roughness exponent (Hurst exponent)*

Rough surfaces, which can be the result of a rupture, may lead to invariance through anisotropic transformation (self-affinity). In this case the surfaces are described by a local roughness exponent or by the Hurst exponent which verifies that: $\forall$ $\mathrm{x} = (x, y) \in \mathrm{R}^2$ in a neighborhood of x_0, $\exists H \in \mathrm{R}$ such that, for all values of $\lambda > 0$:

$$f(x_0 + \lambda x, y_0 + \lambda^{\delta} y) - f(x_0, y_0) \sim \lambda^H [f(x_0 + x, y_0 + y) - f(x_0, y_0)]$$

In the case of a random phenomenon the symbol ~ signifies equality in terms of the laws of probability.

The Hurst exponent always has a value between zero and one and characterizes the fluctuations of height in a given surface. The analyzed surface might possess some properties of self-affinity meaning that it possesses properties of isotropic scale invariance if δ=1 or of anisotropic scale invariance if $\delta \neq 1$.

3.3.1.2. *Local regularity (Hölder exponent)*

Local regularity is introduced with the aim of describing the signals that show signs of local fluctuations, in this case, the Hurst exponent is insufficient when it comes to determining the regularity of the signals. A signal is said to be regular if it can be approached locally by a polynomial.

A signal is said to be a regular Hölder signal $h(x_0) \geq 0$ if there is the presence of a polynomial P_{x_0} of degree $n = \lfloor h(x_0) \rfloor$ and a constant $C > 0$, so that:

$$\left| f(x) - P_{x_0}(x) \right| \leq C \left| x - x_0 \right|^{h(x_0)}$$

If f can be differentiated n times in the neighborhood x_0, then P_{x_0} is equal to the Taylor series of f at x_0.

If $0 \leq \alpha < 1$, then the regular part of $f(x)$ is reduced to $P_{x_0}(x) = f(x_0)$ and therefore $\left| f(x) - f(x_0) \right| \leq C \left| x - x_0 \right|^{h(x_0)}$.

The Hölder exponent provides a measurement of the rugosity of *f(x)*: the closer the value of *f(x)* is to one then the softer its trajectory will be. The closer the value of *f(x)* is to zero, then the more variability the trajectory or the surface will possess. This increased variability corresponds to an increased level of rugosity. Singularities play a fundamental role in the study of signals and they often carry essential and relevant information. In the case of an image, its contours correspond to discontinuities or sudden variations in grayscale. These discontinuities or sudden variations in grayscale are the pieces of information that are recorded by a singularity of scale *h*.

EXAMPLE 3.5.– The calculation of the Hölder exponent of the profile of a rupture surface of an elastomeric material: characteristics of rugosity.

Figure 3.10. *A 3D representation of one of the rupture surfaces of an elastomeric material*

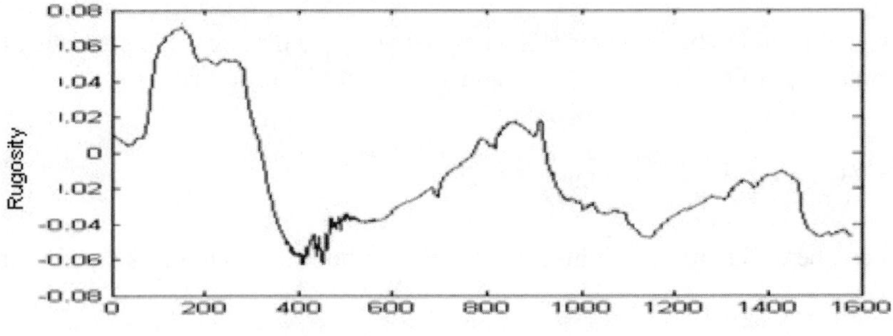

Figure 3.11. *Example of a rupture profile*

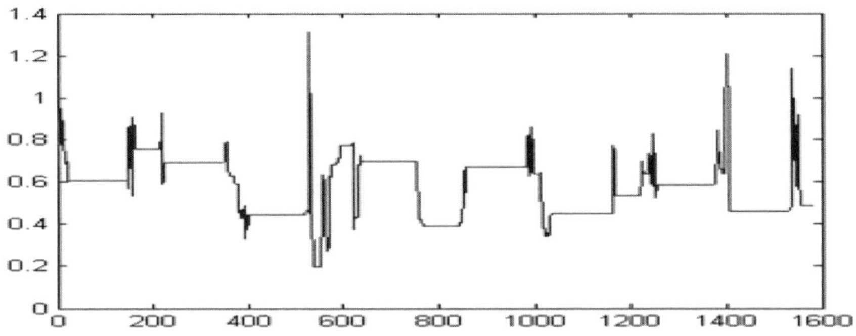

Figure 3.12. *The values of the local Hölder exponent calculated for each point of the profile*

INTERPRETATION 3.1.– Our signal describes a profile (Figure 3.11) which has been extracted from a rupture surface of an elastomeric material that can be seen as a 3D representation in Figure 3.10. The calculation of the local Hölder exponent of this signal was created with the help of the Fraclab toolbox.

Figure 3.12 shows that the majority of the values of the local exponent are between zero and one, with some peak values above one, reaching a maximum of 1.3. The most notable point of this graph peaks at an amplitude of 1.3 for an abscissa of 552. This value of 1.3 is characterized by a sharp increase in h (the Hölder exponent) and is followed by much smaller values of 0.3. The most singular points of the entire profile, i.e. those with the smallest exponent, can be found straight after the maximum value of 1.3. The value of the exponent is approximately 0.25 for an abscissa of between 555 and 570; in other words this means that there is a strong level of rugosity of the rupture's surface. This interval relates to the breaking point of the material and can correspond to a default or inclusion in the surface of the rupture.

Our analysis remains basic and our objective is to trace the causes of the ruptures, for example by carrying out research into the defects which cause the material to break, or by looking at the poor physical and chemical design of the material. This study shows that the evolution of the local Hölder exponent provides clear and relevant information during the propagation analysis of cracks in a given material.

3.3.2. *Multifractal spectrum*

It is important to determine the distribution of the singularities of a multifractal in order to analyze its properties. The singularity spectrum (noted as *D(h)* and defined as the Hausdorff dimension of points where the Hölder exponent carries a particular value (iso-Hölder sets)) provides a geometric description of the singularities and measures the global distribution of the different Hölder exponents.

Mathematically, this can be defined as follows [MAL 98].

DEFINITION 3.4.– *Let A_h be the set of all the points where the punctual regular Hölder signal of a multifractal is worth h. The singularity spectrum D(h) of a multifractal is the fractal dimension (the Hausdorff dimension) of A_h. The support of D(h)is the entire sets of h for which A_h is not empty.*

Consequently, according to the definition of the fractal dimension given in section 3.2, covering the support of a multifractal by unconnected intervals of size *s* gives the number of intervals which intersects A_h:

$$N_h(s) \sim s^{-D(h)}$$

The singularity spectrum thus expresses the proportion of singularities *h* which appear on a given scale *s*. The singularity spectrum of multifractals can easily be measured from the wavelet transform modulus which will be introduced in section 3.4.4.

NOTE 3.2.– A multifractal is said to be homogenous if all of its singularities have the same Hölder exponent $h = \alpha_0$, the support of *D*(*h*) therefore becomes punctual $\{\alpha_0\}$.

The multifractal approach of 2D signals enables scientists to consider light intensity as a measurement of local information held by the image [LEV 00, LEV 97, LEV 94, LEV 92, BER 94].

A Hölder exponent is attributed to each point x on the image $h = h(\mathrm{x})$ according to the following equation:

$$h(\mathrm{x}) = \lim_{\varepsilon \to 0} \frac{\log|f(\mathrm{x}+\varepsilon) - f(\mathrm{x})|}{\log|\varepsilon|}$$

This equation provides information on the local regularity of the image in a neighborhood of *x*. Once this simulation has been carried out on the entire image it is then possible to construct sets of regularities by grouping together the points of the image which have the most similar Hölder exponent values in order to measure the Hausdorff dimension of these sets, noted as d_H. The pair (h, d_H), which is then obtained, can be used as an important piece of information in different applications:

– detection of contours; the contours of an image are identified with the component that is linked to the strongest singularities;

– measurement of the rugosity of a texture;

– denoising.

A large part of the noise (pixels of isolated singularities which do not correspond to any contour) can be eliminated by carrying out threshold techniques which are based on the notion of dimension.

Figure 3.13 shows an example of the multifractal spectrum of a rupture surface image as well as the corresponding contours.

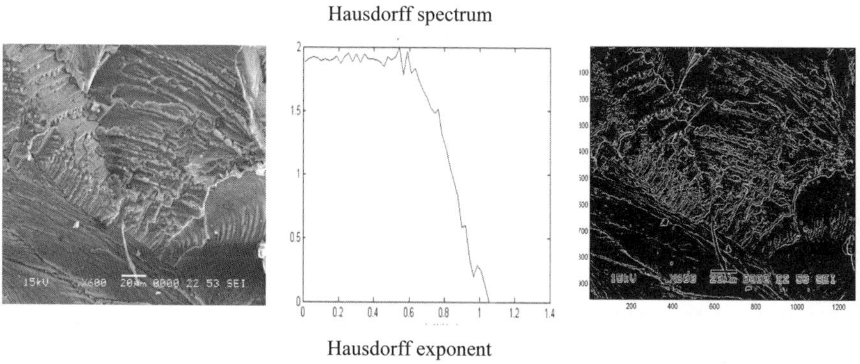

Figure 3.13. *Estimation of the exponents of the singularities of the rupture surface. On the left: the original image of the face. Center: distribution of the values of the image's singularity exponents. On the right: grayscale representation of the singularity exponents: the clearer the pixel, the weaker the exponent of the pixel*

Despite the complexity of the texture of the image, this method ensures the detection of very fine contours.

3.4. Distribution of singularities based on wavelets

3.4.1. *Qualitative approach*

Singularities and irregular structures of a signal carry relevant information. It is very important to detect and characterize the discontinuities in the intensity of pixels which describes an image. This intensity also defines the contours of a scene (Figure 3.13) or the transient of a pathological electro-encephalogram.

In order to characterize singular structures, it is necessary to quantify the regularity of the signal *f(t)*. The Hölder exponent (sometimes also referred to as the Lipchitz exponent) provides the measurements for uniform regularity of intervals as well as for any given point v. If f is singular, i.e. non-derivable, its behavior is described by the Hölder exponent.

The decrease of the wavelet transform modulus according to the respective scale is linked to the global and punctual regularity of the signal. Measuring this asymptotic decrease can be compared to "zooming" into the structures of a signal with a scale that tends towards zero.

3.4.2. *A rough guide to the world of wavelet*

In the case of an image, the "time-scale" or "space-scale" analysis is based on the use of a very extensive range of scales used to analyze a signal. This type of analysis is often referred to as a multiresolution analysis. Multiresolution analysis is based on a variety of different behaviors in terms of the laws that define the scales (e.g. rather large to very fine scales). The different scales are used to "zoom" into the structure of the signal and obtain increasingly precise representations of the signal that is being analyzed. A function $\psi(t)$ or a localized and oscillating wave known as "mother-wavelet" are needed for the operation named above.

The condition of localization implies a rapid decrease when $|t|$ increases indefinitely and the oscillation suggests that $\psi(t)$ vibrates just as a wave does and that $\psi(t)$'s average and its n moments equal zero:

$$\int_{-\infty}^{+\infty} \psi(t)dt = \cdots = \int_{-\infty}^{+\infty} t^k \psi(t)dt = 0 \quad \text{for } 0 \leq k \leq n$$

This property (n vanishing moments) is important in order to analyze the local regularity of a signal. Indeed, the Hölder regularity index increases with increasing the number of vanishing wavelet moments. Furthermore, the wavelet is (a zero average function) centered around zero with a finite energy.

The mother-wavelet $\psi(t)$ – whose scale, based on a convention, is 1 – generates other wavelets $\psi_{u,s}(t)$, $s>0, u \in \mathrm{R}$. This generation of other wavelets is based on the changes of the scale s as well as on temporal translation u:

$$\psi_{u,s}(t) = \frac{1}{\sqrt{s}}\psi\left(\frac{t-u}{s}\right)$$

$\psi_{u,s}(t)$ is therefore centered around u.

The wavelet transform of f, expressed by $Wf(u,s)$, for the scale s and the position u, is calculated by correlating f with the corresponding wavelet:

$$Wf(u,s) = \int_{-\infty}^{+\infty} f(t)\frac{1}{\sqrt{s}}\psi^*\left(\frac{t-u}{s}\right)dt$$

where $\psi^*_{u,s}$ describes the complex conjugate of $\psi_{u,s}$.

In a way, this transformation measures the fluctuations in a signal $f(t)$ around a specific point u and for the scale provided by $s>0$. This type of property has interesting applications and allows for the detection of transients and analysis of fractals. These transients are detected when zooming all along the scales and fractal analysis in determining the distribution of singularities.

It is possible to compare the time-scale analysis (u, s) to a time-frequency representation (u, η/s) where $\psi_{u,s}$ is symbolically represented by rectangles whose dimensions vary according to s. Their surface, however, remains the same (Figure 3.14). This idea therefore also represents the Heisenberg uncertainty principle:

$$\sigma_t \sigma_\omega \geq \frac{1}{2} \quad \text{or} \quad \sigma_t \quad \text{and} \quad \sigma_\omega$$

respectively represents the temporal resolution (or standard deviation) and the frequency resolution.

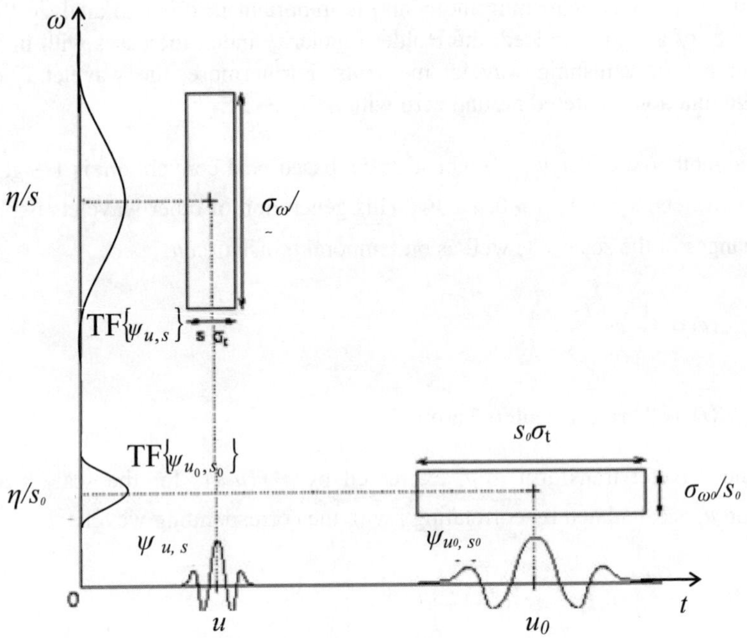

Figure 3.14. *Process of wavelet analysis and the illustration of the Heisenberg uncertainty principle*

Figure 3.14 shows that if s decreases, the frequency support increases and shifts towards higher frequencies. As a result the temporal resolution improves.

Figures 3.15(a) and (b) show some examples of wavelets that are very useful in the analysis of signals.

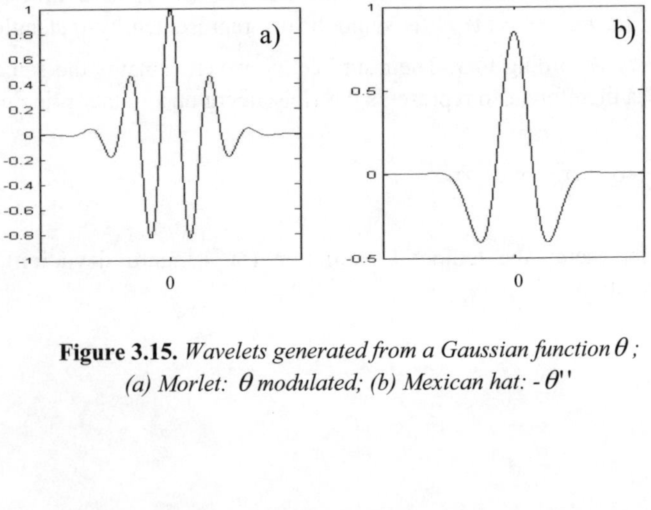

Figure 3.15. *Wavelets generated from a Gaussian function* θ *; (a) Morlet:* θ *modulated; (b) Mexican hat:* $-\theta''$

3.4.3. *Wavelet Transform Modulus Maxima (WTMM) method*

Wavelet ψ is assumed to have n moments equal to zero. This wavelet is also C^n with the derivates of a rapid decrease. This means that for every $0 \leq k \leq n$ and $m \in N$, there is C_m such that:

$$\forall t \in R, \qquad \left|\psi^{(k)}(t)\right| \leq \frac{C_m}{1+|t|^m}$$

Jaffard's necessary and sufficient condition [JAF 91] on wavelet transform, used in order to estimate the Hölder (or Lipchitz) punctual regularity of f at a point v, can be expressed as follows:

if $f \in L^2(R)$ is Hölder[2] $\alpha \leq n$, there exists A such that:

$$\forall (u,s) \in R \times R^+ , \quad |Wf(u,s)| \leq A s^{\alpha + 1/2}\left(1 + \left|\frac{u-v}{s}\right|^{\alpha}\right) \qquad [3.1]$$

Based on what has been discussed above, the Hölder local regularity of f at v depends on the decrease of $|Wf(u,s)|$ at fine scales in the neighborhood of v. The decrease of $|Wf(u,s)|$ can indeed be controlled by the values of its local maxima.

Modulus maximum describes any point (u_0, s_0) such that $|Wf(u_0,s_0)|$ is locally maximum at $u = u_0$ [OUA 02]. This implies:

$$\frac{\partial Wf(u_0,s_0)}{\partial u} = 0$$

NOTE 3.3.– The singularities are detected by searching the abscissa where the wavelet modulus maxima converge at fine scales. Indeed, if ψ has exactly n moments equal to zero and a compact support, there is θ with a compact support such that $\psi = (-1)^n \theta^{(n)}$ with $\int_{-\infty}^{+\infty} \theta(t)dt \neq 0$. The wavelet transform can be expressed

2. In order to conform with the written form used by Jaffard and Mallat [MAL 98], the Hölder (or Lipchitz) exponent will, in the meantime, be expressed as α instead of h.

as a multiscale differential operator of order *n*: $Wf(u,s) = s^n \frac{d^n}{du^n}\left(f * \overline{\theta}_s\right)(u)$ with $\overline{\theta}_s(t) = \frac{1}{\sqrt{s}}\theta\left(-\frac{t}{s}\right)$.

If the wavelet has only one moment which equals zero, wavelet modulus maxima are the maxima of the first order derivate of *f* smoothened by $\overline{\theta}_s$. These multiscale modulus maxima are used for the location of discontinuities as well as when analyzing the contours of an image. If the wavelet has two moments that equal zero, the modulus maxima correspond to large curvatures.

NOTE 3.4.– *A contrario*, if $Wf(u,s)$ does not have a local maximum on the level of fine scales, then *f* is regular in that local area.

NOTE 3.5.– In general nothing guarantees that a modulus maxima is situated at (u_0, s_0) nor that this modulus is part of a line of maxima that propagates finer scales. However, in the case of θ being a Gaussian function, the modulus maxima of $Wf(u,s)$ belong to the related curves that are never interrupted when the scale decreases.

NOTE 3.6.– In the case of an image, the points of the contours are distributed on curves that often correspond to the boundaries of the main structures. The modulus of individual maximum wavelets are linked to form a curve of maxima that follows the outline. For an image, partial derivates of wavelets linked to x_1 and x_2 of a smoothing function θ are also taken into consideration.

$$\psi_1 = \frac{\partial \theta}{\partial x_1} \quad \text{and} \quad \psi_2 = \frac{\partial \theta}{\partial x_2}$$

Function θ is assumed to be localized around $x_1 = x_2 = 0$ and isotropic (does not depend on $|\mathrm{x}|$). The Gaussian function $e^{-(x_1^2 + x_2^2)/2} = e^{-|\mathrm{x}|^2/2}$ and the Mexican hat wavelet $\left(2 - |\mathrm{x}|^2\right)e^{-|\mathrm{x}|^2/2}$ are wavelets that meet these requirements. The corresponding wavelets transform is as follows:

$$Wf(u,s) = \frac{1}{s^2}\nabla\left\{\int f(\mathrm{x})\theta\left(\frac{\mathrm{x} - \mathrm{u}}{s}\right)d^2\mathrm{x}\right\} = \nabla\left\{f * \theta_s(\mathrm{u})\right\}$$

where $*$ describes the operator of the convolution product.

This transformation can equally be expressed by its modulus $|Wf(\mathrm{u},s)|$ and its argument $Arg\{Wf(\mathrm{u},s)\}$.

This signifies that the 2D-wavelet transform defines the gradient field of $f(\mathrm{x})$ smoothened by θ. Note that the gradient $\nabla\{f_* \, \theta_s(\mathrm{u})\}$ indicates the direction of the largest variation of f for a smoothened scale s and that the orthogonal direction is often referred to as the direction of maximum regularity.

Furthermore, in the sense of Canny's detection of the contours, wavelet modulus maxima are defined by the respective points u where $|Wf(\mathrm{u},s)|$ is a local maximum that tends towards the direction of the given gradient described by the angle $Arg\{Wf(\mathrm{u},s)\}$. Its points create a chain that is referred to as the maxima chain. They might also be referred to as a gradient vector that indicates the local direction in which the signal varies the most when compared to the smoothened scale s.

The skeleton of wavelet transformation consists of two lines of maxima that are convergent up to the point of plane (x_1,x_2) in the limits of $s \to 0$. This skeleton carries out the positioning of the space-scale that contains all information concerning the fluctuations of the local regularity of f.

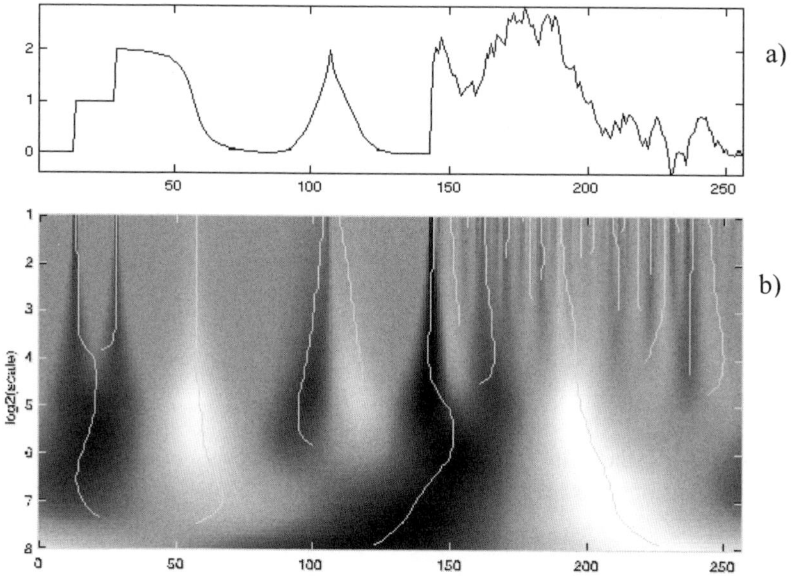

Figure 3.16. *(a) Signal f(t)shows its singularities; (b) the wavelet modulus maxima and the lines of maxima of this modulus*

Figure 3.16 can be accessed on the website http://cas.ensmp.fr/~chaplais/ and shows that the singularities create coefficients of great amplitude in their cone of influence. WTMMs detect the singularities well and allow us to estimate graphically the order of singularity of $f(t)$ at all times in representing the WTMM linked to the function of the scale s. The gradient (based on the hypothesis of linearity) of the curve, e.g. log-log, obtained at a specific point in time, provides an estimation of the Hölder coefficient.

NOTE 3.7.– *Numerical aspects*. It is well known that derivatives of Gaussians are used to guarantee that all maxima lines propagate up to the finest scales. However, the process of chaining maxima must be performed with caution due to the apparition of artefacts in areas where the wavelet transform is close to zero.

Moreover, the finest scale of the wavelet transform is limited by the resolution of data. Then, the sampling period must be sufficiently small so that α is measured precisely.

3.4.4. *Spectrum of singularities and wavelets*

The singularities of multifractals vary from one point to another. It is important to establish the distribution of these singularities to analyze their properties. In practice, the distribution of singularities is estimated by global measurements which use self-similarities of multifractals. In this way, the fractal dimension of the points with the same Hölder regularity is calculated. The function used for this calculation is a function based on global partition calculated on the basis of the WTMM. The self-similarity of the wavelet transform modulus leads to the positions of the values of wavelet transform modulus maxima being equally self-similar.

In practice, the spectrum of singularities written as $D(\alpha)$ or $D(h)$ for multifractals is measured based on the local maxima of the wavelet transformation.

Let ψ be a wavelet with n vanishing moments. If f has pointwise Hölder regularity $\alpha_0 < n$ at v, the wavelet transform $Wf(u,s)$ possesses a set of modulus maxima that are convergent towards v at fine scales.

All maxima at the scale s can be interpreted as a recovery of the singular support of f by wavelets of scale s. For its maxima, the following applies:

$$|Wf(u,s)| \sim s^{(\alpha_0+0.5)}$$

Let $\{u_p(s)\}_{p\in Z}$ represent the positions of all local maxima of $|Wf(u,s)|$ on a fixed scale s, $[\alpha_{\min},\alpha_{\max}]$ is the support of $D(\alpha)$, and ψ a wavelet with $n>\alpha_{\max}$ moments equal to zero. The calculation of the spectrum $D(\alpha)$ for a self-similar signal *f* is carried out as follows:

– calculation of maxima; determine $Wf(u,s)$ and its modulus maxima for each scale s; create chains of the wavelets' maxima across scales;

– calculation of the function of partition (measurement of the sum of power q for these wavelet modulus maxima)

$$Z(q,s)=\sum_p \left|Wf(u_p,s)\right|^q ;$$

– determination of the scaling exponent $\tau(q)$ by linear regression of $\log_2 Z(s,q)$ as function of $\log_2 s$

$$\log_2 Z(s,q) \sim \tau(q)\log_2 s \qquad s\to 0 .$$

Note that the Hölder exponent α and the spectrum of singularities $D(\alpha)$ are conjugated variables of q and $\tau(q)$. This means that $D(\alpha)$ can be obtained by inverting the Legendre transform of $\tau(q)$ under the hypothesis that $D(\alpha)$ is convex;

– determination of the spectrum

$$D(\alpha)=\min_{q\in R}\left(q\left(\alpha+\frac{1}{2}\right)-\tau(q)\right).$$

NOTE 3.8 – The monofractals or the homogenous fractals are characterized by a single Hölder exponent h (or α) = H (the Hurst exponent) of a spectrum τ_q which is linear and whose gradient is given by $h=\dfrac{\partial\tau}{\partial q}=H$. On the other hand, a non-linear behavior of τ_q indicates a multifractal for which the Hölder exponent $h(\mathrm{x})$ is a variable.

3.4.5. *WTMM and some didactic signals*

Highly oscillatory signals

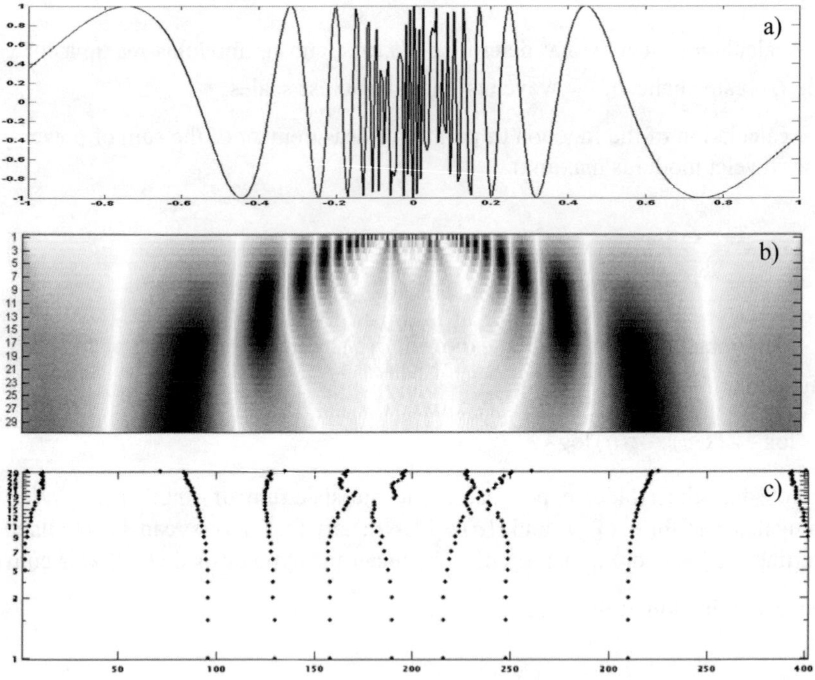

Figure 3.17. *(a) Signal* $f(t) = \sin(a/t)$ *; (b) the modulus of its wavelet transform calculated with* $\psi = -\theta'$ *where* θ *is a Gaussian function; (c) the lines of maxima for this modulus*

Figure 3.17 shows the analysis of a generic signal $f(t) = \sin\left(\frac{a}{t}\right)$ which is highly oscillatory and discontinuous at $t = 0$. If (u, s) is in the cone of influence of 0, condition [3.1] is verified for $\alpha = h = 2$. However, Figure 3.17 shows that there are coefficients for wavelets of high energy outside the cone of influence centered at 0. The instantaneous frequency of f(t) is $-\frac{a}{t^2}$, and if u varies, the set of points (u, s(u)) describes a parabolic curve situated outside the cone of influence centered around 0. The WTTMs confirm that large amplitudes are situated outside the cone of influence.

A signal with a finite number of discontinuities

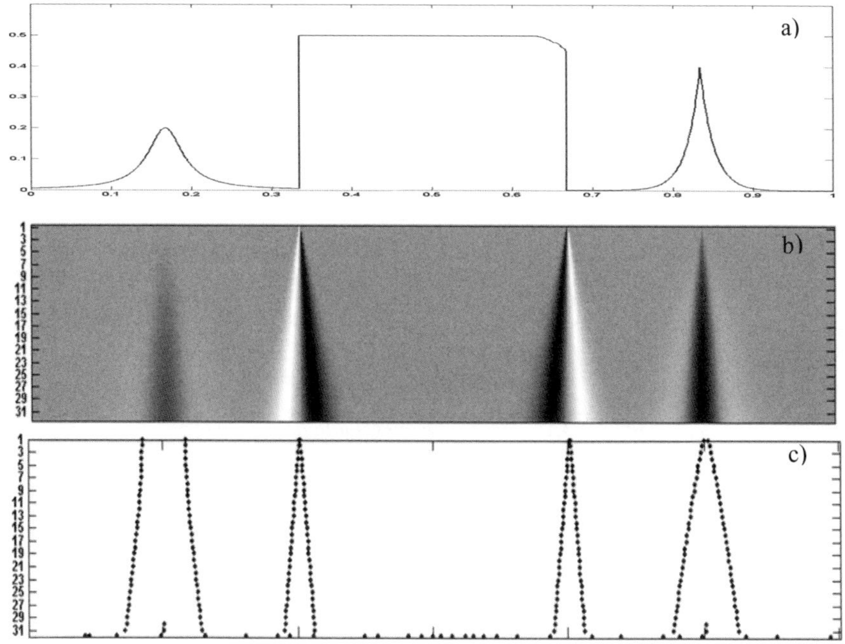

Figure 3.18. *(a) signal f(t) with a finite number of discontinuities; (b) its wavelet transform modulus calculated with* $\psi = -\theta''$ *where* θ *is a Gaussian; (c) the lines of maxima for this modulus*

Figure 3.18 represents the analysis of a generic signal which shows singularities that are marked by discontinuities and non-derivabilities. The WTTMs are calculated with the help of a wavelet which is a second derivative of a Gaussian. The decrease of $|Wf(u,s)|$ described by s is expressed throughout several curves of maxima. These curves correspond to the smoothened and non-smoothened singularities.

3.5. Experiments

3.5.1. *Fractal analysis of structures in images: applications in microbiology*

Fractal analysis can be used in microbiology when determining the number of yeast cells in a digitized image [VES 01].

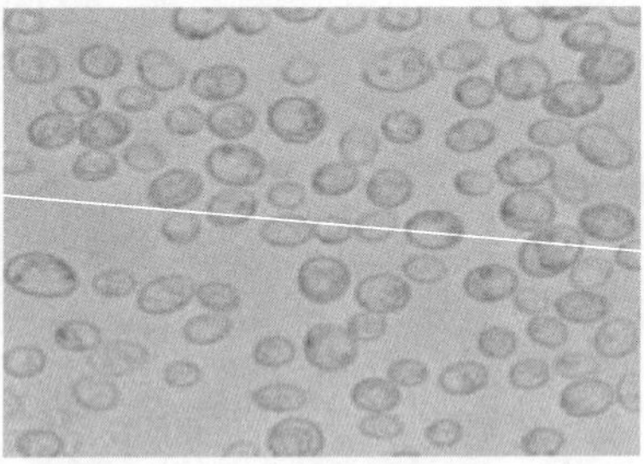

Figure 3.19. *Image of a microbiological sample*

Figure 3.19 comes from an acquisition system that is composed of an SM-6 optical microscope and a digital camera that can be operated with the help of a computer.

The optical level of the microscope and the resolution of the digital camera provide the link between the size of the image and the size of the sample (10 μm/48 pixels) to be studied.

The number of cells can be determined by bearing the following characteristics in mind:

– the cells have a round shape;

– the cells are of a similar size;

– the cells differ from the background in their color and intensity.

The process of determining the number of cells works as follows:

1. Application of a mask to adjust the color: white (W) for the cells and black (B) for the background.

2. Determining the fractal dimension and the fractal measurement (see definition 3.2) of the masked image comprising the interface of the background (K_{WBW}, d_{WBW}) as well as the interface of the cells (K_{BW}, d_{BW}) is done with the help of the following equations that are directly based on the definition of the fractal:

$$N_{BW}(s) = K_{BW} s^{-d_{BW}} \quad N_{WBW}(s) = N_W(s) + N_{BW}(s) = K_{WBW} s^{-d_{WBW}} \qquad [3.2]$$

N_W and N_{BW} respectively represent the number of entirely white boxes as well as the number of partially black boxes.

d_{BW} and d_{WBW} are the fractal dimensions obtained with the help of the box-counting method (see section 3.2.3.1), K_{BW} and K_{WBW} are fractal measurements.

To determine the number of cells n_c as well as their radius r, the following equations are used:

$$N_{BW} = n_c \frac{\pi(2r+\varepsilon)}{s} \sim n_c \frac{2\pi r}{s}, N_{WBW} = n_c \frac{\pi r^2}{s^2} \qquad [3.3]$$

From equations [3.2] and [3.3], the following equation is inferred:

$$n_c = \frac{K_{BW}^2 s_m^{-2d_{BW}}}{4\pi K_{WBW} s_m^{-d_{WBW}}} = \frac{K_{BW}^2}{4\pi K_{WBW}} s_m^{d_{WBW} - 2d_{BW}}$$

s_m is the size of the box corresponding to the maximum fractal dimension. For a radius of r = 38 pixels, the fractal analysis estimates the number of cells which is equal to n_c= 100 cells. This is shown in Figure 3.20.

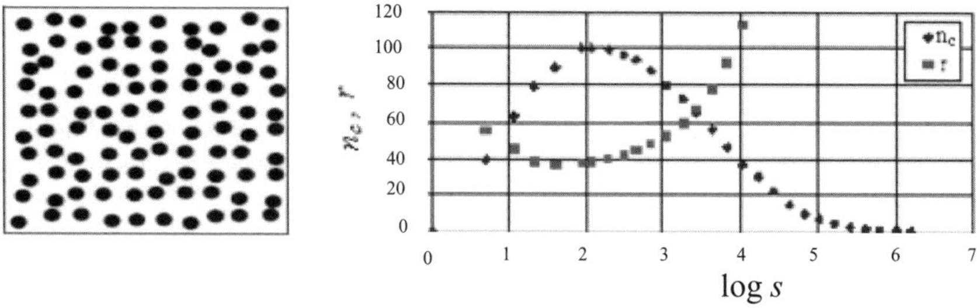

Figure 3.20. *On the left: model of cellular structure, n_c= 100 cells of the radius r = 38 pixels. On the right: Determination of the number of cells n and their radius by fractal analysis (log s = 2 provides n_c = 100 and r = 38 pixels)*

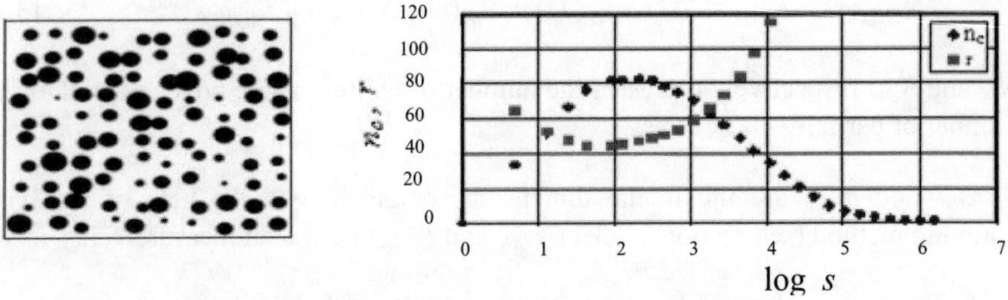

Figure 3.21. *On the left: model of cellular structure* n_c = 100 *cells whose size is distributed according to a Gaussian law on the average radius r* = 38 *pixels of the standard deviation* = 4. *On the right: fractal analysis of the number of cells n and their radius r (log s* = 2 *provides nc* = 82 and *r* = 43 *pixels)*

Note that unless the structures are of a certain size a bias is introduced in the estimation process (Figure 3.21), the number of cells that has been calculated is always lower than the number of cells in reality. The size of the cells is therefore underestimated.

Determining the number and the size of cells with the help of a fractal analysis is therefore only valid if the structures contain cells that are of a similar size.

This experiment shows that fractal analysis is a promising tool when it comes to counting and measuring objects in a digital image.

3.5.2. *Using WTMM for the classification of textures – application in the field of medical imagery*

Numerous works in image processing and pattern recognition are directed to two complex tasks: detection and automated classification of aggregates that might indicate the beginning of breast cancer.

This study, related to medical issues, is based on the application of the WTMM 2D method described in section 3.4.3. The aim of the segmentation of aggregates referred to as accumulation is to dip objects into the rough texture at a specific moment in time, shown in Figure 3.24.

As far as methodology is concerned, this technique is pertinent in discriminating against the groups of singularities that have been sufficiently characterized by the Hölder exponent.

Furthermore, the WTMM method is used to measure the properties of the invariance of the scale of mammography in order to distinguish dense tissue and adipose tissue with the help of the Hurst exponent.

In particular, the invariance of the observed scale is of a monofractal type in which two different groups of tissue are respectively characterized by the Hurst exponent $H=0.3\pm0.05$ and $H=0.65\pm0.05$. This type of result can be interpreted as an adipose feature of nature (fat) or as conjunctive (dense) tissue. The properties of invariance of a scale are associated with two groups of tissue that allow for the segmentation of the image of a breast. This procedure is based on the Hurst exponent of a square size of 256×256 pixels per analysis of around 50 images (see Figure 3.23).

This discrimination of monofractal behavior between adipose and dense tissue is also proven by the calculation of spectra for singularities $D(h)$ (see Figure 3.22).

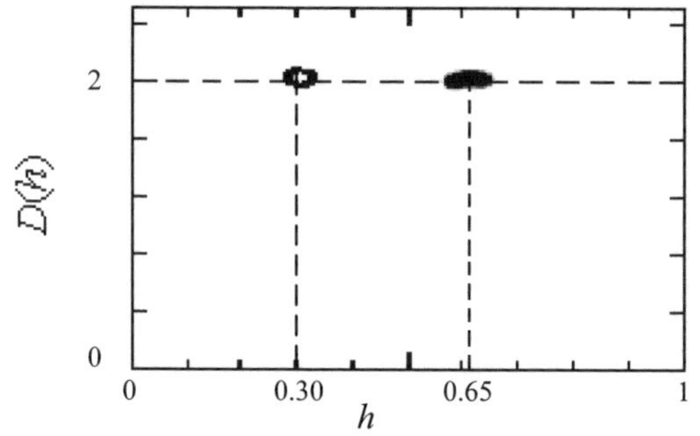

Figure 3.22. *Spectrum of singularities of 49 juxtaposed images (512×512) on the level of scales $21 \leq s \leq 112$*

These spectra are made up of around 50 images and can be reduced to one single point $h=H=0.30$ for adipose tissue and $h=H=0.65$ for dense tissue. This is the proof that the texture represented in the mammogram is of a monofractal nature throughout the entire image. This is why it is possible to distinguish adipose from dense tissue.

For more information and more details on these concepts, please see Kestener's thesis [KES 03] .

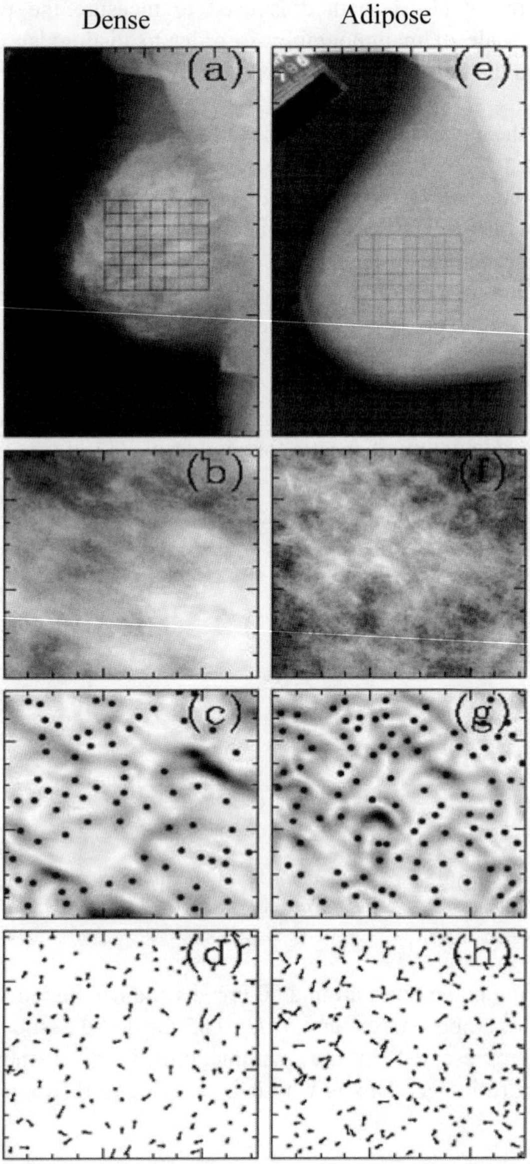

Figure 3.23. *Analysis based on WTMM of two mammograms: (a-d) breast with dense tissue and (e-h) breast with adipose tissue. The wavelet used in the analysis is an isotropic wavelet of the order 1 (derivate of a Gaussian function.) (a) and (e) represent the original mammogram; (b) and (f) represent their respective growth (zoom on the central part:* 256×256 *pixels); (c) and (g) show the WT modulus and the maxima chains at the level of* s = 39 *pixels; (d) and (h) show the chains of local maxima as well as their positions*

Furthermore, this method is used to locate microcalcifications and characterize their disposition in a possible accumulation. In order to do this, the fractal dimension of all objects classed as "microcalcifications" are analyzed.

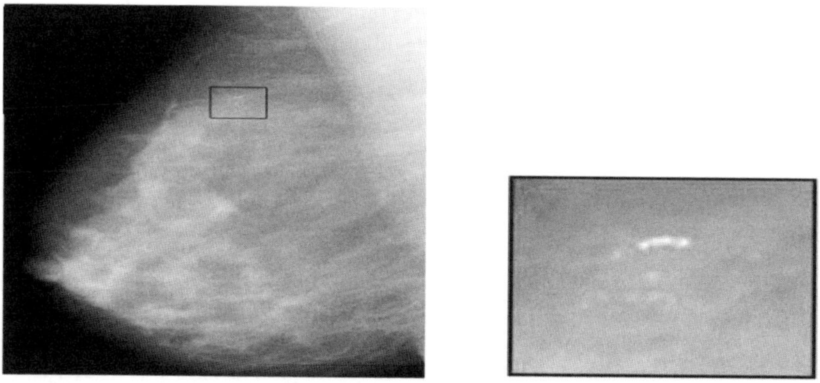

Figure 3.24. On the left*: a mammogram showing microcalcifications. On the right: close-up on the area containing the cluster of microcalcifications*

If the accumulation or cluster is of a linear disposition, the fractal dimension is equal to its Euclidean dimension $d_f = 1$, if a finite surface is filled $d_f = 2$ can be expected. However, if the cluster is of an arborescent structure, the fractal dimension is fractioned $1 < d_f < 2$, which reflects the local complexity of mammary canals as well as the dissemination of the microcalcifications within these canals. An example of detecting microcalcifications in an accumulation or a cluster is shown in Figure 3.25. These images show that it is possible to clearly distinguish the areas of microcalcifications. Consequently, WTMM allows for an analysis and a characterization which still have to prove their effectiveness in clinical tests.

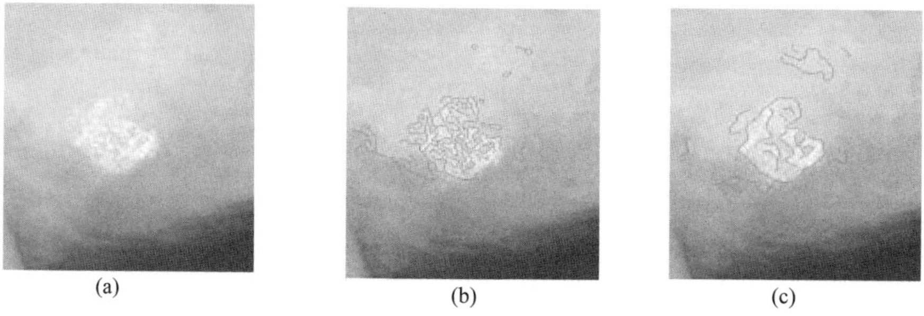

Figure 3.25. *Detection of microcalcifications in a cluster ($d_f = 2 \pm 0.05$). (a) original mammogram. (b) and (c) Chains of maxima that indicate the microcalcifications for the scales s = 14 pixels and s = 24 pixels*

In this study, the WTMM method confirms the interest in a fractal approach and shows that the Hölder exponent (or the Hurst exponent) can be used as an indicator of microcalcifications in the breast (probability of early stages of breast cancer) and, last but not least, this method also establishes the geometric shape of the cluster. This type of information could be used in combination with medical diagnosis and as an automatic system that helps medical staff with the diagnosis.

3.6. Conclusion

This chapter provides a pragmatic approach to the analysis of signals whose regularity varies from one point to another.

The concepts that have been used show multifractal analysis and wavelet analysis. In practice, the distribution of singularities is estimated by global measures using self-similarity or invariance on the scale of multifractals. The criterion of local regularity based on the Hölder exponent can be characterized as a decrease in the wavelet coefficients in the signal that is being analyzed. Furthermore, the spectrum of singularities of the multifractals to be studied is measured by local maxima of wavelet transform.

In the framework of this study, the formalism that was used has been validated in the field of health technology for the detection and characterization of images. Furthermore, the field of mechanical engineering looks at the process in terms of material fatigue.

This promising formalism is currently undergoing a fusion of ideas on its very basic level as well as in the area that aims at developing analysis tools.

3.7. Bibliography

[BEN 00] BENASSI A., COHEN S., DEGUY S., ISTAS J., "Self-similarity and intermittency", in *Wavelets and Time-frequency Signal Analysis*, EPH, Cairo, Egypt, 2000.

[BER 94] BERAN J., *Statistics for Long–Memory Process*, Chapman and Hall, New York, 1994.

[BER 94] BERROIR J., Analyse multifractale d'images. Thesis , Paris University IX, 1994.

[BES 35] BESICOVITCH A.S. , *Mathematische Annalen 110*, 1935.

[CAN 84] CANTOR G., "On the power of perfect sets of points", *Acta Mathematica 2*, 1884.

[GRA 03] GRAZZINI J., Analyse multiéchelle et multifractale d'images météorologiques: application à la détection de zones précipitantes. Thesis, Marne-La-Vallée University, 2003.

[HAU 19] HAUSDORFF F., *Mathematische Annalen 79*, 157, 1919.

[JAF 91] JAFFARD S., "Pointwise smoothness, two-microlocalization and wavelet coefficients", *Publicaciones Matematiques*, vol. 35, p. 155-168, 1991.

[KES 03] KESTENER P., Analyse multifractale 2D et 3D à l'aide de la transformation en ondelettes: application en mammographie et en turbulence développée. Thesis, Bordeaux University I, 2003.

[LEV 00] LEVY VEHEL J., "Analyse fractale: une nouvelle génération d'outils pour le traitement du signal-enjeux, tendances et évolution", *TSI* 19 (1-2-3), pp. 335-350, 2000.

[LEV 97] LEVY VEHEL J., LUTTON E., TRICOT C., *Fractal in Engineering*. Springer Verlag, 1997.

[LEV 94] LEVY VEHEL J., MIGNOT P., "Multifractale segmentation of images", *Fractlas* 2(3), pp. 371-377, 1994.

[LEV 92] LEVY VEHEL J., MIGNOT P., BERROIR J., "Multifractals, texture and image analysis", In: *Proc of Computer Vision and Pattern Recognition, CVPR'92*, p. 661-664, 1992.

[MAN 82] MANDELBROT B.B., *The Fractal Geometry of Nature,* Freeman and Company, San Francisco,1982.

[MAL 98] MALLAT F., *A Wavelet Tour of Signal Processing*, Academic *Press*, San Diego, 1998.

[OUA 02] OUAHABI A., LOPEZ A., BENDERBOUS S., "A wavelet-based method for compression, improving signal-to-noise and contrast in MR Images", *Proc. of the 2nd. IEEE International Conference on Systems, Man & Cybernetics*, Hammamet, Tunisia, 6-9 October 2002.

[SAM 94] SAMORODNITSKY G., TAQQU M.S., *Stable Non-Gaussian Random Processes*, Chapmann and Hall, 1994.

[VES 01] VESLA M., ZMESKAL O., VESELAY M., NEZADAL M., "Fractal analysis of image structures for microbiologic application", *HarFa,* p. 9-10, 2001.

[VON 04] VON KOCH H., "Sur une courbe continue sans tangente, obtenue par une construction géometrique élémentaire", *Arkiv for Matematik* 1, p. 681-704, 1904.

Chapter 4

Information Criteria: Examples of Applications in Signal and Image Processing

4.1. Introduction and context

In this chapter we will focus on the sequence of N observations $x^N = (x_1, \cdots, x_N)$ on a stationary and random process consisting of a family of random variables $X = \{X_n\}_{n \in Z}$ distributed according to the same unknown law θ. A model θ_k based on k free parameters will represent this process X. Determining the optimal estimation $\hat{\theta}_k$ of θ_k in the maximum likelihood (ML) sense enables us to find $\hat{\theta}_k$ which maximizes $f(x^N | \theta_k)$ where f represents the conditional density probability of the observations x^N when choosing the model θ_k. Finding $\hat{\theta}_k$ which will minimize $L(\theta_k) = -\log f(x^N | \theta_k)$ and therefore $\hat{\theta}_k = \arg\min_{\theta_k} L(\theta_k)$ has the same effect.

Even though this criterion of estimation is expressed by a fixed number k, it might be tempting to use this criterion to carry out a simultaneous estimation of the model's parameters and its number of free parameters, which in a written form can be expressed as follows: $\hat{\theta}_k = \arg\min_{\theta_k,\, k=1, \cdots, K} L(\theta_k)$, where K represents the maximum number of free parameters.

Chapter written by Christian OLIVIER and Olivier ALATA.

As a general rule this criterion does not converge; this would lead to an overestimation of the number of free parameters. For sufficient proof see Figure 4.1.

Information criteria (IC) will be referred to as penalized log-likelihood criteria where the penalized term depends on the number k of free parameters in the model θ_k and/or the number of observations N. In some criteria, a third term appears. This term contains Fisher information which is often forgotten about and not very influential in the research focusing on model θ_k. Furthermore, these criteria can be written in a generalized form:

$$IC(k) = -2\log f\left(x^N \middle| \hat{\theta}_k\right) + kC(N) \qquad [4.1]$$

with $\hat{\theta}_k$ as the parametric model which minimizes $L(\theta_k)$ and $C(N)$ as an increasing function of N. The "best" model $\hat{\theta}_k$ is the one that minimizes $IC(k)$. In the case where the hypothesis of independence is issued we could also write:

$$IC(k) = -2\sum_{i=1}^{N} \log f\left(x_i \middle| \hat{\theta}_k\right) + kC(N) \qquad [4.1']$$

where $f\left(x_i \middle| \hat{\theta}_k\right)$, $i = 1, \ldots, N$, describes the conditional probability density of observation x_i while choosing model $\hat{\theta}_k$. Assuming that $\frac{1}{N}L(\hat{\theta}_k)$ tends asymptotically to $E(-\log f(X|\hat{\theta}_k))$ when $N \to +\infty$, the link between information criterion and entropy is created and the reason for the occasional designation of penalized entropic criteria is given to the IC.

Numbers from these criteria are proved or proposed without proof in the context of researches about 1D auto-regressive (AR) models (sections 4.3.1 and 4.3.2). These criteria were then applied to other parameter-based models such as ARMA, 2D AR models (section 4.3.3), mixtures of n-D Gauss's laws (section 4.4) and Markov models (section 4.6.1). They were also taken into consideration for non-parametrical problems such as the approximation of distribution via histograms (section 4.5) and research on features containing a maximum of information about one particular pattern (section 4.6.2).

All of the applications presented in this chapter are, of course, not exhaustive as a large number of recent articles, like those on the processing of images, show.

4.2. Overview of the different criteria

The most common and also the oldest information criterion is Akaike's information criterion (AIC) [AKA 73]. Even though this criterion improves ML estimation, it leads to an over-parameterization in the order of models [SHI 76]. This behavior will be analyzed in the examples which follow. AIC is in itself an improvement of the FPE (Final Prediction Error) criterion which was also created by Akaike (1969). FPE is still used occasionally. FPE is based on the minimization of the Kullback-Leibler (K-L) information between $f(.|\theta)$ and $f(.|\theta_k)$ (see Appendix, section 4.8.1). Through asymptotic approximation of this measure for large N, and leaving out terms that do not specifically depend on the model, the following criterion has been obtained:

$$AIC(k) = -2\log f(x^N \,|\, \hat{\theta}_k) + 2k \qquad [4.2]$$

Even though this criterion has often been criticized and even contested, it still represents major progress in the field of approximation.

Hannan and Quinn [HAN 79] improved this criterion by writing it in the following form:

$$\Phi(k) = -2\log f(x^N \,|\, \hat{\theta}_k) + k \log\log N \qquad [4.3]$$

This criterion is weakly consistent (convergence only in probability when $N \to \infty$; see Appendix, section 4.8.2 for the types of convergence used).

To overcome the inconsistency of AIC, Schwarz [SCH 78] suggested the widely-known BIC criterion based on the Bayesian justification of choosing an order for the AR:

$$BIC(k) = -2\log f(x^N \,|\, \hat{\theta}_k) + k \log N \qquad [4.4]$$

A different approach was introduced by Rissanen [RIS 78]. This approach suggested using the minimization of the length of a code. This code is required to encode observations x^N of the process stated in the number of bits as a criterion of optimization. This criterion is referred to as MDL (*minimum description length*) and can be compared to BIC, even though different terms were added to the penalty (Fisher information was added in [BAR 98], the estimated mean entropy in [BIE 00], etc.)

The BIC and MDL criteria are almost surely convergent (there is a strong consistency) and penalized the term of likelihood more than Φ and AIC. As these two criteria are written in a very similar way, they will no longer be distinguished in the rest of this chapter. A complete theoretical study on MDL can be found in [GRU 04].

Another criterion, referred to as Φ_β, suggested and explored by El Matouat and Hallin [ELM 96], is a generalization of Rissanen's work on stochastic complexity [RIS 89]. It is written in the following form:

$$\Phi_\beta(k) = -2\log f(x^N \,|\, \hat{\theta}_k) + kN^\beta \log\log N \qquad [4.5]$$

with the following condition being required: $0 < \beta < 1$. This criterion is strongly consistent (see Appendix, section 4.8.2) and can be presented as a compromise between Φ and BIC/MDL. It has been shown in [JOU 00] with refined conditions: $0 < \frac{\log\log N}{\log N} \leq \beta \leq 1 - \frac{\log\log N}{\log N} < 1$, which allows for β to be adjusted according to the number of observations N. From now on, $\beta_{\min} = \frac{\log\log N}{\log N}$ and $\beta_{\max} = 1 - \beta_{\min}$ will be referred to as the two bounds regulating β data.

On the basis of the Φ_β penalty it is possible to calculate the penalty of the BIC or AIC criterion. It is sufficient to apply the following values to β: $\beta_{\mathrm{BIC}} = \frac{\log\log N - \log\log\log N}{\log N}$ and $\beta_{\mathrm{AIC}} = \frac{\log 2 - \log\log\log N}{\log N}$. Based on these formulae we can observe that $0 < \beta_{\mathrm{BIC}} < 1$, for $N > 4$, even though $\beta_{\mathrm{AIC}} < 0$ for $N > 1619$, which is coherent with the non-convergence stated in AIC. We can also observe that once $N > 15$, even though $\beta_{\mathrm{AIC}} < \beta_{\mathrm{BIC}} < \beta_{\min} < \beta_{\max} < 1$ and also $\hat{k}_{\mathrm{AIC}} \geq \hat{k}_{\mathrm{BIC}} \geq \hat{k}_{\min} \geq \hat{k}_{\max}$, where $\hat{k}_{\mathrm{IC}}$ represents the estimated number of free parameters with a given IC.

[EM 96] also suggested the AIC* criterion. In this case $C(N)$ is written as follows: $C(N) = 2 + \log N$.

There are also other criteria which are not justified, such as the AIC$_\alpha$ criteria with penalty αk, $\alpha > 2$ [BHA 77]. The penalty $3k$ is part of the AIC$_\alpha$ criteria and features a mixture of laws [BOZ 94] (see section 4.4.1).

Others are similar to the criteria shown above, e.g. the Kashyap criterion [KAS 89], which is a BIC criterion and expressed for the case of an 2D AR model (see section 4.3.3); CAICF (*consistent AIC Fisher*) by Bozdogan [BOZ 94] is equivalent to MDL. Others are specified according to the chosen model and more often an AR model (see section 4.3.1).

As a general rule, acceptable penalties $kC(N)$ need to verify the conditions given in [NIS 88] (see Appendix, section 4.8.2 and Table 4.1) to ensure either almost sure convergence (strong consistency) or only convergence in probability (weak consistency).

Apart from these criteria, there are also other "ICs" such as DIC (deviance IC) [SPI 02] or the GLRT (*generalized likelihood ratio test*) [STO 04], etc. They are used in the estimation of models or in the selection process of models. They will not be discussed further in this chapter as they cannot be classed as a generic form [4.1].

criterion	**author(s)**	**year**	**penalty $kC(N)$**	**cost**	**consistency**
AIC	Akaike	1973	$2k$	K-L	none
Φ	Hannan and Quinn	1978	$k\log\log N$		weak
BIC	Schwarz	1978	$k\log N$	Bayes	strong
Φ_β	El Matouat and Hallin	1994-96	$kN^{\beta}\log\log N$ $0<\beta<1$	Stochasitical complexity + K-L	strong
MDL	Rissanen	1978-87	$k\log N$ $+(k+2)\log(k+2)$	Stochasitical complexity	strong
CAICF	Bozdogan	1987-94	$k(2+\log N)$ $+\log\left\|I(\hat{\theta}_k\right\|$	K-L	strong
AIC*	El Matouat *et al.*	1987-94	$k(2+\log N)$	K-L + Hellinger	strong
AIC_α	Bhansali *et al.*	1977	αk		none

Table 4.1. *Overview of the main criteria*

4.3. The case of auto-regressive (AR) models

We know there is a real interest in using AR models for the modeling of real signals, for example medical signals (ECG, EEG, etc.) or speech signals (encoding by PARCOR (partial correlation) coefficients for example). This fact justifies the existence and development of IC. So, after an overview of the application of IC to an AR model, two applications linked to the analysis of images will be examined in this part.

4.3.1. *Origin, written form and performance of different criteria on simulated examples*

Many of the previous criteria (AIC, BIC, Φ) were initially suggested for research on the order of an 1D auto-regressive (AR) model. Many studies (e.g. [VAN 86], [DIC 94]) have compared these criteria when they are applied within the framework mentioned above.

Note that a random process $X = \{X_n\}_{n \in Z}$ can be modeled by an AR of the order k if:

$$\begin{cases} X_n = -\sum_{i=1}^{k} a_i X_{n-i} + E_n, n \in Z \\ E(E_n) = 0 \quad \text{et} \quad E(E_n E_m) = \sigma_e^2 \delta_{n,m}, (n,m) \in Z^2 \end{cases} \quad [4.6]$$

where $\delta_{n,m}$ is the Kronecker symbol and $E = \{E_n\}_{n \in Z}$ is white noise. Using the previous annotations and the Gaussian hypothesis, θ_k is therefore the model with the following (k+1) parameters: $\{(a_1,...,a_k), \sigma_e\}$ with σ_e being the standard deviation of error. It is widely known that the log-likelihood term can be expressed as follows: $-\frac{N}{2}\log\hat{\sigma}_e^2$, where $\hat{\sigma}_e$ is the estimation in the ML sense of σ_e, and the criteria have the following form:

$$IC(k) = 2N \log \hat{\sigma}_e + kC(N) \quad [4.7]$$

The chosen order $\hat{k}$ and the parameters which have been calculated with the help of methods such as Yule-Walker, Levinson, Burg or lattice, therefore have to verify:

$$\hat{k} = \arg\min_k IC(k) \quad [4.8]$$

In Figure 4.1 the behavior of AIC, BIC/MDL, Φ, Φ_β, ML, is shown for two standard tests:

– a first AR of order 2 with $a_1 = 0.55$ and $a_2 = 0.05$;

– a second AR of order 15 with $a_1 = 0.50$, $a_2 = 0.06$, $a_3 = 0 = \ldots = a_{14}$, $a_{15} = 0.45$.

The research on the respective order is based on 100 experiments and the order is varied from 0 to 20. The number of observations is $N = 1000$ and $\sigma_e = 1$. The curves shown below give an average value of the IC for the 100 experiments. The Φ_β criterion has been calculated for values of β equal to 0.1, 0.2, 0.3 and 0.4. The criteria BIC/MDL and Φ_β (for $\beta = 0.1$, 0.2 or 0.3) give an adequate order. However, if $\Phi_{0.4}$ under-parameterizes because the penalty is too high, even though AIC and Φ are visually confused when it comes to the curves (because loglog1000 = 1.93 ~ 2) and over-parameterize (see Tables 4.2a and 4.2b). The ML criterion therefore systematically over-parameterizes.

Now note the existence of bounds on the value β. For $N = 1000$, $\beta_{\min} = 0.28$, $\beta_{\max} = 0.72$, $\beta_{BIC} = 0.184$ and $\beta_{AIC} = 0.005$ (see section 4.2). These values explain the placement of the curve associated with BIC between the curves associated with $\Phi_{0.1}$ and $\Phi_{0.2}$. The curve $\Phi_{0.3}$ provides the approximate behavior of $\Phi_{\beta_{\min}}$.

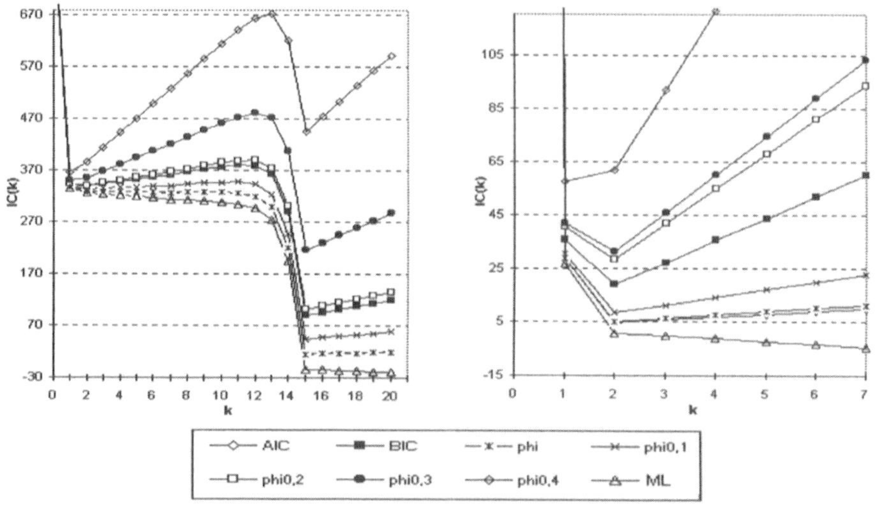

Figure 4.1. *Curves of average values of ICs for the order k for an AR model of order 15 (left) and for an AR model of order 2 (right)*

Tables 4.2 show the performances of the criteria under a different form:

– Tables 4.2a show the influence of N more precisely. This is carried out for the case of an AR order 15, for AIC, BIC/MDL, Φ, $\Phi_{0.2}$. It provides the number of times a value k is estimated. This, of course, applies to the 100 experiments named above (%). The results will have the same orientation as the convergences given in section

4.2. Criterion $\Phi_{0.5}$ is not shown as it systematically under-parameterizes. Note that for N = 10,000, there is β_{min} = 0.241, β_{max} = 0.759, β_{BIC} = 0.154 and β_{AIC} = -0.011. These values explain the poor results of AIC: AIC has the tendency to over-parameterize in comparison to Φ, which corresponds to Φ_0;

– Tables 4.2b give a detailed outline of the results that correspond to an AR order 2. The penalty of $\Phi_{0.5}$ applies to this case even though this criterion systematically under-parameterizes. However, even BIC and $\Phi_{0.2}$ give very contrasting results if the number of observations is insufficient.

order	**AIC**	**BIC**	**Φ**	**$\Phi_{0.2}$**
0	0	0	0	0
1	0	0	0	0
...				
15	***50***	***82***	***49***	***84***
16	9	5	9	3
17	9	7	7	7
18	12	3	14	4
19	14	3	15	2
>19	6	0	6	0
order	**AIC**	**BIC**	**Φ**	**$\Phi_{0.2}$**

N= 1,000

order	**AIC**	**BIC**	**Φ**	**$\Phi_{0.2}$**
0	0	0	0	0
1	0	0	0	0
...				
15	***65***	***100***	***74***	***100***
16	13	0	11	0
17	7	0	6	0
18	6	0	5	0
19	2	0	1	0
>19	7	0	3	0
order	**AIC**	**BIC**	**Φ**	**$\Phi_{0.2}$**

N= 10,000

Tables 4.2a. *Influence of the number of observations on the behavior of criteriafor an AR order 15*

order	**AIC**	**BIC**	**Φ**	**$\Phi_{0.2}$**	**$\Phi_{0.5}$**
0	0	0	0	0	0
1	***47***	***90***	***41***	***91***	***100***
2	35	10	34	9	0
3	8	0	10	0	0
4	5	0	5	0	0
>4	5	0	10	0	0

N= 1,000

order	**AIC**	**BIC**	**Φ**	**$\Phi_{0.2}$**	**$\Phi_{0.5}$**
0	0	0	0	0	0
1	0	1	0	9	***100***
2	***74***	***99***	***80***	***91***	0
3	7	0	7	0	0
4	2	0	2	0	0
>4	17	0	11	0	0

N= 10,000

Tables 4.2b. *Influence of the number of observations on the behavior of criteria for an AR order 2*

Also note that if the number of observations N is insufficient in comparison to the previously established order of the model (e.g. the order of the AR model is greater than $\frac{N}{10}$), Broersen [BRO1 00] suggests two criteria referred to as FSIC and CIC. These criteria compensate for the weaknesses of the log-likelihood and the bias of the criteria. CIC has also been used for a vectorial extension of ARs, as described in [WAE 03]. Other authors such as [SEG 03] base their research on a criterion which stems from the Kullback divergence, abbreviated as KIC (for Kullback IC [CAV 99]), to make up for the insufficient data.

Lastly, for the case of ARMA models, different articles such as [KAS 82] or [BRO2 00] discuss the different approaches in the use of ICs. These articles provide further information. The forms of the criteria are similar to the one given in formula [4.7] but with the order k replaced by $(p+q)$, where (p,q) is the order of the ARMA model.

4.3.2. *AR and the segmentation of images: a first approach*

In the given hypothesis only one line of images of different gray levels is a $\hat{\theta}_c$ mixture of the AR models. Therefore:

$$f(x^N \left| \hat{\theta}_c \right.) = \prod_{i=1}^{c} f(x_{n_{i-1}+1}, \cdots, x_{n_i} / \hat{\theta}_i) \qquad [4.9]$$

where c is the number of AR models (referred to as $\hat{\theta}_i$, i = 1, …, c), of the mixture model $\hat{\theta}_c = \left\{\hat{\theta}_i\right\}_{i=1,\ldots,c}$. The positions n_i, i = 1,…, c-1, correspond to the changes in the models ($n_0 = 0$, $n_c = N$) and can be estimated by using a dynamic programming algorithm. It is suggested that on every segment $[n_{i-1}+1, n_i]$ $\hat{\theta}_i$ is an AR model with order k_i and therefore the parameters are the k_i coefficients and the standard deviation of error. Finally, the retained mixture of AR models which minimizes the IC depends on positions n_i, on their respective number and on the $(k_i + 1)$ parameters of AR models on every segment [JOU 98]. To subdivide a line of images into c AR models there are $\alpha(c) = \left(\sum_{i=1}^{c} k_i\right) + c - 1$ free parameters and the IC criteria take the following written form (see [BOU 91] but only with the AIC criterion and models with fixed order):

$$IC(c) = 2L\left(\hat{\theta}_c\right) + \alpha(c)C(N). \qquad [4.10]$$

First of all, the results obtained from a sequence of synthetic images will be given (Figures 4.2).

A better readability for the order is obtained with the help of the Φ_β and BIC criteria. However, it is clear that in this type of example the interval of validity for β is uncertain as it depends on the variable number N of observations per model that is taken into account.

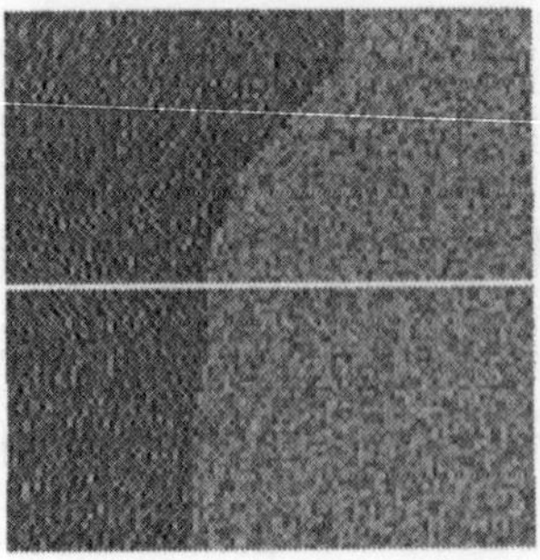

(a)

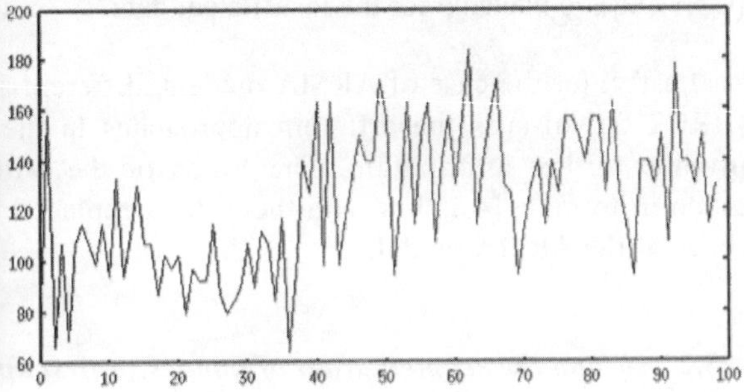

(b)

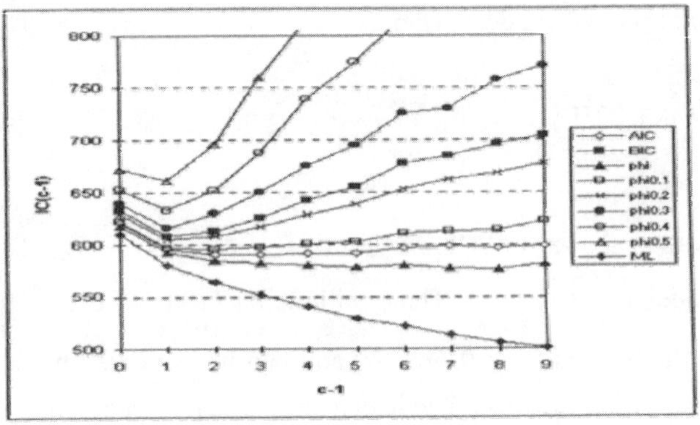

(c)

Figures 4.2. *Synthetic image (a) with the behavior of the ICs (c) on the indicated line (b)*

On a natural image (see Figure 4.3a), the diverse changes obtained according to the criterion are addressed with the help of a dynamic programming algorithm suggested in [THU 97]. Every "white" point corresponds to a change in the model of the respective line. The result of the segmentation is shown using BIC and Φ_β with $\beta = 0.3$; $\beta = 0.3$ corresponds to the best result.

A comparison with the usual techniques of detecting the edges, such as Canny-Derriche or Sobel, shows the high quality of these criteria. However, the computation needs high complexity. Using an 2D AR could refine the segmentation of images even more (see section 4.3.4).

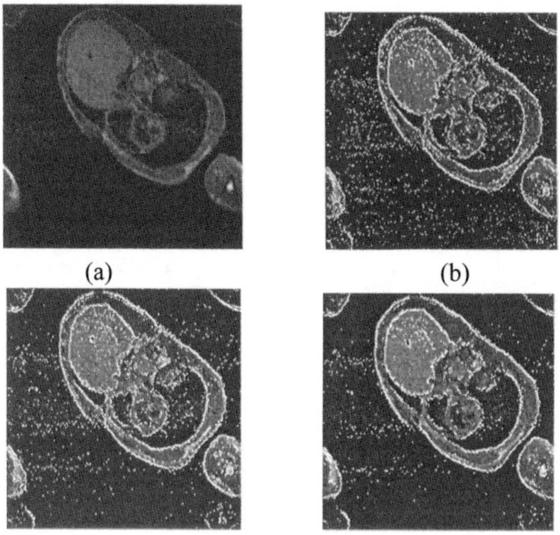

Figures 4.3. *(a) Natural image (heart slice); (b) comparison with the criteria AIC; (c) BIC/MDL and (d)* $\Phi_{0.3}$

4.3.3. *Extension to 2D AR and application to the modeling of textures*

In this section, the focus will lie on the BIC criteria (or Kashyap criteria [KAS 82]) and Φ_β to characterize textures from the basis of an *a priori* hypothesis that they can be represented by a 2D AR. This extension is not only a simple rewriting process of the 1D case since this choice of a support from the "past" is very important. Details on this subject can be found in [ALA 03].

The 2D AR model that has been examined is of a support type *D*, i.e. a Quarter Plane (QP) or a Non-Symmetric Half Plane (NSHP) (Figure 4.4). The model for a process $X = \{X_{n_1,n_2}\}_{(n_1,n_2)\in Z^2}$ is written as follows:

$$X_{n_1,n_2} = -\sum_{(m_1,m_2)\in D} a_{m_1,m_2} X_{n_1-m_1,n_2-m_2} + E_{n_1,n_2} \qquad [4.11]$$

where $E = \{E_{n_1,n_2}\}_{(n_1,n_2)\in Z^2}$ is a white noise with variance σ_e^2.

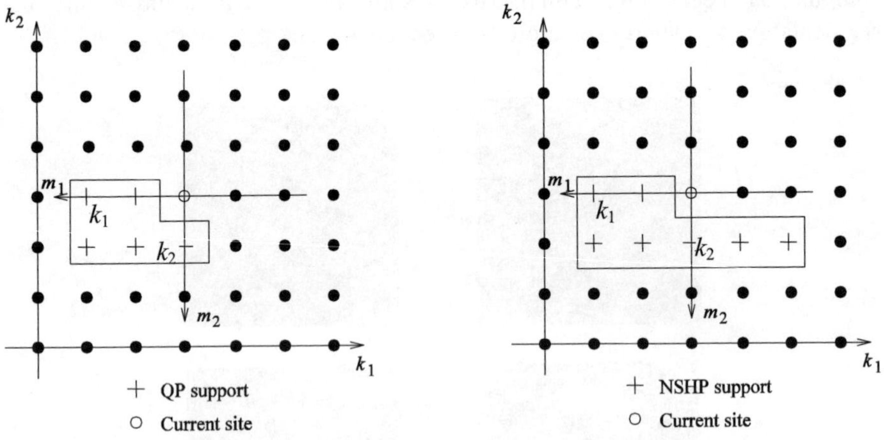

Figure 4.4. *Causal supports QP and NSHP of the order (k_1,k_2)*

In this configuration the IC criteria are written as follows:

$$IC(k_1,k_2) = 2N^2 \log \hat{\sigma}_e + (|D(k_1,k_2)|+1)C(N^2) \quad [4.12]$$

where $N{\times}N$ is here the size of the image and $|D(k_1,k_2)|$ is chosen to be the cardinal number of $D(k_1,k_2)$, i.e. the number of elements in the support of the order (k_1,k_2). For a QP support, $|D(k_1,k_2)| = (k_1+1)(k_2+1) - 1$, and for an NSHP support, $|D(k_1,k_2)| = (2k_1+1)\,k_2+k_1$.

First of all Φ_β will be tested on synthetic textures which are modeled by 2D AR s of the order (i,j) (Figure 4.5). β is still equal to the minimum boundary of the double inequality given in section 4.2, i.e. $\beta_{\min}$. QP is chosen as a support. For every texture, 100 images of the different sizes $N{\times}N$ (1st column of Table 4.3) are synthesized and the rate of correct order estimation is given as a percentage.

The results are very satisfying, regardless of the order or size of the images. However, the results have proven to be very different from BIC/MDL [ALA 03], as this criterion over-parameterizes in comparison to $\Phi_{\beta_{\min}}$ according to section 4.2.

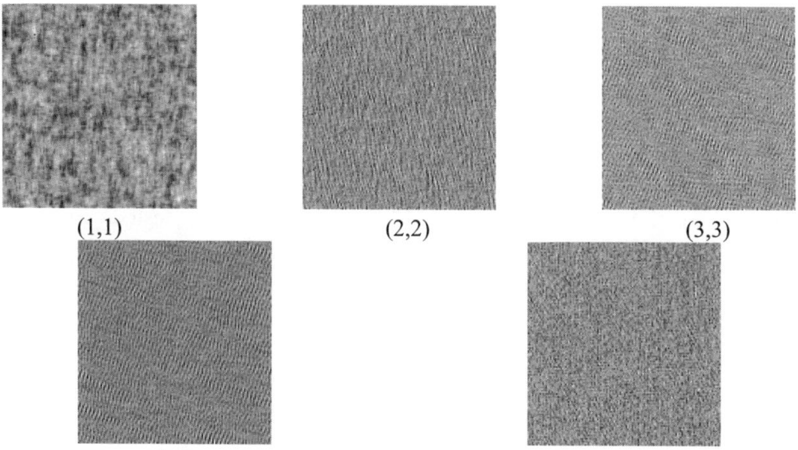

Figure 4.5. *Different synthetic textures and their corresponding orders*

	(1,1)	**(2,2)**	**(3,3)**	**(1,3)**	**(5,2)**
45x45	99	100	100	100	100
55x55	100	100	100	100	100
64x64	100	99	100	100	100
80x80	99	100	100	100	100
90x90	99	100	100	100	100

Table 4.3. *Behavior of the* $\Phi_{\beta_{\min}}$ *criterion in comparison to the size* $N \times N$ *of images. The textures can be identified due to their respective order*

For natural textures the problem of comparison and validity of the criteria is far more complex. To judge the BIC/MDL criteria and $\Phi_{\beta_{\min}}$ for these textures the initial textures T_i are compared to synthetic textures T_s. T_s are generated with a 2D AR model obtained by the two ICs. The comparison measure is the Kullback divergence (see [BAS 96]):

$$J(T_i,T_s)=\frac{1}{2}\left[tr\left(R(T_i)R^{-1}(T_s)\right)+tr\left(R(T_s)R^{-1}(T_i)\right)\right]-121 \qquad [4.13]$$

$R(T)$ is thus the matrix of the covariance associated with the texture T that has been obtained from the QP support of the order (10,10), i.e. $R(T)$ of the size 121×121. "*tr*" describes the trace of a squared matrix.

Figure 4.6a shows Brodatz's four textures which were tested. Figure 4.6b represents the variations of the Kullback divergence, amongst the four textures shown in 4.6a, according to the size $N \times N$ of the squared images, for BIC/MDL and $\Phi_{\beta_{\min}}$ criteria. These are average values of calculated divergences for the different sizes of images given in Table 4.3.

The two criteria are asymptotically comparable. The divergence decreases if N increases. T_s, which stems from $\Phi_{\beta_{\min}}$, is however closer to T_i.

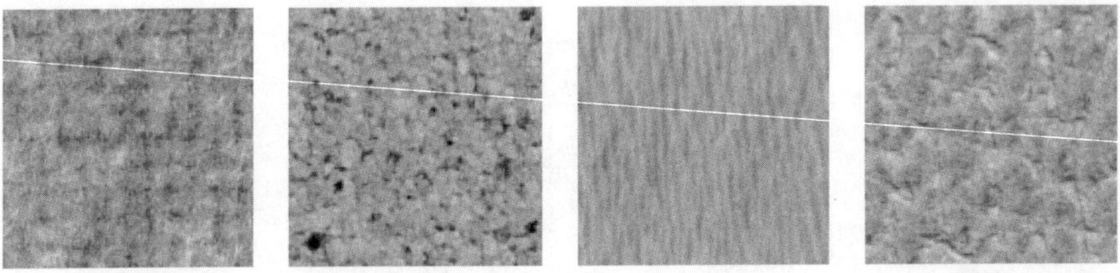

D 19 D 29 D 38 D 92

Figure 4.6a. *Brodatz's textures which were tested*

4.3.4. *AR and the segmentation of images: second approach using 2D AR*

In section 4.3.2, the use of AR in the framework of segmentation of natural images using simple 1D AR models was shown. In section 4.3.3 it was proven that it is possible to obtain a causal 2D AR model associated with a texture which is optimal in the sense of information criteria. This section will show how ICs can be exploited for an unsupervised segmentation of textured images based on a 2D AR model. In order to do this, the order of the associated models of different textures, as well as the number of different textures each image contains, need to be estimated. A parameter-based model needs to be estimated, $\theta = \{\theta_k\}_{k=1,\cdots,K}$, where K is the number of textures in an image and θ_k the AR model with D_k support.

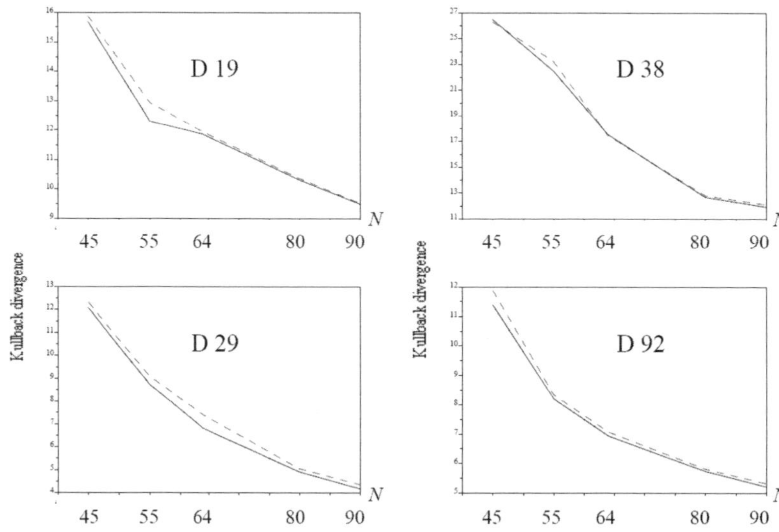

Figure 4.6b. *Values of the Kullback divergence on textures given Figure 4.6a, in comparison to the size N×N of images. The QP support is used. As a line:* $\Phi_{\beta_{\min}}$ *; as a line of dots: BIC/MDL*

The solution to the problem presented here [ALA 05] is based on an article written by Bouman and Liu [BOU 91]. Bouman and Liu have used the AIC criteria to estimate K and fixed the orders of the models AR QP associated to different textures of the image equal to (1,1). Note that in the majority of articles that have been published on unsupervised segmentation, the number of textures within textured images is estimated, while the number of parameters of models has been fixed beforehand. The optimal orders of the models linked to the respective textures and obtained with ICs (section 4.3.3) show to what extent it is insufficient to only use a model with (1,1) order.

The segmentation method can be divided into two steps. The first step consists of estimating the parameter-based model of the textured image. The second step describes the estimation of the segmented image which takes place in a supervised framework. During the first step information criteria come into play. The second step is carried out with the help of a simulated annealing (SA) algorithm based on Gibbs' sampler that optimizes a *maximum a posteriori* (MAP) criterion; the segmented image is modeled by a hidden Markov field.

The first step of the estimation of the parameters of the model associated with the textured image works as follows.

The image is cut into squared blocks (e.g. of the size 8×8 or 16×16) that are assumed to be independent of one another[1]. The distribution of the categories of textures into blocks is supposed to follow a multinomial distribution of unknown parameters $\rho = \{\rho_k\}_{k=1,\cdots,K}$. If now A_k, $k = 1,\ldots, K$, describes all blocks that stem from the texture k, and if $W_k = |A_k|$ describes the number of blocks in A_k, such as $\sum_{k=1}^{K} W_k = W$, with W representing the total number of blocks, it is possible to show that:

$$p(x, A_1, \cdots, A_K | \theta, \rho) = \prod_{k=1}^{K} \exp\left(-\log f\left(x_{A_k} | \theta_k\right) + W_k \log \rho_k\right) \quad [4.14]$$

where S is the set of sites of the image, $x = \{x_s, s \in S\}$ is the textured image, $x_{A_k} = \{x_s, s \in A_k\}$ and $f\left(x_{A_k} | \theta_k\right)$ is the conditional probability law of x_{A_k} subject to the parameter-based model θ_k.

It has been established (section 4.3.2) that [BOU 91] suggests a criterion for the estimation of parameters $(K, \{\theta, \rho\})$ and the distribution of the textures to the blocks. Here it has been given with a nondescript penalty ([ALA 05]):

$$-2\log p(x, A_1, \cdots, A_K | \theta, \rho) + C(|S|) \times \left(\sum_{k=1}^{K} |\theta_k| + K - 1\right) \quad [4.15]$$

The written form of this criterion is analog to formula [4.10] in which the number of free parameters is established $\alpha(K)$. The optimization of this criterion is carried out by successive minimization of $(K, \{\theta, \rho\})$ and the distribution of K textures to the blocks. The two other base qualities remain fixed. Note that, for x_{A_k} the estimation of the order of model θ_k is carried out according to formula [4.12].

Based on the size 256×256 of an image with synthetized textures (Figure 4.7a), the "block" segmentation with the help of the Φ_β criterion, used with $\beta = \beta_{\min}$, has been obtained (Figure 4.7b). Comparisons were also carried out with the AIC and BIC criteria, with mediocre results.

---

1. This first division is necessary because the AR hypothesis implies the use of a correlated process, as opposed to a mixture of independently and infinitely distributed Gaussian processes that implies independence between all elements of the process.

Another example of block segmentation is given in Figure 4.8a. However, this example shows an image that contains natural textures. Figure 4.8b shows that an additional category of texture on the boundaries has been obtained (six categories can be detected). The final segmentation after using SA can give results similar to the type given in Figure 4.8c, which shows how an undesired category of texture has been removed from Figure 4.8b.

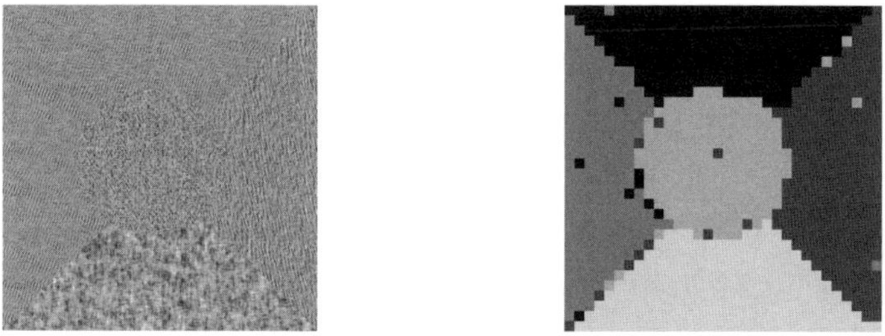

(a) Observed image (b) Image segmented into blocks

Figures 4.7. *Image which contains synthetic textures*

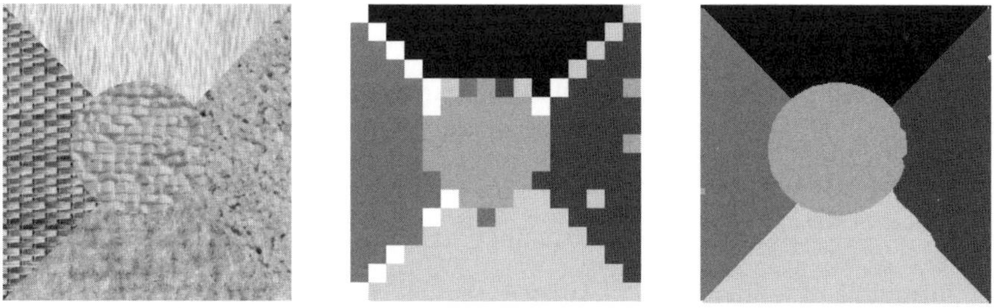

(a) Observed image (b) Image segmented into blocks (c) Correction with help of Simulated Annealing

Figures 4.8. *Image which contains natural textures*

4.4. Applying the process to unsupervised clustering

An application of ICs as presented here is, in fact, very similar to what has already been explained in sections 4.3.2 and 4.3.4. The difference is that the chosen parameter-based models are no longer ARs but n-D parameter-based laws.

Different authors have dealt with this unsupervised clustering problem (the number of classes or categories or clusters is undefined) such as [BOZ 94] (with

$C(N) = 3$), [SAR 96] (dealt with the initial choice of the number of clusters), [LEE 00] (with non-Gaussian laws and the estimation based on ML), etc., to name just a few authors.

Here the focus lies on the choice of the number and form of the components for a Gaussian mixture model. The estimation method used is the EM algorithm which optimizes the function of log-likelihood for a given number of components under the hypothesis of observations x^N being independent of one another.

Population	Expectation vector	Covariance matrix
$n_1 = 300$	$\mu_1 = \begin{bmatrix} 5 \\ 0 \end{bmatrix}$	$\Sigma_1 = \begin{bmatrix} 1 & 0 \\ 0 & 1 \end{bmatrix}$
$n_2 = 300$	$\mu_2 = \begin{bmatrix} 0 \\ 0 \end{bmatrix}$	$\Sigma_2 = \begin{bmatrix} 5 & 2 \\ 2 & 3 \end{bmatrix}$
$n_3 = 300$	$\mu_3 = \begin{bmatrix} -5 \\ 0 \end{bmatrix}$	$\Sigma_3 = \begin{bmatrix} 1 & 0 \\ 0 & 1 \end{bmatrix}$

Table 4.4. *Parameters of the three 2D Gaussian distributions used in Figure 4.9*

If c is the number of components (number of Gaussian laws in the mixture) and n is the dimension of space of observations ($x_i \in \mathrm{IR}^n$, $i = 1, \cdots, N$), the number of free parameters $\alpha(c)$ is equal to (according to [JOU 00]) $\alpha(c) = \frac{c(n+1)(n+2)}{2}$. The optimal number of components still needs to verify (see formula and written form [4.10]): $\hat{c} = \arg\min IC(c)$.

As a first example there is $n = 2$ and the parameters of the three components of the Gaussian mixture are given in Table 4.4 and represented in Figure 4.9a with $N = 900$ samples subdivided into three clusters.

Figures 4.9b to 4.9e show the evolution of different clusters estimated by the EM following the number $c = 1,\ldots,4$. Every ellipsis represents a curve of isodensity for every estimated Gaussian component. Figure 4.9f shows a good readability of $\Phi_{0.3}$ ($\beta_{\min} = 0.282$).

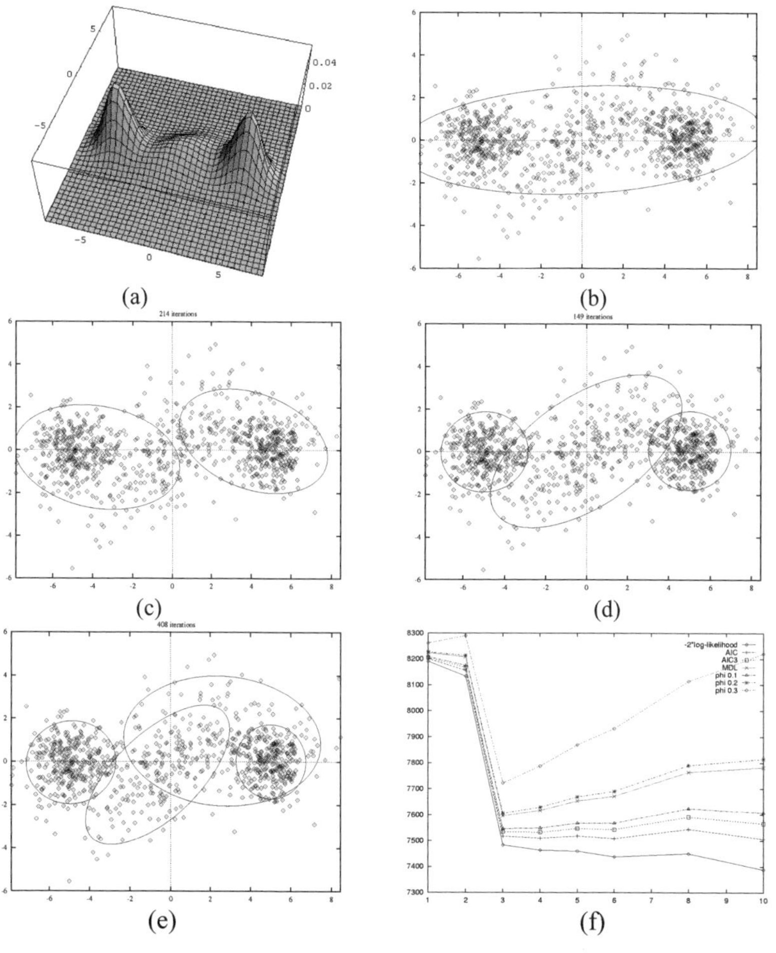

Figures 4.9. *(a) Original model (c = 3); (b) to (e) Gaussian mixture models identified for different values of c; (f) number of components c identified by different ICs. The AIC3 criterion for the penalty C(N) = 3*

Figure 4.10 shows an example of realistic uses of ICs [COU 98] in the case where the space of observations is IR, and therefore with $n = 1$. This sample is part of an envelope with 256 gray levels (Figure 4.10a), and the localization of the threshold (changes of Gauss laws) on the histogram of gray levels is shown (Figure 4.10b) as well as, last but not least, the image which represents the threshold of three different grayscales (Figure 4.10c). The IC criterion is used as a variant of BIC written AIC* (section 4.2) [EM 96]. The threshold T_1 is the result obtained with $c = 2$ components. T_{21} and T_{22} correspond to $c = 3$ components that minimize AIC*.

The use of ICs in this example appears to be an improvement of Kittler's work and the work of the founders of histogram segmentation ([KIT 86], [KUR 92], [WAN 94], etc.). Some authors have also shown to what extent traditional penalties such as AIC or BIC/MDL are insufficient for this type of application. They have therefore suggested empirical values for $C(N)$ of a type that has been mentioned in section 4.2 ([LIA 92] or [THU 97]).

To end this part dealing with unsupervised clustering, problems with the detection of sources that are linked to the ICs can also be mentioned as [FIS 00], where the number of signals which fall in an observation process is estimated using MDL, and the limits which therefore become possible due to the ICs' method [CHE 04]. In [BIE 00], the authors are interested in the number of models that represent a population per integrated completed likelihood (ICL) criterion. ICL should be more robust than BIC/MDL.

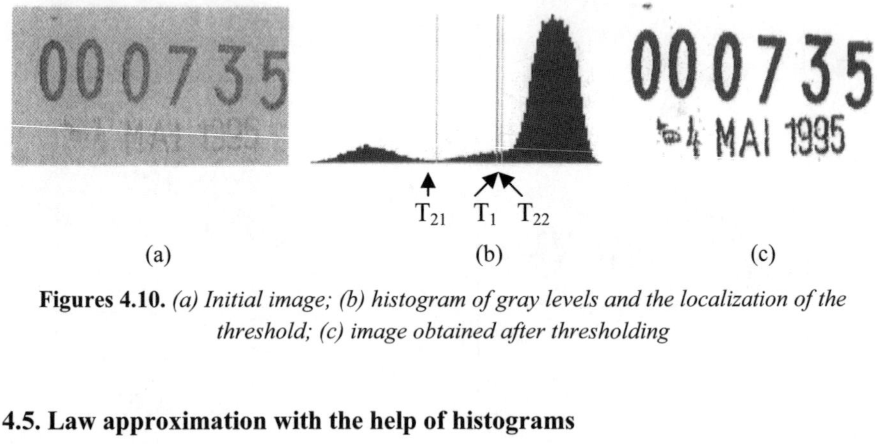

(a) (b) (c)

Figures 4.10. *(a) Initial image; (b) histogram of gray levels and the localization of the threshold; (c) image obtained after thresholding*

4.5. Law approximation with the help of histograms

4.5.1. *Theoretical aspects*

The use of ICs was linked to research on the number of classes in a histogram that represents an approximation of a theoretical law θ followed by a random process X. X is given by N observations x^N and this process is strictly stationary. The aim is to find the histogram that best summarizes the law, i.e. the number and the size of intervals or classes that make up this particular histogram. Note that this no longer deals with the research of parameter-based models. Obtaining the criteria is therefore different. [OLI 94] shows an extension of the AIC criteria to histograms, justified by Hellinger's distance, and demonstrates the AIC* criterion, which has already been mentioned, in the histogram case.

Note $\{B_r, r=1,...,k\}$ which defines the k intervals of the histogram. In this case the IC criteria defining the number of intervals are written in the following form:

$$IC(k) = k(1+C(N)) - 2N\sum_{r=1}^{k}\hat{\theta}_k(B_r)\log\left[\frac{\hat{\theta}_k(B_r)}{\mu(B_r)}\right] \qquad [4.16]$$

where $\hat{\theta}_k$ is the estimation of the law θ in the ML sense, μ is an *a priori* law according to which θ is absolutely continuous with respect to μ.

It is shown that: $\hat{\theta}_k(B_r) = \frac{\textit{number of values } x_i \textit{ in } B_r}{N}$. The desired partition with k intervals is the one that minimizes the IC.

The penalty therefore differs from previous formulae because k is the number of intervals and not the number of free parameters. The k B_r can be obtained using the maximum likelihood. Note that [RIS 92] suggests the method of histogram density estimation based on the MDL principle with B_r intervals of the same size.

4.5.2. *Two applications used for encoding images*

Rissanen's MDL criterion has its origins in the research on the minimal description that is needed to encode a message or information. More precisely, this technique is more refined than the simple entropic encoding of information. This technique consists of arithmetic encoding which does not encode individual symbols that contain information, but instead encodes a sequence of symbols. For more information interested readers can see [GUI 02] or take a look at the texts published on the encoding of fixed images with different gray levels or different colors [HAN 00], [MAD 04], or with video sequences [PAT 98], and references to MPEG 4 and H264 standards.

The two examples that will be given in this section do not deal with binary encoding processes. They are examined with respect to the ICs having a direct impact on the encoding of pixels or the quantization phase.

The first is an encoding problem relating to the different gray levels in an image [COU 98], [COQ 07]. Distribution (Figure 4.11b) of the 256 gray levels of the Lena image with $N = 512{\times}512$ pixels (Figure 4.11a) will be presented below. Figure 4.11c shows a histogram that has been obtained with the help of the AIC* ($k = 40$) and the effect (Figure 4.11d) on the Lena image, which is then encoded in 40 gray levels. With $\Phi_{\beta_{min}}$, we obtain $k = 39$. The result can be interesting in terms of compression of the image and satisfactory in terms of visual quality. Note that PSNR (*peak signal* *to noise ratio*) is equal to 38.52 dB with $k = 39$ and using $\Phi_{\beta_{min}}$. The result becomes even more interesting if the image is made up of homogenous areas of gray levels, which improves the compressing levels after the image has been transformed using discrete cosine transformation (DCT) or discrete wavelet transforms (DWT).

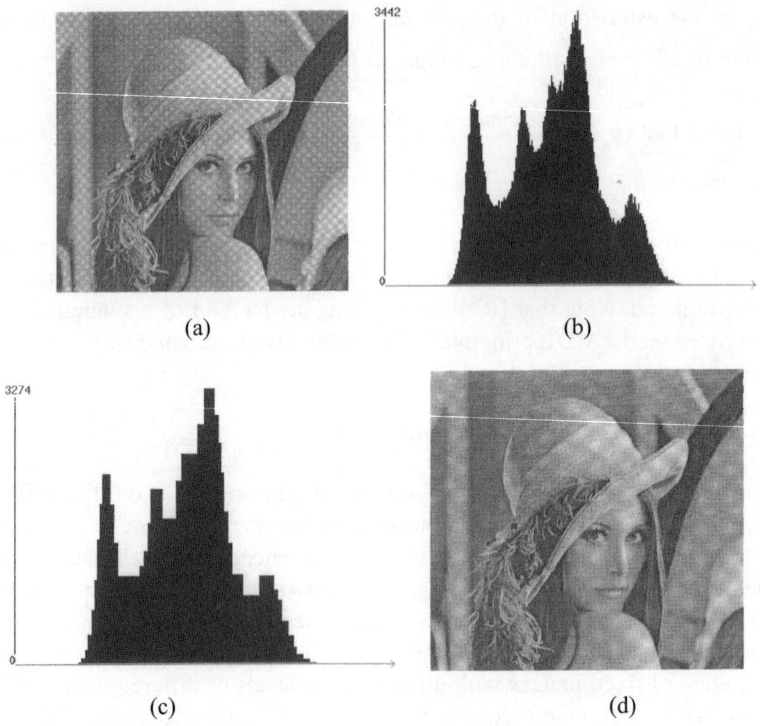

Figures 4.11. *(a)Lena image; (b) distribution of the 256 gray levels; (c) histogram obtained by IC (k = 40 intervals); (d) Lena image encoded on 40 different gray levels*

The second example shows a possible application of ICs in a schema of encoding /quantization of a line of code and transmission (encoding source). DCT is used here, even though the wavelets (DWT) can enter the schema following subband quantization.

Without going into too much detail, it is important to bear in mind that the initial image has been transmitted and split up into blocks of the size 8×8 of DCT coefficients which are decorrelated overall. The first term is called DC and corresponds to the null frequency. Its value is usually high. The 63 other coefficients referred to as AC can be close to zero where such a transformation is of interest.

The distribution of AC values takes place with the position n within the 63 possible positions. For an image of 512 × 512 there is $N = 4{,}096$ DCT coefficients for each of the 63 distributions. The given problem is a problem of non-uniform scalar quantization. Its aim is to determine the threshold of a zone referred to as "dead zone", i.e. the value under which the coefficient AC will be zero. Other coefficients are quantified by other methods that guarantee the minimization of rate-distortion (e.g. RD-OPT method) [RAT 00].

Figure 4.12a shows a test image entitled "boat" and Figure 4.12b shows a distribution of the AC coefficients for the position n, as well as the histogram obtained with $\Phi_{\beta_{\min}}$. The coefficients in the "zero" class will be put to zero. Figure 4.13 compares the standard JPEG method, $\Phi_{\beta_{\min}}$ linked to RD-OPT method (referred to as IC + RD-OPT) [OUL 03] and the RD-OPT method on its own in terms of PSNR. To judge the issue of introducing ICs into the quantization module in an impartial manner, our team has developed a new technique (written down next to the figure as "theory + RD-OPT"). This new technique is based on a quantization that stems from the *a priori* law of coefficients (the law μ of formula [4.16]) which was suggested to be a Laplacian law. In abscissa there is the measured rate of the number of bits per pixel (bpp).

The result is satisfying (gain of several dB) when compared to the JPEG standard at a low rate. This method out-performs RD-OPT by between 0.25 bpp and 0.6 bpp. These results were observed for the entire sequence of tests and reached up to 0.5 bpp. This technique is weaker than the method entitled "theory + RD-OPT" but the IC-based method does not require any parameters to be established beforehand, which is not the case for the other two methods. The explanation of this weakness is that adjacent classes of a similar number, i.e. of AC with high values (the "extremities" of Laplacian law in Figure 4.12b) are merged, contrary to the aim of this research. Indeed, if the rate is increasing (above to 0.7 bpp) the IC method is no longer of any interest.

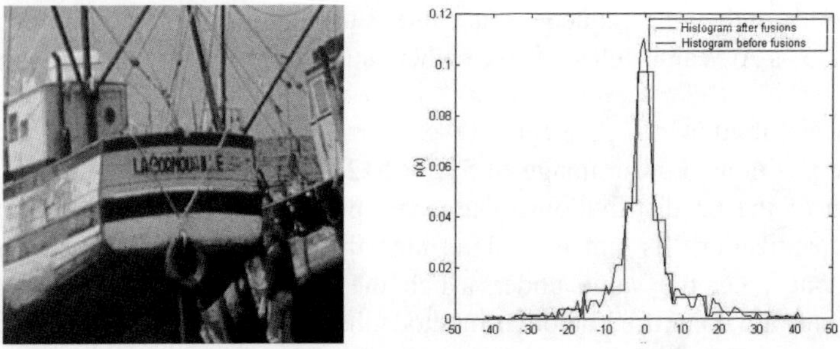

(a) (b)

Figures 4.12. *(a) Test image: boat; (b) distribution of an AC coefficient (n = 37) and merging of the intervals by* $\Phi_{\beta_{\min}}$ *on image (a)*

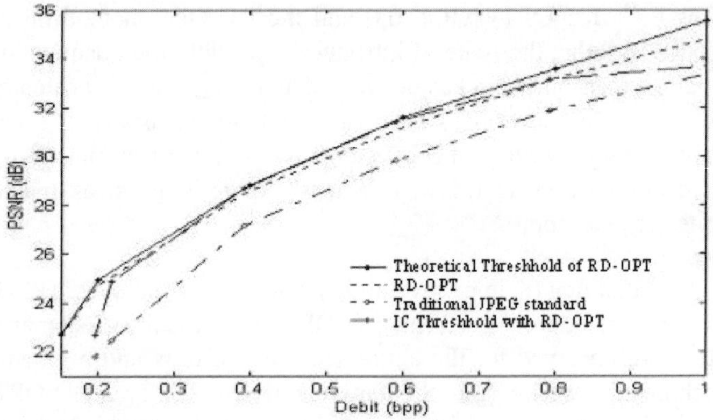

Figure 4.13. *Behavior of the IC-based method compared to other techniques*

Note that when using ICs for high frequency subbands of wavelet-based compression methods (JPEG2000), the *a priori* law of formula [4.16] works in the same way as a generalized Gaussian distribution.

In conclusion, IC for the creation of histograms can be interesting due to the unsupervised choice of parameters which determine quantization. These parameters determine the number of intervals in the quantization process and the length of the intervals. As for the use of ICs on the arithmetic encoding (MDL), it has been mentioned in the introduction of this section.

4.6. Other applications

4.6.1. *Estimation of the order of Markov chains*

For a homogenous and ergodic Markov chain with M states and length N, the order k is to be found.

Noting $n_{i_1,\ldots,i_k,i}$ as the number of occurrences of $(k + 1)$ in successive states $(i_1, \ldots, i_k, i)$ observed in the chain. In the sense of maximum likelihood the following applies:

$$P(X_t = i / X_{t-1} = i_k, \ldots, X_{t-k} = i_1) \overset{ML}{=} \frac{n_{i_1,\ldots,i_k,i}}{\sum_i n_{i_1,\ldots,i_k,i}},$$

and the criteria are written in the usual from:

$$IC(k) = -2 \sum_{i_1,\ldots,i_k,i} n_{i_1,\ldots,i_k,i} \log \frac{n_{i_1,\ldots,i_k,i}}{\sum_i n_{i_1,\ldots,i_k,i}} + M^k (M-1) C(N) \qquad [4.17]$$

To validate this method, a binary chain is simulated with $M = 2$, order $k = 2$, $N =$ 10,000, with M^{k+1} parameters which are the following probabilities:

P(1/11) = 0.3; P(1/01) = 0.4; P(1/10) = 0.8; P(1/00) = 0.1;

P(0/11) = 0.7; P(0/01) = 0.6; P(0/10) = 0.2; P(0/00) = 0.9.

When ignoring the initial probabilities, we observe 4 free parameters can be obtained: $M^{k+1} - M^k = M^k(M\text{-}1) = 4$. These free parameters are represented in factor $C(N)$ in formula [4.17].

Table 4.5 is based on the results of principal criteria indicated as a percentage. This example confirms the information on types of convergence and predictable orders of a model according to the chosen criterion stated in section 4.2.

order	AIC	BIC/MDL	Φ	$\Phi_{\beta_{\min}}$
0	0	0	0	0
1	0	0	0	0
2	***85***	***100***	***91***	***100***
3	12	0	9	0
>3	3	0	0	0

Table 4.5. *Searching for the order of a Markov model*

[OLI 97] describes an application of these criteria in the context of the handwritten information on checks. All words in the dictionary make up a Markov model. The aim is to obtain the optimal order k of this Markov model MM(k) to determine the handwritten amount on the check by reading the words (e.g. 35 dollars/euros).

A tool is available in the form of an alphabet of M = 15 states (known as graphemes), which represents letters or parts of letters. In this case, we dispose of N = 63,000 observations for a learning process of the model parameters, such as probabilities of initial states or the probability matrix for transition of order k. Only ICs of the order 0 (independent states) up to order 3 have been tested – for reasons of complexity and validity of the calculation of different parameters.

These tests have shown that the obtained order (k = 2) and the associated Markov model MM(2) effectively correspond to the best rate when it comes to identifying the words written on a check.

The application of ICs to Markov fields where the parameters (choosing potentials, choosing the environment) could be optimized seems natural. To our knowledge, this task has yet to be carried out successfully due to the challenge of determining the number of free parameters and obtaining different laws in the ML sense in a learning process.

4.6.2. *Data fusion*

Last but not least, two applications which have been tested in the field of pattern recognition will be presented.

1 – The choice of attributes or features as well as deciding on how many of those should be used in an identification process is a widely recorded issue. In a book published in 1986, Sakamoto [SAK 86] showed how the AIC criterion could be applied to the selection process of the most informative features within a respective population. This idea was taken up once more [OLI 96] for the identification of handwritten Arabic numbers (basis of $N = 10{,}000$ samples per number) with the help of Hu's seven invariant moments. The results can be questioned, again due to the complexity of the calculation. Despite the low number of attributes, the quantity of models to be tested has been high due to the possible combination of moments (in this case 126 combinations). These combinations have to undergo a significant learning process, which is, however, impossible for the case of 10,000 samples per number.

2 – A theoretical study in [LEF 00] has integrated AIC into an algorithm aiming at the fusion of information taken from a set of features. The criteria applied to histograms allow for a more "realistic" calculation of reliability coefficients could be assigned to every primitive. They also refined a Dempster-type operator of fusion. We have applied the method by [LEF 00] to the identification process of inner walls situated very close to one another by taking $\Phi_{\beta_{\min}}$ as a criterion instead of AIC. The context is the learning process in an indoor environment for wireless communication. These inner walls should be recognized according to their different radio-electric qualities that stem from differences in their surface. Three geometric and frequential measures are chosen [XIA 04] as features. We study two types of roughness (Figure 4.14) using these three measurements, which are at least considered to be decorrelated. Two primitives are badly processed (reliability coefficients not connected to 1 – see Dempster-Shaffer theory). The tests are carried out on 108 samples size 64×64 of every texture. As a decision-making tool, pignistic probability is used [LEF 00]. This method delivers the best result if the tool is of a square measure weighted by inversing the variances (subject to the hypothesis of effective decorrelation of attribute measures (otherwise see Mahalanobis or Kullback-Leibler measures)).

These two simple examples confirm that the use of ICs in a large number of applications aimed at the pattern recognition is not only possible, but could also help to make algorithms used in these processes less complex while increasing their significance.

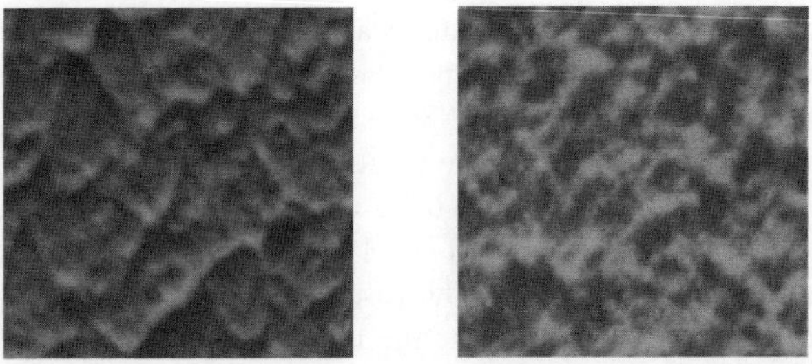

Figure 4.14. *Examples of the coating on indoor walls which are to be identified*

4.7. Conclusion

Finally, and without going back to the applications mentioned in this chapter, information criteria have the advantage of providing justification of the choice of parameters in several problems linked to the processing of signals and images. They require a high amount of observations similar to stochastic approaches, if these statistics are to make sense. However, as has been mentioned for the case of ARs, penalties can eventually be adapted if there is a low number of observations. The complexity of the calculation might be very high, especially in an unsupervised context, but dynamic programming algorithms can reduce the level of complexity for certain applications.

4.8. Appendix

4.8.1. *Kullback (-Leibler) information*

Kullback (-Leibler) information between two probability laws $f(./\theta)$ and $f(./\lambda)$ is defined as follows:

$$K(\theta,\lambda) = \int f(X|\theta)\log\frac{f(X|\theta)}{f(X|\lambda)}dX = E_\theta\left[\log f\left(X|\theta\right)\right] - E_\theta\left[\log f\left(X|\lambda\right)\right]$$

The Shannon entropy can be observed in the first term of the difference as well as mutual information in the second term.

Kullback divergence, which is symmetric, is equal to: $K(\theta,\lambda) + K(\lambda,\theta)$.

4.8.2. *Nishii's convergence criteria [NIS 88]*

Nishii's convergence criteria set up the conditions under which the IC(k) criteria are consistent, i.e. respectively strongly or weakly. Therefore the following applies:

if: $\lim_{+\infty} \frac{C(N)}{N} = 0$ and $\lim_{+\infty} \frac{C(N)}{\log \log N} = +\infty$ so $\lim_{+\infty} \hat{k}=k$ (almost sure convergence);

if: $\lim_{+\infty} \frac{C(N)}{N} = 0$ and $\lim_{+\infty} C(N) = +\infty$ so $\lim_{+\infty} P(\hat{k} = k) = 1$ (convergence in probability).

4.9. Bibliography

[AKA 73] AKAIKE H., "Information theory and an extension of the maximum likelihood principle", *2nd Int. Symposium on Information Theory*, Budapest, p. 267-281, 1973.

[ALA 03] ALATA O., OLIVIER C., "Choice of a 2D causal AR Texture Model using Information Criteria", *Pattern Recognition Letters*, 24, no.9-10, p. 1191-1201, 2003.

[ALA 05] ALATA O., RAMANANJARASOA C., "Unsupervised Texture Image Segmentation using 2-D Quarter Plane Autoregressive Model with Four Prediction Supports", *Pattern Recognition Letters*, 26, p. 1069-1081, 2005.

[BAR 98] BARRON A., RISSANEN J., YU B., "The MDL Principle in Coding and Modeling", *IEEE Trans. On Information Theory*, 44, no.6, p. 2743-2760, October 1998.

[BAS 96] BASSEVILLE M., "Information: entropies, divergences et moyennes" *Publication Interne IRISA*, no.1020, May 1996.

[BHA 77] BHANSALI R., DOWHAM D., "Some properties of the order of an AR Model selected by a Generalization of Akaike's FPE Criterion", *Biometrika*, 64, p. 547-551, 1977.

[BIE 00] BIERNACKI C., CELEUX G., GOVAERT G., "Assessing a mixture Model for Clustering with the integrated completed Likelihood", *IEEE Trans. on PAMI*, 22, no.7, p. 719-725, July 2000.

[BOU 91] BOUMAN C.A., LIU B., "Multiple resolution segmentation of textured images", *IEEE Trans. on PAMI*, vol. PAMI-13, no. 2, p. 99-113, February 1991.

[BOZ 94] BOZDOGAN H., "Mixture-Model cluster analysis using model selection criteria and a new informational measure of complexity", *Proc. of the First US/Japan Conf. on the Frontiers of Statistical Modeling: An Informational Approach,* Kluwer Academic Publishers, p. 69-113, 1994.

[BRO1 00] BROERSEN P.M., "Finite sample Criteria for AR order Selection", *IEEE Trans. on Signal Processing*, 48, no.12, p. 3550-3558, December 2000.

[BRO2 00] BROERSEN P.M., "AR Model Order for Durbin's MA and ARMA Estimators", *IEEE Trans. on Signal Processing*, 48, no.8, p. 2454-2457, August 2000.

[CAV 99] CAVANAUGH J.E., "The Kullback Information Criterion", *Statistics and Probability Letters*, 42, p. 333-343, 1999.

[COQ 07] COQ G., OLIVIER C., ALATA O., ARNAUDON M., "Information criteria and arithmetic codings: an illustration on raw images", *15th EUSIPCO – EURASIP*, Poznan (Poland), pp 634-638, September 2007.

[COU 98] COURTELLEMONT P., OLIVIER C., JOUZEL F., "Information criteria for histogram thresholding techniques", *EUSIPCO'98, Signal Processing IX*, Rhodes (Greece), 4, p. 2509-2512, September 1998.

[DIC 94] DICKIE J., NANDI A., "A comparative Study of AR-order selection methods", *Signal Processing*, 40, no.2, p. 239-256, 1994.

[ELM 96] EL MATOUAT A., HALLIN M., "Order selection, stochastic complexity and Kullback-Leibler information", Springer Verlag, *Time Series Analysis*, 2, p. 291-299, New York, 1996.

[FIS 96] FISCHLER E., MESSER H., "On the use of order statistics for improved detection of signal by the MDL criterion", *IEEE Trans. on Signal Processing*, 48, no.8, p. 2242-2247, August 2000.

[GRU 04] GRÜNWALD P., "A tutorial introduction to the minimum description length principle", *Advances in Minimum Description Length: Theory and Applications,* MITT Press, 2004.

[GUI 02] GUILLEMOT C., PATHEUX S., "Eléments de théorie de l'information et de communication", *Compression et codage des images et des vidéos*, ed. M. Barlaud and C. Labit, Hermes, p. 22-43, 2002.

[HAN 79] HANNAN E.J., QUINN B.G., "The determination of the order of an autoregression", *Journal of the Royal Statistic Society*, 41, no.2, p. 190-195, 1979.

[HAN 00] HANSEN M., BIN Y., "Wavelet thresholding via MDL for natural images", *IEEE Trans. on Information Theory*, 46, no.5, p. 1178-1188, August 2000.

[JOU 98] JOUZEL F., OLIVIER C., EL MATOUAT A., "Information criteria based edge detection", *EUSIPCO'98, Signal Processing IX*, Rhodes (Greece), 2, p. 997-1000, September 1998.

[JOU 00] JOUZEL F., OLIVIER C., EL MATOUAT A., "Choix du nombre de composantes d'un modèle de mélange gaussien par critères d'information", *12ème Congrès RFIA*, Paris (France), 1, p. 149-156 , Feburary 2000.

[KAS 82] KASHYAP R., "Optimal choice of AR and MA parts in autoregressive moving average models", *IEEE Trans. on PAMI*, 4, p. 99-104, 1982.

[KAS 83] KASHYAP R., CHELAPPA R., "Estimation and choice of neighbors in spatial-interaction models of images" , *IEEE Trans. on Information Theory*, 29, no.1, p. 60-71, 1983.

[KIT 86] KITTLER J., ILLINGWORTH J., "Minimum error thresholding", *Pattern Recognition*, 19, p. 41-47, 1986.

[KUN 91] KUNDU A. HE Y., "On optimal order in modeling sequence of letters in words of common language as a Markov chain", *Pattern Recognition*, 22, no.7, p. 603-608, 1991.

[KUR 92] KURITA T., OTSU N., ABDELMALEK N., "Maximum likelihood thresholding based on population mixture models", *Pattern Recognition*, 25, no.10, p. 1231-1340, 1992.

[LEE 00] LEE T., LEWICKI M., SEJNOWSKI T., "ICA mixture models for unsupervised classification of non-Gaussian classes and automatic context switching in blind signal separation", *IEEE Trans. on PAMI*, 22, no.10, p. 1078-1089, Octobre 2000.

[LEF 00] LEFEBVRE E., COLOT O., VANNOONRENBERGHE P., "Contribution des mesures d'information à la modélisation crédibiliste de connaissance", *Traitement du Signal*, 17, no.2, p. 87-97, 2000.

[LIA 92] LIANG Z., JASZACK J., COLEMAN R., "Parameter estimation of finite mixture using the EM algorithm and IC with application to medical image processing", *IEEE Trans. on Nuc. Science*, 39, no.4, p. 1126-1131, 1992.

[MAD 04] MADIMAN M., HARRISON M., KONTOYIANNIS I., "MDL vs maximun likelihood in lossy data compression ", *IEEE-Int. Symposium on Information Theory,* 461, July 2004.

[NIS 88] NISHII R., "Maximum likelihood principle and model selection when the true model is unspecified", *Journal of Multivariate Analysis*, 27, p. 392-403, 1988.

[OLI 94] OLIVIER C., COURTELLEMONT P., COLOT O., DE BRUCQ D., EL MATOUAT A., "Comparison of histograms: a tool of detection", *European Journal of Diagnosis and safety in Automation,* 4, no.3, p. 335-355, 1994.

[OLI 96] OLIVIER C., COURTELLEMONT P., LECOURTIER Y., "Histogrammes et critères d'information en reconnaissance de formes", *10ème Congrès RFIA*, Rennes (France), 2, p. 1033-1042, January 1996.

[OLI 97] OLIVIER C., PAQUET T., AVILA M., LECOURTIER Y., "Optimal order of Markov models applied to bank checks" *Int. Journal of Pattern Recognition and Artificial Intelligence*, 11, no.8, p. 789-800, 1997.

[OUL 01] OULED-ZAÏD A., OLIVIER C., ALATA O., MARMOITON F., "Optimisation du codage d'images par les critères d'information", *18ème GRETSI*, paper # 316 – Toulouse (France), September 2001.

[OUL 03] OULED-ZAÏD A., OLIVIER C., ALATA O., MARMOITON F., "Transform image coding with global thresholding: application to baseline JPEG", *Pattern Recognition Letters*, 24, no.3, p. 959-964, April 2003.

[PAT 98] PATEUX S., Segmentation spatio-temporelle et codage orienté-régions de séquences vidéo basés sur le formalisme MDL, PhD thesis, University of Rennes 1, September 1998.

[RAT 00] RATNAKAR V., LIVNY M., "An efficient algorithm for optimizing DCT quantization", *IEEE Trans. Image Processing*, 9, no.2, p. 267-270, Feburary 2000.

[RIS 78] RISSANEN J., "Modeling by shortest data description", *Automatica*, 14, p. 465-471, 1978.

[RIS 89] RISSANEN J., *Stochastic Complexity in Statistical Inquiry*, World Scientific ed., New Jersey, 1989.

[RIS 92] RISSANEN J., SPEED T.P., YU B., "Density estimation by stochastic complexity", *IEEE Trans. on Information Theory*, 38, no.2, p. 315-323, March 1992.

[SAK 86] SAKAMOTO Y., ISHIGURO M., KITAGAWA G., *AIC in Statistics Mathematics and Applications*, KTK Scientific Publishers, Tokyo, 1986.

[SAR 96] SARDO L., KITTLER J., "Minimum complexity PDF estimation for correlated data", *ICPR'96*, p. 750-754, 1996.

[SCH 78] SCHWARZ G., "Estimating the dimension of a model", *The Annals of Statistics*, 6, p. 461-464, 1978.

[SHI 76] SHIBATA R., "Selection of the order of an AR model by AIC", *Biometrika*, 63, p. 117-126, 1976.

[SEG 03] SEGHOUANE A.K., BEKARA M., FLEURY G., "A small sample model selection criterion based on Kullback symmetric divergence", *IEEE-ICASSP'03*, Hong-Kong, 6, p. 145-148, April 2003.

[SPI 02] SPIEGELHALTER J., BEST N.G., CARLIN B.P., VAN DER LINDE A., "Bayesian measures of model complexity and fit", *Journal of the Royal Statistic Society*, Series B, 64, p. 583-640, 2002.

[STO 04] STOICA P., SELEN Y., LI J., "On IC and the GLRT of model order selection", *IEEE Signal Processing Letters*, 11, no.10, pp 794-797, October 2004.

[THU 97] THUNE M., OLSTAD B., THUNE N., "Edge detection in noisy data using finite mixture distribution analysis", *Pattern Recognition*, 3, no.5, p. 685-699, 1997.

[VAN 86] VAN ECK, "On objective AR model testing", *Signal Processing*, 10, p. 185-191, 1986.

[WAE 03] DE WAELE S, BROERSEN P.M., "Order selection for vector autoregressive Models", *IEEE Trans. on Signal Processing*, 51, no.2, p. 427-433, Feburary 2003.

[WAN 94] WANG Y., LEI T., MORRIS J.M., "Detection of the number of image regions by minimum bias variance criterion", *SPIE-VCIP'94*, 2308, Chicago, p. 2020-2029, September 1994.

[XIA 04] XIA F., OLIVIER C., KHOUDEIR M., "Utilisation d'une modélisation crédibiliste pour une aide à la décision dans un environnement indoor", *SETIT'04*, CD-ROM, Vol. Image, Sousse (Tunisia), March 2004.

Chapter 5

Quadratic Programming and Machine Learning – Large Scale Problems and Sparsity

5.1. Introduction

For a child, learning is a complex process that consists of acquiring or developing certain competencies on the basis of multiple experiences. For a machine, this learning process can be reduced to examples or observations that are used to improve performance. Machine learning can be seen as the optimization of criteria defined on examples. The higher the number of examples, the better the learning process. In terms of optimization, this learning process includes several specific problems. How are the criteria to be optimized defined? How can we manage large amounts of data? Which algorithm is efficient in this context?

When dealing with those problems, neural network approaches suggest the usage of non-convex criteria associated with gradient descent methods. This procedure leads to several difficulties linked to the non-convex criteria. The key to the success of kernel-based methods (obtained about a decade after the introduction of neural networks) is their capacity to express the learning problem as a large scale quadratic programming problem (convex). Kernel-based methods often lead to sparse solutions, i.e. a large number of their components equal zero. Based on this particularity, learning algorithms can solve large scale problems in a reasonable time. Solving this type of problem currently takes about one day of calculating when using a mono-processor for 8 million unknowns. Among these 8 million unknowns only 8,000 to 20,000 variables do not equal zero depending on the complexity of the problem.

Chapter written by Gaëlle LOOSLI and Stéphane CANU.

This chapter deals with quadratic programming for machine learning. In the general case of quadratic programming, "interior points" methods are the most efficient. On the other hand, when it comes to quadratic programming for machine learning, "active set" or "active constraint" methods are of a higher performance. This is due to the fact that "active set" methods take advantage of the geometric particularities of the problem and have sparse solutions.

Algorithms which are based on "active sets" have to know all training points in advance to provide the exact solution to a problem. This functioning is known as "off-line" or "batch". Not all machine learning problems, however, are covered by this type of algorithm. It may occur that data are revealed during learning. This type of learning process is referred to as "on-line". On-line learning processes must also be used if the database is too large for all examples to be processed simultaneously. This even applies if all examples are known right at the beginning. The on-line framework has led to iterative stochastic optimization methods that are adapted to large scale and sparse problems. It is widely accepted that no method of polynomial convergence (especially quadratic programming) can be used to solve problems of large dimensions [NEM 05]. However, if this is true when looking for precise solutions, it is possible to develop methods that provide non-optimal solutions (which can be neglected given the incertitude of examples), but much faster. An on-line stochastic approach will be presented that can both solve large scale and/or on-line problems and thus outperform off-line methods.

This chapter is organized as follows. First the general framework of learning processes and tools for convex quadratic programming will be explained. Secondly "active sets" and their usage of geometrics and sparsity will be shown. A very efficient method of iterative stochastic optimization, LASVM, will also be explained. Last but not least, some results of experiments will underline the possibilities offered by the method described above.

5.2. Learning processes and optimization

Statistical learning based on data use form vectors x of dimension d. For supervised learning, every individual x comes with its label y. This label characterizes the individual (its group, value, structure etc.). The aim of learning is to enable a machine to find the label of a specific piece of data.

5.2.1. *General framework*

Define the family $\mathcal{P}$, of all laws on $(\mathbb{R}^d \times \{-1, 1\})$. Training points are generated from a unknown probability law $\mathbb{P}(x, y)$. Within this framework the result of the learning process is defined as the function f (for more information on this function see section 5.2.2) which for all $\mathbb{P} \in \mathcal{P}$ is able to classify well (on average) points drawn from $\mathbb{P}$ with a high probability, depending on the sample.

DEFINITION 5.1. The training set $\{\mathcal{X}_n^d, \mathcal{Y}_n\}$ *contains all known examples drawn from the underlying law of* $\mathbb{P}(x, y) \in \mathcal{P}$. *The size of this database is* n *and every example (or individual) is* x_i, *with* $i \in [1, n]$. *The examples lie in a space of dimension* d *(usually* $\mathbb{R}^d$*). The labels associated with each example are mainly scalars. The most frequent cases are* $y_i \in \mathbb{R}$ *for regression and* $y_i \in \{-1, 1\}$ *for discrimination.*

DEFINITION 5.2. The test set $\{\mathcal{X}_t^d, \mathcal{Y}_t\}$ *represents a group of unseen examples generated by the same law* $\mathbb{P}(x, y) \in \mathcal{P}$.

An algorithm is efficient if it is able to provide similar results (in terms of errors) for the training set as well as the test set. This type of performance is referred to as the ability to generalize.

DEFINITION 5.3. The empirical risk *measures the algorithm's efficiency on the training set depending on a given cost function* C.

$$R_{emp}[f] = \frac{1}{n} \sum_{i=1}^{n} C(f, x_i, y_i) \quad \textit{with, for example} \quad C(f, x_i, y_i) = \frac{1}{2} |f(x_i) - y_i|.$$

The empirical risk is to be minimized during the learning process.

Nothing guarantees that if the function f provides low empirical risks, it also obtains good results for unknown data. Generalization has to be checked and the *risk* has to be taken into account during the learning process.

DEFINITION 5.4. The risk *is a similar idea to the empirical risk but concerns test data. As this type of data is unknown,* $\mathbb{P}(x, y)$ *is required to express the risk:*

$$R[f] = \mathbb{E}\big(C(f, X, Y)\big) = \int C(f, x, y) d\mathbb{P}(x, y).$$

In order to do this, the empirical risk and a quantity that controls generalization error are minimized together. The statistical theory of learning (see [VAP 95, SCH 02] for more details) uses the notion of capacity to express this type of quantity.

DEFINITION 5.5. The capacity *of an algorithm is defined as the cardinal of the space of hypothesis. In the case of infinite dimension, the capacity is defined by VC-dimension.*

VC-dimension provides a notion of capacity for a group of functions. VC-dimension is defined by the number of points that can be separated in every possible manner by the functions of this group. On the basis of VC-dimension, the Vapnik-Chervonenkis theory provides bounds on the risk. The risk can therefore be bounded in terms of the empirical risk, the capacity of functions to be used (their VC-dimension h) and the number of training examples with a probability $1 - \eta$ for $C = \frac{1}{2}|f(x_i) - y_i|$, [BUR 98, equation 3, page 3]:

$$R[f] \leq R_{\text{emp}}[f] + \sqrt{\frac{h\big(\log(2n/h) + 1\big) - \log(\eta/4)}{n}}.$$

The idea consists of obtaining sufficiently *flexible* functions to be able to learn data but also sufficiently regular to generalize well. The learning process is therefore seen as an optimization problem of two criteria. These criteria are finding the function of the weakest capacity and minimizing the empirical risk.

5.2.2. *Functional framework*

A learning algorithm chooses the best solution from all hypotheses $\mathcal{H}$ depending on the training set and the two criteria to be optimized. Kernel-based methods build the pool of hypotheses $\mathcal{H}$ using the kernel. When it comes to classifying the points, it makes sense to know which objects are close to each other or are similar to each other. If it is possible to establish the distance between two objects, no matter what their form is, this information is sufficient to carry out the process of discrimination. The *kernel* is defined on the basis of this idea.

DEFINITION 5.6. A kernel *is a symmetric function with two variables that returns a scalar to express the distance between two variables. If $s, t \in \mathbb{R}^d$, k is defined as:*

$$\begin{aligned} k : \mathbb{R}^d \times \mathbb{R}^d &\longrightarrow \mathbb{R} \\ s, t &\longmapsto k(s,t) \end{aligned}$$

On the basis of the kernel (if it positive definite[1]) the space of hypothesis can be constructed via the pre-Hilbertian space.

DEFINITION 5.7 (Pre-Hilbertian space). *Let k be a positive definite kernel. $\mathcal{H}_0$ is defined as the vector space based on the linear combinations of the kernel:*

$$\mathcal{H}_0 = \left\{ f : \mathbb{R}^d \longmapsto \mathbb{R} \mid \ell < \infty,\ \{\alpha_i\}_1^\ell \in \mathbb{R},\ \{s_i\}_1^\ell \in \mathbb{R}^d,\ f(x) = \sum_{i=1}^{\ell} \alpha_i k(s_i, x) \right\}$$

1. Only cases for which the kernel is positive definite will be taken into consideration, i.e. $\forall \ell < \infty, \forall \alpha_i \in \mathbb{R}, \forall x_i \in \mathbb{R}^d, i = 1 \ldots \ell; \sum_i \sum_j \alpha_i \alpha_j k(x_i, x_j) \geq 0$.

This space is provided with a scalar product defined as a bilinear form such that for all $f, g \in \mathcal{H}_0$, $f(x) = \sum_{i=1}^{\ell} \alpha_i k(s_i, x)$ *and* $g(x) = \sum_{j=1}^{m} \beta_j k(t_j, x)$,

$$\langle f, g \rangle_{\mathcal{H}_0} = \sum_{i=1}^{\ell} \sum_{j=1}^{m} \alpha_i \beta_j k(s_i, t_j)$$

If the kernel is positive, this also guarantees that the scalar product is positive and definite. The induced space with this scalar product is a pre-Hilbertian space.

Completing the pre-Hilbertian space $\mathcal{H}_0$ according to the norms introduced by the scalar product turns it into a Hilbert space $\mathcal{H} = \overline{\mathcal{H}_0}$.

PROPERTIES. The norm induced in this space equals $\|f\|_{\mathcal{H}}^2 = \langle f, f \rangle_{\mathcal{H}}$. Also note that:

$$\langle f(\cdot), k(\cdot, x) \rangle_{\mathcal{H}} = \left\langle \sum_{i=1}^{\ell} \alpha_i k(s_i, \cdot), k(\cdot, x) \right\rangle_{\mathcal{H}} = f(x)$$

This is the property of reproduction. The space $\mathcal{H}$ is thus a *Reproducing Kernel Hilbert.*

Due to the usage of this property of reproduction, the problem of finding a general solution f in the space of functions $\mathcal{H}$ comes down to the problem of finding a vector $\boldsymbol{\alpha} \in \mathbb{R}^n$ for which n describes the number of available examples for the learning process. The link between the function f and the vector $\boldsymbol{\alpha}$ is provided by the representer theorem ([SCH 02, Theorem 4.2, page 90]):

$$f(x) = \sum_{i=1}^{n} \alpha_i k(x_i, x).$$

5.2.3. *Cost and regularization*

Now all hypotheses have been defined, the two criteria to be minimized during the learning process will be explained in detail. These are the empirical risk and the capacity. They are referred to as the minimization of regularized cost. The *regularization* is a form of capacity control. The penalization term is to be used. This term grows if the decision function increases in complexity. For an exact solution to the problem, at least one of the two criteria has to be of the type $\mathcal{L}_1$ (as opposed to the cost type $\mathcal{L}_2$) [NIK 00][2].

2. See the article *"When does sparsity occur?"* on O. Bousquet's blog http://ml.typepad.com/machine_learning_thoughts/.

DEFINITION 5.8 (Cost of the type $\mathcal{L}_2$). *The cost is known as type $\mathcal{L}_2$ if it takes a quadratic form.*

For example, this is the case for the cost $C_2 = (f(x_i) - y_i)^2$ used for regression, the quadratic hinge cost $C_h^2 = \max(0, y_i(f(x_i) + b - 1))^2$ used for discrimination and terms of quadratic penalization type $\Omega_2 = \|f\|_{\mathcal{H}}^2$. The cost linked to logistical regression $C_\ell = y_i \log f(x_i) + (1 - y_i)\log(1 - f(x_i))$ does not strictly refer to a cost type $\mathcal{L}_2$ but possesses the following characteristics: regularity and convexity.

Costs of the type $\mathcal{L}_1$ are more difficult to define formally. This type of costs leads to sparsity. These are often built on the basis of absolute values and are singular at origin [NIK 00]. For example, this is the case of cost $C_1 = |f(x_i) - y_i|$ and $C_1(\varepsilon) = \max(0, |f(x_i) - y_i| - \varepsilon)$ used for the regression, for the cost $C_h = \max(0, y_i(f(x_i) + b - 1))$ used as a form of discrimination, and the terms of penalization type $\Omega_1 = \sum_{i=1}^n |\alpha_i|$ when the solution f we are looking for verifies $f(x) = \sum_{i=1}^n \alpha_i k(x_i, x)$.

The choice of cost functions and regularization leads to different types of algorithms (see Table 5.1 for examples). These algorithms minimize the cost $\mathcal{L}_2$. They are of a Gaussian type and very popular due to their properties of derivation and their simple calculations. Algorithms that minimize the term $\mathcal{L}_1$ are also often used due to their capacity to provide sparse solutions. On the other hand, the usage in terms of $\mathcal{L}_1$ implies rather slow solutions (see simplex). In [POR 97] the authors show how the $\mathcal{L}_1$ method may compete with the $\mathcal{L}_2$ method in terms of algorithms. This chapter will show how, in practice, it makes more sense to benefit from sparsity.

Cost	Ω	Discrimination	Regression
$\mathcal{L}_1$	$\mathcal{L}_1$	LP SVM [MAN 98]	LP SVR: $C_1(\varepsilon)$ and Ω_1
$\mathcal{L}_2$	$\mathcal{L}_1$	HLAR: C_h^2 and Ω_1 [KEE 05]	LARS: C_2 and Ω_1 [EFR 04]
$\mathcal{L}_1$	$\mathcal{L}_2$	SVM: C_h and Ω_2	SVR: $C_1(\varepsilon)$ and Ω_2
$\mathcal{L}_2$	$\mathcal{L}_2$	Logistical K-regression: C_ℓ and Ω_2 Lagrangian SVM: C_h^2 and Ω_2 [MAN 01]	Splines: C_2 and Ω_2

Table 5.1. *Overview of different learning algorithms according to the nature of objective functions*

5.2.4. *The aims of realistic learning processes*

Learning processes have to be a tool for the processing of data taken from the real world. Particularly when processing signals there is a large amount of signals which are affected by ambient noise and variables. The challenge in learning processes does therefore not only affect development of high performing methods and their primary tasks (classification, regression, etc.), but also has an impact on the available amount of computer storage. This is why on-line algorithms and those of a lower complexity (less than $\mathcal{O}(n^2)$) are likely to be used in the development of realistic applications.

5.3. From learning methods to quadratic programming

One of the most widely used algorithms that requires a quadratic program is probably the SVM (Support Vector Machine) as well as a variety of declinations. This chapter will show how a binary SVM, or one class SVM or SVM for regression (SVR – Support Vector Regression machine) may use identical forms for problem solving. Several algorithms will be shown as well as their capacity to solve a particular system.

5.3.1. *Primal and dual forms*

5.3.1.1. *Binary classification: SVMs*

For a problem of discriminating two classes, the separating functions have to be found between the examples of each class. This separation is based on a $\mathcal{RKHS}$ known as $\mathcal{H}$. The decision function is $D(x) = \text{sign}(f(x) + b)$. The primal expression of a binary SVM is as follows:

$$\begin{cases} \displaystyle\min_{f\in\mathcal{H},\boldsymbol{\xi},b} \frac{1}{2}\|f\|^2_{\mathcal{H}} + C\sum_{i=1}^{n}\xi_i \\ \quad y_i\big(f(x_i)+b\big) \geq 1-\xi_i \quad i \in [1,\dots,n] \\ \quad \xi_i \geq 0 \quad i \in [1,\dots,n]. \end{cases} \qquad [5.1]$$

This system finds the function of $\mathcal{H}$ of the smallest norm $\mathcal{L}_2$ (regularization). This function correctly subdivides the learning points into different classes. The constraints of correct classification (the cost, here described as $\mathcal{L}_1$) are linked to a margin of 1 that obliges the separating function to be situated far from training points. Furthermore, it is possible to infringe the constraints of correct classification (i.e. loosen up those limits) due to variables ξ_i. C represents the maximum influence of a point within this solution. Solving this kind of system is based on the usage of a Lagrangian:

$$\begin{aligned} \mathcal{L}(f,\boldsymbol{\xi},b) = &\frac{1}{2}\|f\|^2_{\mathcal{H}} + C\sum_{i=1}^{n}\xi_i \\ &- \sum_{i=1}^{n}\alpha_i\big(y_i\big(f(x_i)+b\big) - 1 + \xi_i\big) - \sum_{i=1}^{n}\beta_i\xi_i \end{aligned} \qquad [5.2]$$

with $i \in [1, \ldots, n]$, $\xi_i \geq 0$, $\alpha_i \geq 0$ and $\beta_i \geq 0$. Minimizing [5.1] comes down to canceling the derivatives of $\mathcal{L}$ in combination with every variable:

$$\begin{cases} \Delta_f(\mathcal{L}) = 0 \\ \Delta_\xi(\mathcal{L}) = 0 \\ \Delta_b(\mathcal{L}) = 0 \end{cases} \iff \begin{cases} f(\cdot) = \sum_{i=1}^{n} \alpha_i y_i k(x_i, \cdot) \\ C - \beta_i - \alpha_i = 0 \\ \sum_{i=1}^{n} y_i \alpha_i = 0 \end{cases} \quad [5.3]$$

When replacing these two relations in [5.2] a dual problem can be obtained which has been described in a matrix form:

$$\begin{cases} \max_{\boldsymbol{\alpha} \in \mathbb{R}^n} -\frac{1}{2} \boldsymbol{\alpha}^\top G \boldsymbol{\alpha} + \mathbf{e}^\top \boldsymbol{\alpha} \\ \mathbf{y}^\top \boldsymbol{\alpha} = 0 \\ 0 \leq \alpha_i \leq C \quad i \in [1, \ldots, n]. \end{cases} \quad [5.4]$$

$\mathbf{e}$ is a vector of ones and G describes the kernel matrix weighted by the labels of the points such as $G_{ij} = y_i K(x_i, x_j) y_j$. This dual system consists of convex quadratic programming with the particularity that it is confronted with a problem of n unknowns with $2n + 1$ constraints. Furthermore, G is a positive, semi-definite matrix.

5.3.1.2. *One class classification: OC-SVM*

For a problem of one class classification, the border line around the training class has to be found. Primal expression of *one class SVM* is expressed as:

$$\begin{cases} \min_{f \in \mathcal{H}, b, \xi} \frac{1}{2} \|f\|^2_{\mathcal{H}} + C \sum_{i=1}^{n} \xi_i - b \\ f(x_i) \geq b - \xi_i \quad i \in [1, \ldots, n] \\ \xi_i \geq 0 \quad i \in [1, \ldots, n] \end{cases} \quad [5.5]$$

As for binary cases, the Lagrangian is also applied and the following dual expression can be obtained:

$$\begin{cases} \max_{\boldsymbol{\alpha} \in \mathbb{R}^n} -\frac{1}{2} \boldsymbol{\alpha}^\top K \boldsymbol{\alpha} \\ \mathbf{e}^\top \boldsymbol{\alpha} = 0 \\ 0 \leq \alpha_i \leq C \quad i \in [1, \ldots, n] \end{cases} \quad [5.6]$$

This system is also a convex quadratic form of programming.

5.3.1.3. *Regression: SVR*

When dealing with the problem of regression, expected outputs are reals. Here the data is not to be separated but the aim is to obtain a function that represents the best of the training points. The idea is for the solution to fit in a tube of the size 2ϵ based on training examples. The primal formula of this problem is expressed as follows:

$$\begin{cases} \min_{f\in\mathcal{H},\boldsymbol{\xi},\boldsymbol{\xi}^*,b} \frac{1}{2}\|f\|^2_{\mathcal{H}} + C\left(\sum_{i=1}^{n}\xi_i^* + \sum_{i=1}^{n}\xi_i\right) \\ \quad y_i - f(x_i) - b \leq \epsilon + \xi_i \quad i \in [1,\ldots,n] \\ \quad -y_i + f(x_i) + b \leq \epsilon + \xi_i^* \quad i \in [1,\ldots,n] \\ \quad \xi_i^* \geq 0 \quad i \in [1,\ldots,n] \\ \quad \xi_i \geq 0 \quad i \in [1,\ldots,n] \end{cases} \qquad [5.7]$$

According to the same method, a Lagrangian is always used for dual formulae:

$$\begin{cases} \max_{\boldsymbol{\alpha}\in\mathbb{R}^n} -\frac{1}{2}(\boldsymbol{\alpha}^{*\top}K\boldsymbol{\alpha}^* - \boldsymbol{\alpha}^{*\top}G\boldsymbol{\alpha} - \boldsymbol{\alpha}^\top G\boldsymbol{\alpha}^* + \boldsymbol{\alpha}^\top G\boldsymbol{\alpha}) \\ \quad -\epsilon\mathbf{e}^\top\boldsymbol{\alpha} - \epsilon\mathbf{e}^\top\boldsymbol{\alpha}^* - \mathbf{y}^\top\boldsymbol{\alpha} + \mathbf{y}^\top\boldsymbol{\alpha}^* \\ \quad \mathbf{e}^\top\boldsymbol{\alpha} = \mathbf{e}^\top\boldsymbol{\alpha}^* \\ \quad 0 \leq \alpha_i \leq C \quad i \in [1,\ldots,n] \\ \quad 0 \leq \alpha_i^* \leq C \quad i \in [1,\ldots,n] \end{cases} \qquad [5.8]$$

This formula is similar to convex quadratic programming and it is sufficient to adapt the data as follows:

$$\begin{cases} \max_{\boldsymbol{\beta}\in\mathbb{R}^{2n}} -\frac{1}{2}\boldsymbol{\beta}^\top G\boldsymbol{\beta} + \mathbf{c}^\top\boldsymbol{\beta} \\ \quad \mathbf{u}^\top\boldsymbol{\beta} = 0 \\ \quad 0 \leq \beta_i \leq C \quad i \in [1,\ldots,2n] \end{cases} \qquad [5.9]$$

G describes the matrix which consists of four kernel matrices $G = [K, -K; -K, K]$. $\boldsymbol{\beta}$ is a vector that contains both $\boldsymbol{\alpha}$ and $\boldsymbol{\alpha}^*$, i.e. $\boldsymbol{\beta} = [\boldsymbol{\alpha}, \boldsymbol{\alpha}^*]$. $\mathbf{c}$ is the vector which represents $[-y-\epsilon, y-\epsilon]$, while the vector $\mathbf{u}$ represents $[\mathbf{e}, -\mathbf{e}]$.

5.4. Methods and resolution

A certain number of problems in learning processes can be expressed in a convex quadratic program. This part of the chapter will show how to efficiently obtain a solution for minimization under "box" constraints.

5.4.1. *Properties to be used: sparsity*

First of all note that the formula to be optimized is convex and only produces a single optimal solution. Also note that the constraints are linear and of a finite number. The best solution α is verified for each of its components α_i according to the Karuch Kuhn Tucker (KKT) conditions, i.e.

$$\begin{aligned} \alpha_i = 0 &\longrightarrow f(x_i)y_i > 1 \\ 0 < \alpha_i < C &\longrightarrow f(x_i)y_i = 1 \\ \alpha_i = C &\longrightarrow f(x_i)y_i < 1. \end{aligned} \qquad [5.10]$$

The general principle of methods used for subdivision and/or active sets is based on the observation that only the points that are not affected by the constraints in the solution require a calculation of their coefficient. This general principle is known as sparsity. Other methods have fixed values as coefficients for each particular problem. They are bounded (by 0 and C for the case of SVM for example). Based on this statement, it is sufficient to know the distribution of points in three groups (constrained at 0, constrained at C, not constrained) for the solution of the problem. These groups will from now on be referred to as I_0, I_C and I_w, for the respective points of 0, C or not constrained. This statement promotes different techniques of distribution and different algorithms that will be explained after presenting some tools that are shared by several methods.

5.4.2. *Tools to be used*

ADMISSIBLE DIRECTIONS. The geometry of a dual and convex quadratic problem is explained in Figure 5.1. The box constraints $0 \leq \alpha_i \leq C$ limit the solution to a hypercube of the dimension n. The equality constraint $\sum y_i \alpha_i = 0$ also limits the solution to a polytope $\mathcal{F}$ of dimension $n - 1$. Let us consider an admissible point $\boldsymbol{\alpha} \in \mathcal{F}$ and a straight line containing this point. This line indicates the admissible direction if its intersection with the polytope contains other points apart from $\boldsymbol{\alpha}$.

An algorithm of admissible direction updates the position of the admissible point $\boldsymbol{\alpha}_t$ in an iterative way. First of all a direction u_t has to be chosen. The algorithm then follows this direction and searches for an admissible point $\boldsymbol{\alpha}_{t+1}$ which maximizes the cost function. The optimum is reached once improvements are no longer possible [ZOU 60]. There are two possible configurations when searching for an admissible point. The quadratic cost function is limited to the direction and reaches its maximum either inside or outside the polytope.

SHRINKING. *Shrinking* [JOA 99] is a form of heuristics that determines, while the algorithm is being carried out, which points are excluded from the solution or are

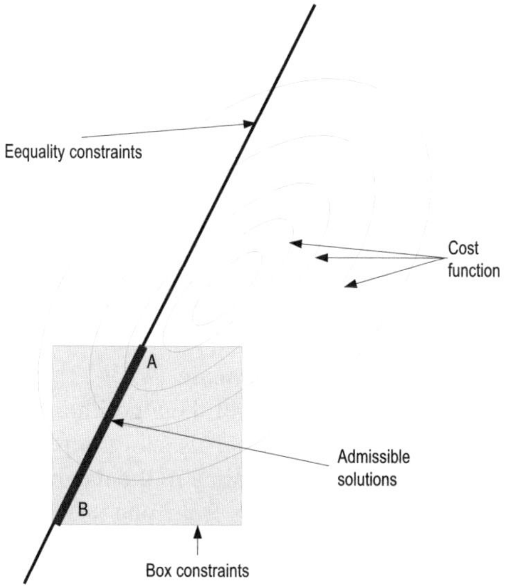

Figure 5.1. *The geometry of a dual and quadratic problem. The box constraints* $0 \leq \alpha_i \leq C$ *limit the solution to a hypercube of the dimension* n. *The equality constraint* $\sum \mathbf{y}_i \alpha_i = 0$ *also limits the solution to a polytope of the dimension* $n - 1$. *The function for quadratic cost is limited to the line that is being searched for. This line has its maximum at the inside or at the outside of the constraint box*

bounded. The value of these points is therefore known without having to calculate their coefficient α. This is why the points can no longer be taken into consideration and the size of the problem is reduced mechanically so that the problem can be solved. As this type of heuristic might lead to errors, it is sufficient to check the algorithm once it has stopped and verify whether the points to be excluded are in group I_0 or group I_C. An optimization step might be required at this point.

5.4.3. *Structures of resolution algorithms*

Iterative algorithms for resolution share a common structure (Algorithm 5.1). The methods are distinguished by their method of distributing the points and the way in which they calculate α.

5.4.4. *Decomposition methods*

As their name indicates, the decomposition methods use a sub-group of the training database at every stage of problem solving. The action of problem solving is therefore subdivided into smaller dimensions. The results of these sub-problems are combined in order to obtain the best global solution.

Initialization
While (current solution is not optimal) **do**
  update groups
  compute the coefficients α_i which correspond to the changes
done

Algorithm 5.1. *General schema showing algorithms of an iterative resolution*

CHUNKING. The method known as *chunking* [VAP 82] eliminates the points for which α becomes equal to 0 in the course of the algorithm. The main idea is to solve the QP problem for a sub-group of points. This sub-group is made up out of support vectors of the previous solution and M points in the database. These points infringe KKT more than any other point. Moving from one sub-group to the next ensures that the problem is eventually solved. The limit of this method is the size of the final solution when the entire kernel matrix cannot be stored in the memory.

DECOMPOSITION. The technique of decomposition [OSU 97] is similar to the principle of *chunking*. In this approach the size of the sub-groups remains the same and allows for vector points to be removed. This is how all quadratic problems can be solved regardless of their size.

SMO (sequential minimal optimization). The extreme case only considers two points in every stage. This method is referred to as SMO [PLA 99]. The main advantage is that every quadratic sub-problem of a limited size can be solved analytically. Digital resolution, which is often very cost intensive, can therefore be avoided.

SMO only considers admissible solutions that modify only two coefficients α_i and α_j on the basis of opposed values. The most widely used version of SMO is based on the first order criteria which are used to select the pairs (i, j) which define the following directions:

$$
\begin{aligned}
i &= \underset{\{s|\alpha_s<\max(0,y_sC)\}}{\arg\max} \frac{\partial W}{\partial \alpha_s} \\
j &= \underset{\{s|\alpha_s>\min(0,y_sC)\}}{\arg\min} \frac{\partial W}{\partial \alpha_s}.
\end{aligned}
\qquad [5.11]
$$

Furthermore, the majority of SMO implementations (iterative schema of Algorithm 5.2) use the *shrinking* technique in order to limit the space of searching α. New criteria based on second order information have been published recently [FAN 05]. These criteria ensure that the results can be used in theory and practice for a better convergence.

chose two initial points i and j
calculate their coefficients α_i and α_j
While (the current solution is not optimal) **do**
  select two new points
  calculate the associated coefficients
done

Algorithm 5.2. *Iterative schema of an SMO algorithm*

The use of SMO does not require the usage of QP resolving tools. For other decomposition methods every sub-problem has to be solved in the form of a quadratic problem.

5.4.5. *Solving quadratic problems*

5.4.5.1. *Methods based on interior points*

Methods based on interior points, suggested by [KAR 84], solve linear programming in a polynomial amount of time. Their derivation for semi-definite, positive and non-linear programming has led to algorithms such as LOQO and OOQP [VAN 99, GER 01]. Even though these methods are very efficient when it comes to optimization, they have been outperformed by methods that will be presented later on. The methods of interior points work on the complete training set. Thus, they are not dedicated to large databases whatever the efficiency of the algorithm. The main challenge in learning processes is sparsity. This property is the key to solving large quadratic problems. Even when working on a sub-group of points taken from the learning process's database, as is the case for decomposition methods, it still saves time and effort to use methods which rely on sparsity and reduce complexity.

5.4.5.2. *Active constraint methods*

The principle of these methods is an efficient distribution of the points into three groups I_0, I_C and I_w. These groups obtain the values of $\alpha_i, i \in I_w$ by solving a linear system. For SVM, the algorithm known as SimpleSVM [VIS 03] (it can, however, be applied to all convex quadratic problems with box constraints) distributes points iteratively into three groups.

At every stage the minimization is solved on group I_w without constraints. If one of the values represented in α violates the constraints (i.e. the solution is situated outside the box), α is projected into the admissible space (i.e. in practice the indicated point changes group via the value that infringes the constraints.). This procedure will be explained in detail.

The iterative schema of this type of algorithms works as follows:

choose initial distribution
calculate initial solution
While (the current solution is not the best one) **do**
  update the distribution into groups
  calculate the associated coefficients for group I_w
done

Algorithm 5.3. *Iterative schema of a SimpleSVM*

The choice of a particular distribution into groups, calculating coefficients and the evaluation of performances for a solution still have to be defined.

SOLVING A DUAL PROBLEM. The problem to be solved: if the points with coefficients equal to zero are removed in equation [5.4], the system can be expressed as follows:

$$\begin{cases} \max\limits_{\boldsymbol{\alpha}_w \in \mathbb{R}^{|I_w|}} & -\frac{1}{2}\boldsymbol{\alpha}_w^\top G_{ww}\boldsymbol{\alpha}_w + \mathbf{e}_w^\top \boldsymbol{\alpha}_w - C e_c^\top G_{cw}\alpha_w \\ & \mathbf{y}_w^\top \boldsymbol{\alpha}_w = -C y_C^\top \mathbf{e}_C \\ \max\limits_{\boldsymbol{\alpha}_w \in \mathbb{R}^{|I_w|}} & 0 < \alpha_i < C \quad i \in I_w \end{cases} \quad [5.12]$$

G_{ww} describes the kernel between the points of I_w and G_{cw} between I_C and I_w. The Lagrangian is used to obtain the solution with the multipliers λ. With $e_v = e_w - CG_{wc}e_c$, $\boldsymbol{\alpha}_w = G_{ww}^{-1}(\mathbf{e}_v - \lambda \mathbf{y}_w)$ can be obtained. With $N = G_{ww}^{-1}\mathbf{e}_{wc}$ and $M = G_{ww}^{-1}\mathbf{y}_w$,

$$\boldsymbol{\alpha}_w = N - \lambda M. \quad [5.13]$$

Given the constraints in problem [5.12], the following relationship can be obtained: $\mathbf{y}_w^\top(N - \lambda M) = -Cy_c^\top \mathbf{e}_C$. The values of the Lagrange multipliers (which correspond, in the primal, to the bias indicated as b in different primal expressions) is deduced:

$$\lambda = \frac{\mathbf{y}_w^\top N + C y_c^\top \mathbf{e}_c}{\mathbf{y}_w^\top M}. \quad [5.14]$$

For a given distribution of data between the three groups of points, the coefficients of the decision function are calculated. It the distribution is correct, all elements of $\boldsymbol{\alpha}_w$ are between 0 and C.

DISTRIBUTION OF GROUPS OF POINTS. In order to attribute every point to a group the corresponding values of α will be used. The straightforward (but not very delicate) strategy is to place the entire training set into the group I_w at the first step. After solving the quadratic program in I_w once, (all negative multipliers send their point to group I_0 and all multipliers larger than C send their points to group I_C) The problem of this method is the machine aspect, since solving the entire system directly from a large database is very cost intensive in terms of time and computer storage. However, this notion is used when producing an iterative version. Let's have another look at the iterative schema given for Algorithm 5.3.

```
choose initial distribution: (random)
calculate initial solution: eq. [5.13] and [5.14]
While (the current solution is not optimal: there are violated constraints) do
    update groups:
    If (one of the calculated α is negative or higher than C) then
        Transfer this point to I_0 or I_C
    else
        If (constraints in the primal is violated by a point from I_C or I_0) then
            Transfer this point to I_w
        else
            no constraints are violated anymore
        end If
    end If
    calculate associated coefficients for group I_w: eq. [5.13] and [5.14]
done
```

Algorithm 5.4. *Detailed SimpleSVM algorithm*

Initialization can be carried out randomly (making a good choice for the first point can accelerate the convergence of the algorithm). Checking whether a solution actually is the best one is carried out in two stages. The box's constraints have to be checked ($0 < \alpha < C$) as well as the classification constraints (or regression, or aim of the algorithm) given by the primal expression. These constraints also have to be respected.

SOLVING LINEAR SYSTEMS. Calculating Lagrange multiplications α requires solving a linear equation. This is usually a complex operation $\mathcal{O}(n^3)$. This operation is very cost intensive and one of the main issues in the algorithm. It therefore makes sense to take a more detailed look at this topic. To solve the equation $\boldsymbol{\alpha}_w = G_w^{-1}(\mathbf{e}_{wc} - \lambda \mathbf{y}_w)$, the inverse of the matrix does not have to be calculated. Generally G_w is subdivided. When doing so, G_w's properties (symmetric, semi-defined and positive) play an important role and can be used when subdividing

QR. Q is an orthogonal matrix and R a triangular matrix. This subdivision ($G_w = QR$) leads to $QR\boldsymbol{\alpha}_w = \mathbf{e}_{wc} - \lambda\mathbf{y}_w \Leftrightarrow R\boldsymbol{\alpha}_w = Q^\top(\mathbf{e}_w - \lambda\mathbf{y}_w) \Leftrightarrow \boldsymbol{\alpha}_w = R^{-1}(Q^\top(\mathbf{e}_{wc} - \lambda\mathbf{y}_w))$. The latter operation is less cost intensive as R is triangular. Subdividing a matrix, no matter whether this is done on the basis of QR or other methods, is complex $\mathcal{O}(n^3)$ and is therefore the most cost intensive operation in the solving process. However, for every iteration one single point is transferred from one group to another. First range updating techniques can be used for different methods of subdividing matrixes. Every iteration is therefore linked to a complex $\mathcal{O}(n^2)$. There are techniques to prove convergence for SMO [KEE 02] and SimpleSVM [LOO 05].

5.4.6. *Online and non-optimized methods*

Online methods are very interesting as they allow for realistic applications. Their aim is to combine performance of results with SVM type methods. This type of usage can be adapted to different contexts and used on a large scale. LASVM is an algorithm that combines the advantages of SMO and SimpleSVM to reach this aim.

LASVM [BOR 05] is an online SVM that increases incrementally with respect to the dual objective. LASVM maintains a vector of current coefficients $\boldsymbol{\alpha}$ and the indexes for support vectors that correspond to I_v (here, I_v is made up out of I_w and I_c). Every iteration of LASVM receives a new example from I_0 and updates the vector of coefficients $\boldsymbol{\alpha}$ by using two stages of SMO referred to as *process* and *reprocess*:

– *Process* means searching for an admissible point. The direction is defined by a pair of points created with the index of the current example σ and another example chosen from I_v by using the criterion of the first order (equation [5.11]). This operation provides a new vector for the coefficients $\boldsymbol{\alpha}'_t$ and can insert σ into all corresponding I_v.

– *Reprocess* also means searching for an admissible point. The direction is, however, defined by a pair of points (i, j) both taken from I_v by using a criterion of the first order (e.g. equation [5.11]). This operation provides new vector coefficients $\boldsymbol{\alpha}$ and may delete i, j from the group of indexes I_v.

Repeating the LASVM operations (Algorithm 5.5) on a random choice of examples leads to the convergence of SVM solutions with an arbitrary level of precision. However, depending on the size of n, the experiment shows that LASVM also provides good results after a single presentation of every example. After every example has been presented, the final coefficients are refined and carry out a *reprocess* up to the convergence of the dual function.

Every iteration of an online LAVSM randomly chooses examples for the learning process $(x_{\sigma(t)}, y_{\sigma(t)})$. The most sophisticated strategies of choosing examples lead to a better performance on the level of a scale. There are four strategies in total:

– *random selection*: randomly choose an example;

```
Initialize
While (there are points which still have to be covered) do
    | select point which has to change group
    | optimize α of I_v
done
finalize
```

Algorithm 5.5. *Schema of the LASVM algorithm for online functioning*

– *gradient-based selection*: choose the example with the lowest classification (lowest value for $y_k f(x_k)$) amongst the points which have not been dealt with before. This criterion is close to what happens during a SimpleSVM;

– *active selection*: choose an example that is close to the limitations of decision-making, i.e. the smallest value of $|f(x_k)|$ amongst a random group of points. This criterion automatically chooses the example without considering its label;

– *self-active selection*: choose from 100 points which have not been dealt with before but stop as soon as five of them are outside the margin. From those five, this type of selection chooses the closest one to the limits of decision making.

There is empirical proof that the active or self-active selection leads to similar or better performances if the number of support vectors to be used is limited. This fact is based on the linear increase of the number of support vectors as every example which has been wrongly classified is automatically a support vector in the traditional form of SVM. Choosing the points purely on the basis of which one is close to the limit excludes a high number of points from a possible solution.

On the basis of the structure, LAVSM exploits large databases. Of course, a problem for the learning process is not automatically a problem of optimization. It does not make any sense to optimize the function of the primal objective with a higher precision than when working with a finite number of examples. Formal results exist for the online learning process and use an estimation of the stochastic gradient. These algorithms have been designed for online functioning and are much faster than direct optimization of a function with a primal objective [BOT 04]. This type of algorithm does not optimize the function as precisely as traditional optimization, but leads to the same number or errors in tests. A specific stage is also reached more rapidly by this algorithm than by traditional methods.

5.4.7. *Comparisons*

For a better illustration of the links between SimpleSVM, SMO and LASVM, the essential functions of these algorithms will be shown below:

– initialize: correspond to what has to be done before entering the iterative phase;

– select: choose points to be moved from one group to another;

– optimize: calculate the Lagrange multipliers;

– finalize: guarantee local and global optima.

Table 5.2 compares these different stages to one another. If these three algorithms are similar, the major difference takes place during the optimization stage α. SimpleSVM maintains the same group of coefficients at optimized values throughout the entire process, while LASVM and SMO update those values. Due to its online functioning LASVM is faster than SMO.

General	SimpleSVM [VIS 03]	LASVM [BOR 05]	SMO [PLA 99]
Selection	Search groups I_0 and I_C for a point that violate constraints	Choose one point from the following points	Search for a point in I_w and a point in the entire database according to SMO's heuristics
Optimization	Carries out a complete optimization for I_w	Two stages of SMO, one between the new point and the point of I_w (**Process**) and one stage between two points of I_w (**Reprocess**)	One SMO stage between two chosen points (optimizes the two α correspondents)
Final	Stops if constraints are no longer violated	Stops when all points have been dealt with and carries out a complete optimization for I_w	Stops if constraints are no longer violated

Table 5.2. *Comparison between the different functions used by SimpleSVM, LASVM and SMO*

Amongst non-optimized methods, processing large dimensions also note (even though this method will not be explained in detail here) that the CVM (core virtual machines [TSA 05]) use statistical properties for the selection of the candidates and therefore decrease the time required for the learning process at relatively high efficiency in terms of performance. A rather small quantity of points guarantees to 95% that 95% of the points are distributed correctly.

5.5. Experiments

In order to illustrate the capacities of the methods which have been presented, some results of experiments will be shown. First of all a problem whose size can be changed will be used to show the empirical complexity, sparsity and efficiency of different algorithms. Now an experiment will be explained which shows the behavior of LASVM for a database with up to 8 million points.

5.5.1. *Comparison of empirical complexity*

For a problem of separable complexity it is important to be familiar with a 4×4 draughtsboard (Figure 5.2) as it shows developments in terms of time required for calculations and number of points used in the learning process.

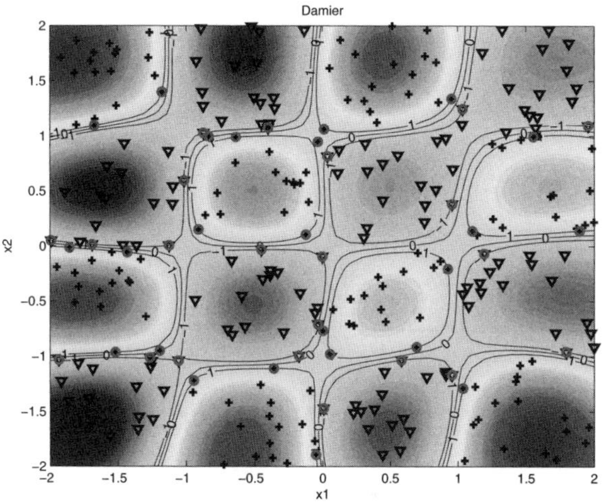

Figure 5.2. *Example of a draughtsboard and an SVM associated solution. Note sparsity of the result (round points in the figure). The other points are not used in the solution*

The draughtsboard is an example of sparsity of the solution. No matter the rate of filling the squares, only the points on the border lines are useful for the decision-making process. In an extreme case, a point on every angle of a square is sufficient to obtain the correct border lines, i.e. 54 points are sufficient. Support vector points might, of course, be amongst those points but sparsity of the solution shows the efficiency of the methods presented earlier.

With the implementation of SMO, libSVM [CHA 01] is used in version 2.7 and version 2.81. Version 2.7 uses first order criteria for the selection of points and version 2.81 uses criteria of the second order. For second order criteria, versions are also compared to each other with or without *shrinking*. For SimpleSVM the tool box is used [LOO 04]. For LASVM the code is derived from this tool box and obtains a Matlab implementation. The code for CVM is a variation of libSVM 2.7.

Figure 5.3 shows the exact methods which provide the level of acknowledgement. Empirical complexity is given by the gradient of the straight lines obtained for the period of the learning process (above); the bias of these straight lines partially depends on the language used during implementation (C is compiled, Matlab interpreted). An improvement can be observed in terms of the time required for calculations due to the *shrinking* heuristics and the superiority of second order criteria above those of the first order for SMO. In terms of sparsity, the two non-optimized methods retain fewer points than the other methods. Increasing sparsity means that the time required for calculations can be reduced. However, care needs to be taken in order not to deteriorate the quality of the solution. If LASVM is convergent to similar solutions compared to that obtained by optimized methods, CVM loses some of its quality.

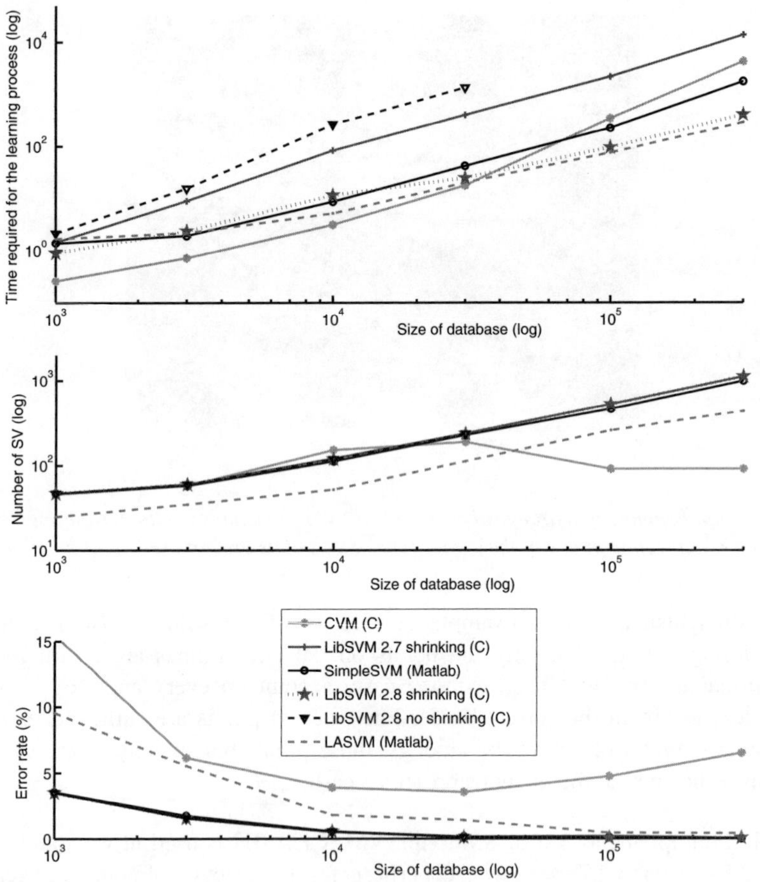

Figure 5.3. *Comparison of different algorithms on the draughtsboard problem. The first figure shows the development in terms of learning process. The figure in the middle shows the number of support vectors. The figure below indicates the error rate*

5.5.2. *Very large databases*

For further illustration of the online method, an experiment on recognizing handwritten characters (numbers 0 to 9) will be shown. MNIST [LEC 98] is the database which is being used and which has been created on the basis of invariance [LOO 06]. Without going into any detail on the generation of the points, the experiment and its results will be presented. The main idea is to solve the problem of invariance (enable the algorithm to recognize numbers independently from transitions, rotations, or minor deformations). In order to do this, a database with a virtually infinite number of examples, different from the original, has been created. It is sufficient to use a generic parameter-based formula to create an example which does not take the same shape. As soon as a new example is required the parameters are chosen randomly and calculate a virtual example. An online algorithm is adapted to increasing levels of demand. The more the examples are out of shape, the more robust the algorithm is when it comes to variations, but this also increases the size of the problem to be solved. If an off-line method is used (e.g. SMO or SimpleSVM) the entire database has to be available at the beginning. This limits the possibility of introducing a new invariance.

MNIST contains 60,000 training examples of 10 classes. The test database contains 10,000 examples. The dimension of an example is 784 (an image of 28×28 pixels of shades of gray). During the learning process up to 130 changes in shape were generated for every example. These changes in shape make up to 8 million points per classifier (every classifier has a binary learning process of one class against all other classes, i.e. 9 against 1).

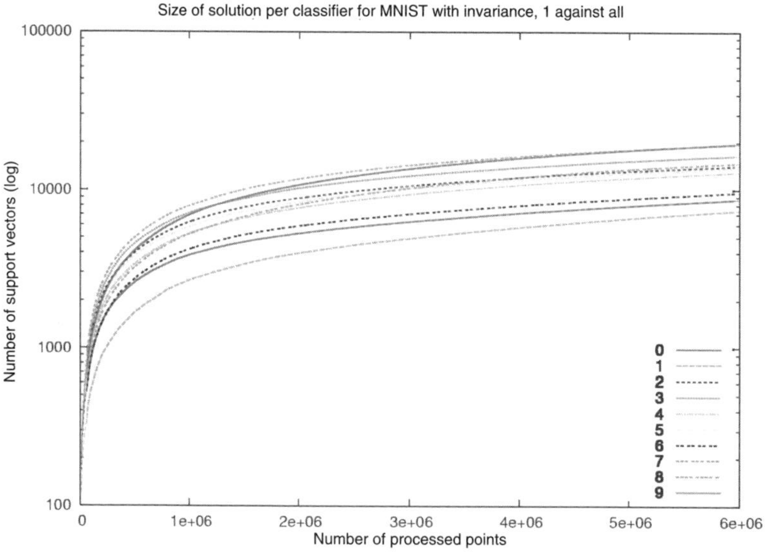

Figure 5.4. *Development of the size of solution per classifier during the learning process. These results were obtained during preliminary studies of up to 6 million points*

In this example LASVM shows its capacity to process large databases. These results were obtained after eight days of learning and testing carried out by one single machine for all 10 classifiers. The number of support vectors retained by the classifier varies between 8,000 and 20,000 (see Figure 5.4). This variation depends on how complex the task is (distinguishing 1 from other figures is easier than distinguishing 3, 5 and 8. Fewer examples are therefore needed when distinguishing 1). These results were obtained due to the combination of two essential elements. These are sparsity and the online mode. Sparsity allows for a high number of examples to be considered and the online mode ensures that every point is dealt with only once. In terms of performance, this learning process has an error rate of 0.67%, which is the best possible result for kernel-based methods that use invariance and do not modify the examples used in tests.

5.6. Conclusion

This chapter has shown the link between learning processes and convex quadratic programming. One of the major challenges for the learning process is the capacity to process large databases. An overview on the best methods of resolution has been given, as well as an explanation on how to increase development towards non-optimized methods which are more efficient.

Defining the learning process as a problem in convex quadratic programming has improved efficiency and reliability. The currently available algorithms are able to solve very large problems (8 million variables and 16 million constraints) within a reasonable timescale. This is linked to a particular problem in the learning process the level of sparsity within the solution. If in the general case of quadratic programming the methods of an interior point are more efficient, this no longer applies to the specific problem of a learning process. The methods of active constraints benefit from the problem's particularities. The result given by these programs is reliable as the problem has a unique solution and because of the convergence of these algorithms. These methods are very robust to small disturbances of data that lead to changes in the solution. These solutions then move on to a high number of operational systems which are able to recognize previously learnt forms. In the near future it will be possible to integrate these algorithms into independent systems in order to provide them with the capacity to learn.

This chapter has focused on *support vector* algorithms. There are, however, other interesting approaches. For regression, the LAR (*least angle regression* [EFR 04]) algorithm suggests a very efficient method of quadratic programming for problems with a penalization lower than $\mathcal{L}_1$ (i.e. sparsity). One of the main advantages of this method (apart from its efficiency) is how regularization is carried out on the basis of minimization. This method therefore has not just one, but of a group of potentially useful solutions. Finding the right algorithms is based on regularizations and the main challenge when optimizing learning processes. This method is a sophisticated way of studying optimization problems of two or three criteria for a solution.

When it comes to the learning process, the limits of algorithms presented above are linked to their own strengths. Once an algorithm has chosen many points (up to 10,000 in the example which has been presented) a mechanism is required to compile this solution, i.e. find an equivalent form which is, however, more economical when using resources. This can currently only be done via non-convex approaches such as neural networks. These solutions are, however, not very satisfactory. Choosing the right algorithms according to the size and the complexity of the problem is important. If needed, the algorithm may also be changed during learning.

5.7. Bibliography

[BOR 05] BORDES A., ERTEKIN S., WESTON J. and BOTTOU L., "Fast kernel classifiers with online and active learning", *Journal of Machine Learning Research*, vol. 6, pp. 1579–1619, 2005.

[BOT 04] BOTTOU L. and LECUN Y., "Large scale online learning", in THRUN S., SAUL L. and SCHÖLKOPF B. (Eds.), *Advances in Neural Information Processing Systems 16*, Cambridge, MA, MIT Press, 2004.

[BOY 02] BOYD S. and VANDENBERGHE L., "Advances in convex optimization: interior-point methods, cone programming and applications", *IEEE Conference on Decision and Control*, 2002.

[BUR 98] BURGES C.J.C., "A tutorial on support vector machines for pattern recognition", *Data Mining and Knowledge Discovery*, vol. 2, no. 2, pp. 121–167, 1998.

[CHA 01] CHANG C.-C. and LIN C.-J., *LIBSVM: a Library for Support Vector Machines*, 2001 (software available at http://www.csie.ntu.edu.tw/~cjlin/libsvm/).

[EFR 04] EFRON B., HASTIE T., JOHNSTONE I. and TIBSHIRANI R., "Least angle regression", *Annals of Statistics*, vol. 32, no. 2, pp. 407–499, 2004.

[FAN 05] FAN R.-E., CHEN P.-H. and LIN C.-J., "Working set selection using second order information for training support vector machines", *Journal of Machine Learning Research*, vol. 6, pp. 1889–1918, 2005.

[GER 01] GERTZ M. and WRIGHT S., *Object-oriented Software for Quadratic Programming*, 2001 (http://pages.cs.wisc.edu/~swright/ooqp/ooqp-paper.pdf), *ACM Transactions on Mathematical Software*, vol. 29, pp. 58–81, 2003.

[JOA 99] JOACHIMS T., "Making large-scale SVM learning practical", in SCHOLKOPF B., BURGES C. and SMOLA A. (Eds.), *Advanced in Kernel Methods – Support Vector Learning*, MIT Press, pp. 169–184, 1999.

[KAR 84] KARMARKAR N.K., "A new polynomial-time algorithm for linear programming", *Combinatorica*, vol. 4, pp. 373–395, 1984.

[KAU 99] KAUFMAN L., "Solving the quadratic programming problem arising in support vector classification", *Advances in Kernel Methods: Support Vector Learning*, pp. 147–167, MIT Press, 1999.

[KEE 02] KEERTHI S. S., GILBERT E.G., "Convergence of a generalized SMO algorithm for SVMClassifier design", *Machine Learning*, vol. 46, no. 1-3, pp. 351–360, 2002.

[KEE 05] KEERTHI S. S., "Generalized LARS as an effective feature selection tool for text classification with SVMs", *ICML'05: Proceedings of the 22nd International Conference on Machine learning*, New York, NY, USA, ACM Press, pp. 417–424, 2005.

[LEC 98] LECUN Y., BOTTOU L., BENGIO Y. and HAFFNER P., "Gradient-based learning applied to document recognition", *Proceedings of the IEEE*, vol. 86, no. 11, pp. 2278–2324, 1998, http://yann.lecun.com/exdb/mnist/.

[LOO 04] LOOSLI G., "Fast SVM Toolbox in Matlab based on SimpleSVM algorithm", 2004, http://asi.insa-rouen.fr/~gloosli/simpleSVM.html.

[LOO 05] LOOSLI G., CANU S., VISHWANATHAN S., SMOLA A.J. and CHATTOPADHYAY M., "Boîte à outils SVM simple et rapide", *Revue d'Intelligence Artificielle*, vol. 19, pp. 741–767, 2005.

[LOO 06] LOOSLI G., BOTTOU L. and CANU S., *Training Invariant SupportVectorMachines using Selective Sampling*, Rapport, LITIS – NEC Laboratories of America, 2006.

[MAN 98] MANGASARIAN O., "Generalized support vector machines", *NIPS Workshop on Large Margin Classifiers*, 1998.

[MAN 01] MANGASARIAN O.L., MUSICANT D.R., "Lagrangian support vector machines", *Journal of Machine Learning Research*, vol. 1, pp. 161–177, 2001.

[NEM 05] NEMIROVSKI A., "Introduction to convex programming, interior point methods, and semi-definite programming", *Machine Learning, Support Vector Machines, and Large-Scale Optimization Pascal Workshop*, March 2005.

[NIK 00] NIKOLOVA M., "Local strong homogeneity of a regularized estimator", *SIAM Journal on Applied Mathematics*, vol. 61, no. 2, pp. 633–658, 2000.

[OSB 05] OSBORNE M.R., "Polyhedral function constrained optimization problems", in MAY R., ROBERTS A.J. (Eds.), *Proc. of 12th Computational Techniques and Applications Conference CTAC-2004*, vol. 46, pp. C196–C209, April 2005, http://anziamj.austms.org.au/V46/CTAC2004/Osbo [April 22, 2005].

[OSU 97] OSUNA E., FREUND R. and GIROSI F., "An improved training algorithm of Support Vector Machines", in PRINCIPE J., GILE L., MORGAN N. and WILSON E. (Eds.), *Neural Networks for Signal Processing VII – Proceedings of the 1997 IEEEWorkshop*, pp. 276–285, 1997.

[PLA 99] PLATT J., "Fast training of support vector machines using sequential minimal optimization", in SCHOLKOPF B., BURGES C. and SMOLA A. (Eds.), *Advanced in Kernel Methods – Support Vector Learning*, MIT Press, pp. 185–208, 1999.

[POR 97] PORTNOY S. and KOENKER R., "The Gaussian hare and the Laplacian tortoise: computability of squared-error versus absolute-error estimators", *Statistical Sciences*, 1997.

[SCH 02] SCHÖLKOPF B. and SMOLA A.J., *Learning with Kernels*, MIT Press, 2002.

[TSA 05] TSANG I.W., KWOK J.T. and CHEUNG P.-M., "Core vector machines: fast SVM training on very large data sets", *Journal of Machine Learning Research*, vol. 6, pp. 363–392, 2005.

[VAN 99] VANDERBEI R.J., "LOQO: an interior point code for quadratic programming", *Optimization Methods and Software*, vol. 11, pp. 451–484, 1999.

[VAP 82] VAPNIK V.N., *Estimation of Dependences Based on Empirical Data*, Springer-Verlag, 1982.

[VAP 95] VAPNIK V., *The Nature of Statistical Learning Theory*, Springer, NY, 1995.

[VIS 03] VISHWANATHAN S.V.N., SMOLA A.J. and MURTY M.N., "SimpleSVM", *Proceedings of the Twentieth International Conference on Machine Learning*, 2003.

[ZOU 60] ZOUTENDIJK G., *Methods of Feasible Directions*, Elsevier, 1960.

Chapter 6

Probabilistic Modeling of Policies and Application to Optimal Sensor Management

6.1. Continuum, a path toward oblivion

In the domain of operational research several practical questions often lead on to combinatorial issues. This chapter covers planning and exploring problems. Generally speaking, this chapter will deal with the path planning or placement of sensors. Specific tasks like the detection of intrusion or surveillance are possible applications. For such a task, large scale planning is often required and/or different levels of decision making are considered. For some applications, an uncertain area might be explored by mobile sensors. In such cases, the observations are involved in the whole planning process, and feedback on the sensors orientations is necessary. For such problems, the optimal planning of the complete mission is very difficult to solve and various simplification techniques are addressed by the literature.

Main difficulties in optimization arise when manipulating discrete data. Even for continuous non-convex optimization, the implication of discrete data is implicit and associated to local minima. The combinatorics implied by integer data not only impact the optimization but also have logical implications (e.g. calculability and logical complexity). Various approaches have been investigated in order to reduce the combinatorics. Systematic enumeration methods like Branch and Bound can be used in combination with the computation of branching bounds of the enumeration tree. The point here is to approximate the combinatorics using a relaxed problem, and, on this basis, to explore the interesting branches of the combinatorial tree. Typically, integer linear programs could be solved by relaxing to a linear program. Now, even

---

Chapter written by Frédéric DAMBREVILLE, Francis CELESTE and Cécile SIMONIN.

a Branch and Bround approach cannot apply when the combinatorics become too complex. In practice, obtaining solutions which are close to the optimum is sufficient. Deterministic approaches could provide near-optimal solutions.

However, stochastic methods and metaheuristics offer better versatility. Metaheuristics share two principles: exploring the optimization space by generating solution proposals and focusing gradually toward the best or promising proposals. This focus corresponds to a degressive relaxation in the optimization process. In the case of simulated annealing, high initial temperatures enable the particles to make big jumps, which somehow is equivalent to smoothing down the functional to be optimized. In the case of cross-entropy (CE) the solutions are approximated using a family of parametric probabilistic laws, which gradually sharpen toward the best populations. These methods imply a progressive migration from a continuous and weak informational content toward a sharp and strong informational content.

This chapter introduces the CE method [DEB 03, RUB 04], a metaheuristic based on rare events simulation, and its applications to some optimal exploration problems. The method relies on a relaxed modeling of the policies to be optimized by means of a family of probabilistic laws. It is underlined here that the capability of probability for logical representation is not only instrumental and necessary, but it also endows the CE method with a powerful asset. Practically, the choice for the modeling family is implied by the logical properties of the optimization problem. Throughout this chapter, different kinds of probabilistic models are considered in accordance with the nature of the problems to be solved.

Even though metaheuristics can provide near-optimal solutions for complex combinatorics, there are cases where the solution itself cannot be formalized in a simple way. In the case of dynamic control under partial observation, the global solution is a huge decision tree, which depends on all potentially earned observations. Such huge dimension variables cannot be manipulated. Under some conditions, dynamic programming answers this problem theoretically by providing a recursive construction of the optimal solution. In practice, this method takes far too much time and computer storage; approximation strategies are necessary. The final section of this chapter investigates the approximation of control strategies under partial observation by means of a cross-entropic approach. As has already been mentioned, an exact modeling of the decision trees is intractable. The decision tree itself will be approximated by means of a family of probabilistic laws. The chapter will show that a *double-approximation*, i.e. using the metaheuristics and the modeling, still leads to acceptable solutions. These solutions are stochastic processes with bounded memories and constitute continuous approximations of the optimal decision tree.

6.2. The cross-entropy (CE) method

The CE method was developed by Reuven Rubinstein [DEB 03] and was initially intended for rare event simulation. A rare event is an event of weak probability. The need for evaluating the probability of rare events arises when analyzing reliability, e.g. what is the probability of a breakdown? A mathematical expression of such weak probabilities is not always possible. In the worst case scenario, the only possible way to evaluate a rare event is by simulation. However, simulating the rare event directly using the law under study will not work: too many samples are necessary to accurately evaluate the probability of the rare event. Importance sampling is a method usually implemented in order to overcome this difficulty. Its principle consists of changing the law so as to favor the sampling around the rare event. However, adjusting the parameters for this change in laws is generally difficult. The CE method provides an iterative estimation of these parameters. Subsequently, the CE method for rare event simulation is introduced at first.

Then, it is explained how this method could be adapted to optimization tasks. Since an event maximizing a function is typically a rare event, such usage of CE is a corollary. This section of the chapter is very succinct. Please see [RUB 04] or the tutorial [DEB 03] for an improved introduction. Furthermore, a large amount of documentation is available at http://www.cemethod.org.

Notations:

– [Kronecker] Given a logical proposition A, the quantity $\delta[A]$ has to be 1 if A is true and requires 0 in the opposite case.

– The quantity $\mathbf{E}_p f(x) = \int_X f(x)p(x)\,dx$ describes the expectancy of the value $f(x)$ relative to the density of probability p of the variable $x \in X$.

– Given the measurable sub-set $Y \subset X$ and a density of probability p over X, the quantity $\mathbf{P}_p(Y) = \mathbf{P}_p(x \in Y) = \mathbf{E}_p\delta[x \in Y] = \int_Y p(x)\,dx$ is the probability of the event $x \in Y$.

– Given the finite set E, the quantity $\sharp E$ is the number of elements, i.e. the cardinality, of this set.

– The object $\mathcal{N}(\mu, \Sigma)$ refers to the Gaussian distribution with mean μ and covariance matrix Σ.

6.2.1. *Probability of rare events*

Let us assume a measurable space X with density of probability p and a real-valued measurable mapping f over X. We have to evaluate the probability that function f exceeds a certain threshold $\gamma \in \mathbb{R}$. This event $F_\gamma = \{x \in X / f(x) \geq \gamma\}$ is a *rare event*, when its probability is very weak, typically when f shows the malfunctioning of a system.

6.2.1.1. *Sampling methods*

Monte-Carlo. Let $x_1, \ldots, x_N$ be samples of the variable x, generated according to the law p. The probability $P_\gamma = \mathbf{P}_p(f(x) \geq \gamma) = \mathbf{P}_p(F_\gamma)$ can be estimated by the empirical average:

$$\hat{P}_\gamma = \frac{1}{N}\sum_{n=1}^{N} \delta\big[f(x_n) \geq \gamma\big].$$

The variance in this unbiased estimate is given by:

$$\mathbf{E}_p\big(\hat{P}_\gamma - P_\gamma\big)^2 = \frac{1}{N}P_\gamma\big(1 - P_\gamma\big).$$

In the case where P_γ is very weak, the variance is not the correct criterion for evaluating this estimate. The relative standard deviation would be more appropriate:

$$\frac{\sigma_\gamma}{P_\gamma} = \sqrt{\frac{1 - P_\gamma}{NP_\gamma}}.$$

Note that this standard deviation is inversely proportional to $\sqrt{NP_\gamma}$. Thus, the empirical averaging cannot be used to estimate a rare event F_γ.

Importance sampling. The law p being inadequate for an estimation of P_γ, an alternative approach consists of using an auxiliary sampling law q, with a sampling scope focused around the rare event. Let $x_1, \ldots, x_N$ be samples of the variable x, generated according to the law q. Then, the probability P_γ is estimated by the weighted empirical average:

$$\hat{P}_\gamma = \frac{1}{N}\sum_{n=1}^{N} \delta\big[f(x_n) \geq \gamma\big]\frac{p(x_n)}{q(x_n)}.$$

The variance for this estimate is:

$$\mathbf{E}_q\big(\hat{P}_\gamma - P_\gamma\big)^2 = \frac{\mathbf{E}_p\big(\delta\big[f(x) \geq \gamma\big]p(x)q(x)^{-1}\big) - P_\gamma^2}{N}.$$

This variance is zeroed for an optimal probabilistic density $q^*(x) = \delta[f(x) \geq \gamma]p(x)P_\gamma^{-1}$. Unfortunately the law q^* is inaccessible, since the probability P_γ is unknown. Actually, a wise definition of a sampling law q is not easy. The method explained subsequently consists of choosing the sampling law from a family of laws $\pi(\cdot;\lambda) \mid \lambda \in \Lambda$. The objective is to optimize the parameter λ in order to minimize the distance $\mathcal{D}(q^*, \pi(\cdot;\lambda))$ with the optimal sampling law. As a criterion for distance between two probabilities, the *Kullback-Leibler divergence is* *considered*. This divergence is the CE between the laws *diminished* by the entropy of the reference law:

$$\mathcal{D}(q,p) = \mathbf{E}_q \ln \frac{q(x)}{p(x)} = \int_X q(x) \ln q(x)\, dx - \int_X q(x) \ln p(x)\, dx.$$

Sampling by means of a family of laws. Now, the sampling law is optimized within the family of laws $\pi(\cdot;\lambda) \mid \lambda \in \Lambda$. Minimizing the divergence $\mathcal{D}(q^*, \pi(\cdot;\lambda))$ means maximizing the CE $\int_X q^*(x) \ln \pi(x;\lambda)\, dx$. Taking into consideration the definition of q^*, the following optimization is implied:

$$\lambda_* \in \arg\max_{\lambda\in\Lambda} \mathbf{E}_p\big(\delta\big[f(x) \geq \gamma\big] \ln \pi(x;\lambda)\big).$$

At this stage, a rare event estimation is again instrumental. The calculation of the optimal parameter λ_* thus requires an importance sampling, say by the law q:

$$\lambda_* \in \arg\max_{\lambda\in\Lambda} \mathbf{E}_q\left(\delta\big[f(x) \geq \gamma\big]\frac{p(x)}{q(x)} \ln \pi(x;\lambda)\right).$$

Let $x_1, \ldots, x_N$ be samples of the variable x generated according to law q. The optimal parameter λ_* is approximated by the estimation $\hat{\lambda}_*$ defined as:

$$\hat{\lambda}_* \in \arg\max_{\lambda\in\Lambda} \sum_{n=1}^{N} \left(\delta\big[f(x_n) \geq \gamma\big]\frac{p(x_n)}{q(x_n)} \ln \pi(x_n;\lambda)\right). \qquad [6.1]$$

The family of law $\pi(\cdot;\lambda) \mid \lambda \in \Lambda$ has been chosen in order to guarantee a solution for this maximization. This chapter will show that there is no problem in doing so. The family of natural exponential laws is used in this case [RUB 04].

Now, the estimation of λ_* based on formula [6.1] is still not satisfactory. This estimation process is based on an importance sampling, in which law q remains unknown. Indeed, estimating λ_* is the basis for creating such a law. This is a typical framework, where a fixed-point process may be used; by the way, such an iterative approach is the essence of the CE method.

6.2.1.2. *CE-based sampling*

The previous section showed us how to optimize a parameterized law in order to address the importance sampling. However, estimating this parameter via equation [6.1] also requires an importance sampling. To avoid this problem, Rubinstein suggests an iterative method on the basis of a degressive relaxation of the rare event.

For the sake of simplicity, we assume the existence of parameter $\lambda_0 \in \Lambda$ such that $\pi(\cdot;\lambda_0) = p$.

Basic principle. Let $(\gamma_t \mid t \geq 1)$ be a sequence such that:

– event $f(x) \geq \gamma_1$ is not rare;

– event $f(x) \geq \gamma_{t+1}$ is not rare in regards to the law $\pi(\cdot;\lambda_t)$;

– $\lim_{t\to\infty} \gamma_t = \gamma$.

Then, a sequence of parameters $(\lambda_t \mid t \in \mathbb{N})$, with $\pi(\cdot;\lambda_0) = p$, may be constructed on the basis of an iteration of estimation process [6.1]:

– generate $x_1, \ldots, x_N$ according to law $\pi(\cdot;\lambda_t)$;

– choose $\lambda_{t+1} \in \arg\max_{\lambda\in\Lambda} \sum_{n=1}^{N} (\delta[f(x_n) \geq \gamma_{t+1}] \frac{p(x_n)}{\pi(x_n;\lambda_t)} \ln \pi(x_n;\lambda))$.

Rubinstein proposed an adaptive algorithm for constructing sequence $(\gamma_t \mid t \geq 1)$.

Adaptive construction of the bounds sequence. The construction process introduced now implies the use of a selection parameter $\rho \in \,]0,1[$. This parameter impacts the convergence speed of the sequence $(\gamma_t \mid t \geq 1)$. Specifically, the bound γ_t is defined as the $(1-\rho)$ quantile of $f(x)$ for the law $\pi(\cdot;\lambda_{t-1})$. More precisely, when $x_1, \ldots, x_N$ are samples generated by $\pi(\cdot;\lambda_{t-1})$ and are ordered so that $f(x_n) \leq f(x_{n+1})$ for $1 \leq n < N$, then γ_t is defined by $\gamma_t = \min\{\gamma, f(x_{\lceil(1-\rho)N\rceil})\}$. Parameter ρ has to be large enough, so as to provide a fair estimation of the probability of event $f(x) \geq \gamma_t$ for the law $\pi(\cdot;\lambda_{t-1})$. Note that the sequence $(\gamma_t \mid t \geq 1)$ increases except for the noise. When combining the constructions of $(\lambda_t \mid t \in \mathbb{N})$ and $(\gamma_t \mid t \geq 1)$, the CE algorithm, as described in [DEB 03], is then defined.

Integrated algorithms. Let $\rho \in \,]0,1[$. The CE algorithm for simulating rare events is expressed as follows:

1) choose $\lambda_0 \in \Lambda$ such that $\pi(\cdot;\lambda_0) = p$. Set $t = 1$;

2) generate N samples $x_1, \ldots, x_N$ by the law $\pi(\cdot;\lambda_{t-1})$. Put these samples into an increasing order according to f. Calculate the $(1-\rho)$ quantile γ_t:

$$\gamma_t = \min\left\{\gamma, f\left(x_{\lceil(1-\rho)N\rceil}\right)\right\}, \qquad [6.2]$$

3) compute λ_t by the maximization:

$$\lambda_t \in \arg\max_{\lambda\in\Lambda} \sum_{n=1}^{N} \left(\delta\left[f(x_n) \geq \gamma_t\right] \frac{p(x_n)}{\pi(x_n;\lambda_{t-1})} \ln \pi(x_n;\lambda)\right); \qquad [6.3]$$

4) if $\gamma_t < \gamma$, iterate from stage 2) again, while setting $t \leftarrow t+1$;

5) estimate the probability P_γ for rare events by:

$$\hat{P}_\gamma = \frac{1}{N} \sum_{n=1}^{N} \delta\left[f(x_n) \geq \gamma\right] \frac{p(x_n)}{\pi(x_n;\lambda_t)}. \qquad [6.4]$$

Note. There are different evolutions of the CE method. In particular parameter ρ itself may be adaptive. Theoretical convergence results [HOM 04, MAR 04] have been based on hypotheses of adaptive ρ.

6.2.2. *CE applied to optimization*

It was quickly noticed that the CE-based sampling process could be adapted and applied to optimization. This idea makes sense, since parameters which optimize an objective are hard to find, and actually constitute a rare event. Let us have a more detailed look at the algorithm presented in section 6.2.1.2. This algorithm is based on the construction of a sequence of bounds $(\gamma_t \mid t \geq 1)$, which might be considered as increasing except for the noise. These bounds define the events $F_{\gamma_t} = \{x \in X/f(x) \geq \gamma_t\}$ which become increasingly *rare*. During the process of construction, $(\gamma_t \mid t \geq 1)$ is truncated by γ, the parameter of the rare event to be evaluated. However, it is possible to carry on with the process. This leads to a sequence of increasingly rare events, which converge towards the event $F_{\gamma_\infty} = \{x \in X/f(x) \geq \max_{y \in X} f(y)\}$. This event is almost empty in general. If the process works long enough, it leads to the construction of a parameter λ, such that law $\pi(\cdot; \lambda)$ is very close to F_{γ_∞}. This is a way to reach the optimum of f.

While not subsequently addressed in detail, certain implied difficulties have to be discussed since they influence the convergence of the method:

– choosing the selection parameter ρ, or its evolution throughout the process;

– defining a criteria that stops the convergence;

– *choosing a family of laws π well adapted to the optimization problem.*

As previously discussed, some theoretical results answer the first two points [HOM 04]. It is not the purpose of this chapter to investigate these theoretical aspects. In addition, empirical criteria are sufficient and functional for our applications. The latter point is not easy to formalize; it is mostly driven by the experiments. However, the examples show that the laws naturally follow from the structure of the problem.

The next section presents the CE algorithm for optimization in the initial version and in its smoothed version.

6.2.2.1. *Algorithms*

f is a real-valued function of the variables $x \in X$. Variable x has to be optimized with $f(x)$. A family of laws $(\pi(\cdot; \lambda) \mid \lambda \in \Lambda)$ is defined over X.

Original algorithm. Let us take a selection parameter $\rho \in]0, 1[$. The CE algorithm for optimization works as follows:

1) initialize $\lambda_0 \in \Lambda$; This parameter is generally chosen so that $\pi(\cdot; \lambda_0)$ is close to the uniform law. Set $t = 1$;

2) generate a set E_t of N samples, according to the law $\pi(\cdot;\lambda_{t-1})$. Select the $\lfloor \rho N \rfloor$ best samples in regards to the function f. Let $S_t \subset E_t$ be the subset of samples such that:

$$\sharp S_t = \lfloor \rho N \rfloor \quad \text{and} \quad \forall x \in S_t,\ \forall y \in E_t \setminus S_t,\ f(x) \geq f(y);$$

3) compute λ_t by the maximization:

$$\lambda_t \in \arg\max_{\lambda \in \Lambda} \sum_{x \in S_t} \ln \pi(x;\lambda); \qquad [6.5]$$

4) repeat from stage 2) while setting $t \leftarrow t+1$ until the stop criteria is achieved;

5) the final stage being at $t = t_f$, an optimal solution for f can be sampled by the law $\pi(\cdot;\lambda_{t_f})$.

A simple stop criterion could be obtained by evaluating the stationarity of $\gamma_t = \inf_{x \in S_t} f(x)$.

Notes. Since the aim of the algorithm is no longer to evaluate a probability, some simplifications appear in the sampling algorithm. In particular, the update equation [6.5] no longer contains the term of correction $\frac{p(x)}{\pi(x;\lambda_{t-1})}$.

When the variable x is discrete, update [6.5] may cancel the probability of some states: as a result, theses states become inaccessible. An exponential smoothing of the update avoids this problem.

Smoothed algorithm. In addition to the selection parameter $\rho \in]0,1[$, a smoothing parameter α is defined. The set of law parameters Λ is assumed to be convex (this is often the case in practice). The smoothed algorithm differs from the previous algorithm at its update stage:

3) compute κ_t by the maximization:

$$\kappa_t \in \arg\max_{\kappa \in \Lambda} \sum_{x \in S_t} \ln \pi(x;\kappa),$$

[Smoothed update] Then, set $\lambda_t = \alpha\kappa_t + (1-\alpha)\lambda_{t-1}$.

The CE algorithm for optimization is now applied to a simple example: *the traveling salesman.*

6.2.2.2. *A simple example*

A salesman has to visit n cities and then come back to the starting point. The distance between city i and city j is described by the value $d(i,j) = d(j,i)$. The aim is to optimize the travel by the definition of a cycle $(v_1, v_2, \ldots, v_n, v_1)$, in which all cities are visited only once, and which minimizes the distance to be traveled, i.e.:

$$f(v_1, v_2, \ldots, v_n) = d(v_n, v_1) + \sum_{k=1}^{n-1} d(v_k, v_{k+1}).$$

The CE method is applied to this problem. It is assumed without loss of generality that $v_1 = 1$.

While applying the CE method, the first stage is the definition of a family of laws which is adapted to the problem. This preliminary stage is fundamental, as all other stages take place automatically. For this problem, *Markov chain laws are elected* [RUB 04]. Within a Markov chain, each new destination is sampled relatively to the current location of the salesman[1]. This choice is not at all random, but is led by the cinematic nature of the problem. Somehow, it locates certain difficulties of the optimization: it is likely that the cities close to the current position would be favored by the best solution.

These laws have to be formalized now. Let $p(v)$ denote the law for sampling a circuit v. Then, the construction of $p(v)$ is that of a Markov chain:

$$p(v) = p(v_2 \mid 1) \prod_{k=2}^{n} p(v_{k+1} \mid v_k).$$

As a consequence, the law p is characterized by the parameters $p(j \mid i)$, that is the probability of moving form city i to city j. The following constraints are applied to $p(j \mid i)$:

$$p(j \mid i) \geq 0, \quad \sum_{j=1}^{n} p(j \mid i) = 1 \quad \text{and} \quad p(i \mid i) = 0 \text{ (non-stationarity).} \qquad [6.6]$$

Note that the Markov chain does not comply with the circuit constraint $n = 1$. Some additional adaptation of the sampling process will be implemented.

Family of laws. Let $P = (p(j \mid i))_{ij}$ denote the matrix of components $p(j \mid i)$. The laws family, π, is naturally defined as follows:

– Λ is the set of matrices P compliant with constraint [6.6]; note that Λ is convex;

– $\pi(v; P) = p(v_2 \mid 1) \prod_{k=2}^{n} p(v_{k+1} \mid v_k)$, for all $P \in \Lambda$.

1. Another possibility would have been to map each city at a travel stage. However, such a law is insufficient for reasons explained hereafter.

As previously discussed, the Markov chain does not always generate samples for valid circuits. Actually, the following constraints are not managed by the laws:

$$v_{n+1} = 1 \quad \text{and} \quad v_i \neq v_j \text{ for all } i, j \leq n \text{ such that } i \neq j. \qquad [6.7]$$

It is shown at the very end of this section, how to handle this difficulty. However, before that, updating stage [6.5] is now expressed in detail.

Update. The update of a parameter $P \in \Lambda$ is performed by maximizing:

$$\max_P \sum_{v \in S_t} \sum_{k=1}^{n} \ln p\big(v_{k+1} \mid v_k\big),$$

under constraint [6.6]. This problem is convex and separable. It is easily solved by the Kuhn-Tucker conditions. The following formula is valid for $i \neq j$:

$$p(j \mid i) = \frac{\sharp\big\{(v, k)/v \in S_t,\ 1 \leq k \leq n,\ v_k = i \text{ and } v_{k+1} = j\big\}}{\sharp\big\{(v, k)/v \in S_t,\ 1 \leq k \leq n \text{ and } v_k = i\big\}}. \qquad [6.8]$$

Constraints on the itinerary. Constraints [6.7] are generally unsatisfied by the sampled circuits. Then, the approach in use consists of sampling with rejection. The sampling process is thus repeated until the quota (i.e. N) of valid samples has been reached. Samples which have not been validated by [6.7] are simply rejected. This procedure attributes implicitly the evaluation $-\infty$ to invalid samples and adapts the coefficient of the selection rate ρ so as to remove systematically the invalid cases. Now, a literal implementation of the sampling with rejection is intrinsically inefficient, since the probability of rejection is very high. The rejections have to be simulated, by reprocessing the law $\pi(\cdot; P)$ along the circuit:

– after each Markovian choice for a city, remove this choice from matrix P, by canceling the column of P *linked to that city; renormalize all lines of* P,

– always choose 1 for the last transition.

This selection method is equivalent to $\pi(\cdot; P)$ with rejection, and remains compatible with updating formula [6.8].

The results of detailed experiments are available in the tutorial on CE [DEB 03]. CE is of a high performance for these problems.

6.3. Examples of implementation of CE for surveillance

The suggested example is inspired by the research carried out on behalf of intelligence services working for military institutions. The aim is to find clues in the theater of operations with the help of different tools. The territory to be explored is often very large in comparison to the capacities of surveillance sensors. It is therefore necessary to place the sensors in a way which maximizes the chances of finding the clue. Two main levels of optimization have been considered:

– a global level which decides upon the distribution of sensors to subdivisions of the territory;

– a local level: the search efforts are optimized within each subdivisions of the territory.

The levels of optimization are mutually dependent. In particular, the global distribution of the sensors is subject to the results of local optimizations of the search efforts. The CE method is used for the global optimization.

6.3.1. *Introducing the problem*

The probability of detecting a hidden target in a space has to be maximized. The target is non-moving and some prior information is known by means of a localization probability. The sensors are characterized by their detection capability and by their autonomy; their operation radii are limited.

The space to be searched is denoted E and divided into zones E_z, $z \in \{1, \ldots, Z\}$. These zones are limited areas and have been chosen so as to take into account the autonomy of the sensors. A sensor will investigate one zone. Each zone z is split up into cells Unit_{zu}, $u \in \{1, \ldots, U_z\}$. These cells are homogenous, i.e. all points in one cell have the same characteristics (same surface, same kind of vegetation etc. refer to Figure 6.1). These subdivisions are subject to the following constraints:

$$E = \bigcup_{z=1}^{Z} E_z \quad \text{with: } E_z \cap E_{z'} = \emptyset \text{ for } z \neq z',$$

$$E_z = \bigcup_{u=1}^{U_z} \text{Unit}_{zu} \quad \text{with: } \text{Unit}_{zu} \cap \text{Unit}_{zu'} = \emptyset \text{ for } u \neq u'.$$

The target is positioned somewhere in this space. The presence of the target is characterized by the probabilistic prior α. Typically, $\alpha(\text{Unit}_{zu})$ is the prior probability that the target is in the cell Unit_{zu}. The target may be detected by sensors s, $s \in \{1, \ldots, S\}$. These sensors only deal with one zone and their autonomy is limited. This limited autonomy is characterized by an available amount of resources Φ_s Exploring a cell Unit_{zu} requires a certain amount of resources $\varphi_s(\text{Unit}_{zu})$. The optimization problem thus has to respect the following constraint:

$$\sum_{u=1}^{U_z} \varphi_s(\text{Unit}_{zu}) \leq \Phi_s.$$

The chances of detecting a target located in cell Unit_{zu} depends on the quantity of resources that are being used (e.g. time used to explore the cell). These probabilities also depend on the performance of the sensor within the cell Unit_{zu}. Classically, an exponential function of detection is employed [DEG 61, KOO 57]. Thus

$$P_s(\varphi_s) = \exp\big(- w^s_{zu}\varphi_s(\text{Unit}_{zu})\big)$$

represents the conditional probability that sensor s will not detect a target hidden in Unit_{zu}, taking into account the search effort provided $\varphi_s(\text{Unit}_{zu})$. Coefficient $-w^s_{zu}$ characterizes the visibility of the Sensor_s in cell Unit_{zu}.

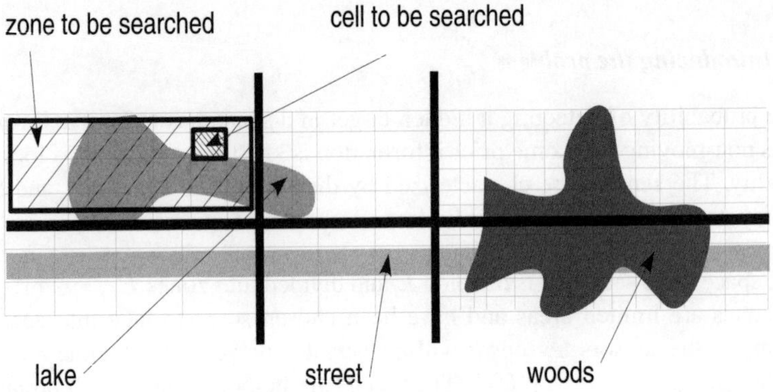

Figure 6.1. *Space to be searched*

In addition to the search functions φ, a mapping $m : s \mapsto z$ is defined, which allocates each sensor to a zone. The aim is to optimize φ and m conjointly, so as to maximize the chances of detecting the target:

$$F(m,\varphi) = \sum_{z=1}^{Z}\sum_{u=1}^{U_z} \alpha(\text{Unit}_{zu}) \prod_{s\in m^{-1}(z)} P_s\big(\varphi_s(\text{Unit}_{zu})\big).$$

The entire problem is therefore expressed as follows:

$$\min_{\varphi,m} F(m,\varphi), \text{ under the constraints: } \sum_{u=1}^{U_z} \varphi_s\big(\text{Unit}_{zu}\big) \leq \Phi_s \text{ and } \varphi \geq 0. \qquad [6.9]$$

This problem is subdivided into two mutually dependent levels:

– find the best possible allocation of the sensors to the zones;

– find the best possible distribution of resources to the sensors.

The problem of allocating the sensors to the respective zones can be solved with the help of the CE method. The underlying problem is, however, convex and therefore needs to be optimized by more traditional methods [DEG 61].

Note that the sensor allocation can be solved by linear programming (LP) when the mapping m is one-to-one. A comparative study between CE and LP has been carried out. Well adjusted parameters in CE enable us to provide optimal solutions. When LP is a possible approach, it provides optimal solutions to this allocation problem much more quickly.

However, in most cases m is not one-to-one and cannot be optimized by LP. When LP-based optimization is no longer possible, integer programming could be employed.

Optimizing the allocation of sensors to search zones by integer programming is, however, impossible for large dimensions (combinatorics). CE, on the other hand, is often able to find the optimal solution in a reasonable time. The following sections present the CE method for the optimization of the sensor allocation.

6.3.2. *Optimizing the distribution of resources*

This section will show how to optimize the distribution of sensor resources.

When only one sensor s is allocated to zone z, the distribution of the resources is solved as follows:

$$\min_{\varphi_s} \sum_{u=1}^{U_z} \alpha(\text{Unit}_{zu}) \exp\big(-w_{zu}^s \varphi_s(Cel_{zu})\big)$$

under the constraints:

$$\sum_{u=1}^{U_z} \varphi_s(\text{Unit}_{zu}) \leq \Phi_s \text{ and } \varphi \geq 0.$$

This is a *de Guenin* problem [DEG 61], solved by *Kuhn and Tucker*:

$$\varphi_s(\text{Unit}_{zu}) = \max\left\{0, -\frac{\ln\left(\frac{\lambda}{w_{zu}^s \alpha(\text{Unit}_{zu})}\right)}{w_{zu}^s}\right\}, \qquad [6.10]$$

where the parameter λ is a dual variable to be adjusted. As the objective function is convex, λ is obtained by dichotomy in $]0, \max_u(w_{zu}^s \alpha(\text{Unit}_{zu}))]$ so as to fulfill the condition $\sum_{u=1}^{U_z} \varphi_s(\text{Unit}_{zu}) = \Phi_s$.

When no assumption is made about the mapping m, the set of sensors $m^{-1}(z)$ is affected to zone z. A recursive method is employed in order to optimize the usage of the sensors of $m^{-1}(z)$, based on the de Guenin optimization scheme. Optimization of the variables φ_s takes place recursively in the form of a cycle (all other variables remain unchanged).

Algorithm. 1) Initialize $\varphi_s(\text{Unit}_{zu}) = 0$ for each sensor $s \in m^{-1}(z)$,

2) Compute the probability p_1 of not detecting the target in zone E_z, denoted p_1

3) Carry out the following operations for each sensor $s \in m^{-1}(z)$:

- compute the best distribution of resources according to the optimization of *de Guenin* for the sensor s in concern (for every other sensor i, φ_i is kept unchanged),

- memorize the distribution φ_s, previously computed.

4) Compute the new probability p_2 of not detecting the target,

5) If $p_1 \approx p_2$ then stop; otherwise, set $p_1 := p_2$ and reiterate from step 3).

Figure 6.2 provides a simple illustration of this algorithm.

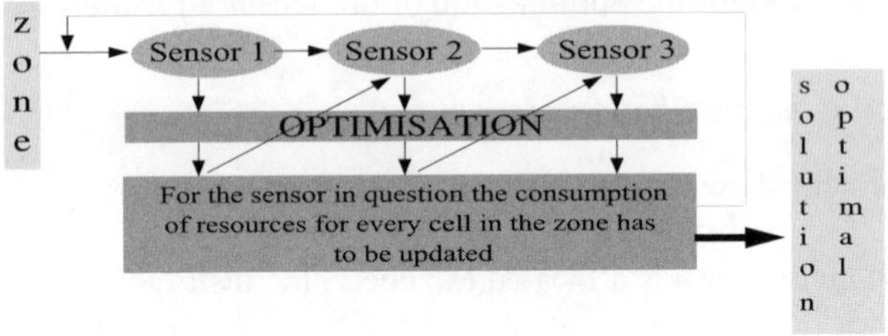

Figure 6.2. *Optimizing the distribution of resources for several sensors*

6.3.3. *Allocating sensors to zones*

An exhaustive research of the best possible allocation of sensors to zones requires the computation of the best resource distribution for each sensor configuration in each zone. For a given zone z, this means that φ is optimized for any possible choice of $m^{-1}(z) \subset [\![1, S]\!]$. The total number of sub-problems to be solved by local optimization is $Z \times 2^S$. This is, of course, impossible. CE on the other hand, which is adaptive with the data, does well with this problem. Subsequently, a family of laws for sampling the allocation functions m is defined.

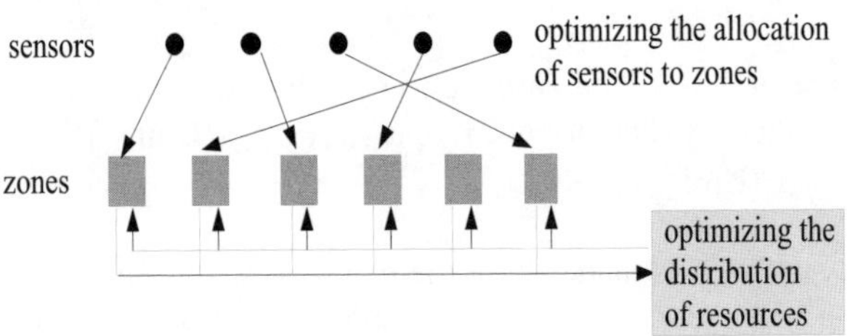

Figure 6.3. *Allocating sensors*

Family of laws. A discrete probabilistic law $p(\cdot \mid s)$ is associated with each sensor s and is intended for sampling the allocation of sensor s to a particular zone z. The matrix $P = (p(j \mid i))_{ij}$ is constructed from the components $p(z \mid s)$. On the basis of P, it is possible to define a sample prototype for the allocation functions m:

$$\pi(m; P) = \prod_{s=1}^{S} p\big(m(s) \mid s\big).$$

The construction of a family of laws π for the CE is naturally implied:

– Λ is the set of matrices P which satisfy $p(j \mid i) \geq 0$ and $\sum_{j=1}^{Z} p(j \mid i) = 1$,

– $\pi(m; P) = \prod_{s=1}^{S} p(m(s) \mid s)$, for $P \in \Lambda$.

The update of P is similar to section 6.2.2.2:

$$p(j \mid i) = \frac{\sharp\{m/m \in S_t,\ m(i) = j\}}{\sharp S}. \qquad [6.11]$$

Injective allocation. The previous sampling method will generate any possible sensor-to-zone mapping. In general, such mapping is not injective. In order to provide one-to-one mapping, a sampling with a reject will be implemented as shown in section 6.2.2.2. Thus, the sampling process is iterated until N injective maps are obtained. The remaining non-injective maps are simply rejected. As in section 6.2.2.2, this process is accelerated by making the rejections virtual. This can be carried out via a normalization process.

In the case of injective maps, a computation of the best resource allocation is possible for all configurations. Only $Z \times S$ cases have to be taken into consideration. In such a case, LP is a better approach.

6.3.4. *Implementation*

This section will focus on optimizing the allocation of sensors to zones in a general case (i.e. several sensors can be allotted to the same zone). In order to do so, an example (for an easy example, see Figure 6.4) of a territory divided into 20 zones of research is used. Every zone is made up of 25 cells. 10 sensors are available for the exploration of the territory.

Parameters of the CE algorithm. For each iteration of the algorithm, $N = 500$ samples are generated and evaluated. The coefficient of the selection rate is $\rho = 0.05$. The smoothing parameter α is adaptive, depending on the number nb of elapsed iterations:

- For $nb \leq 4$, $\alpha = 0.1$ - For $nb > 4$, $\alpha = \left(1 - \frac{1}{\sqrt[4]{nb}}\right)$.

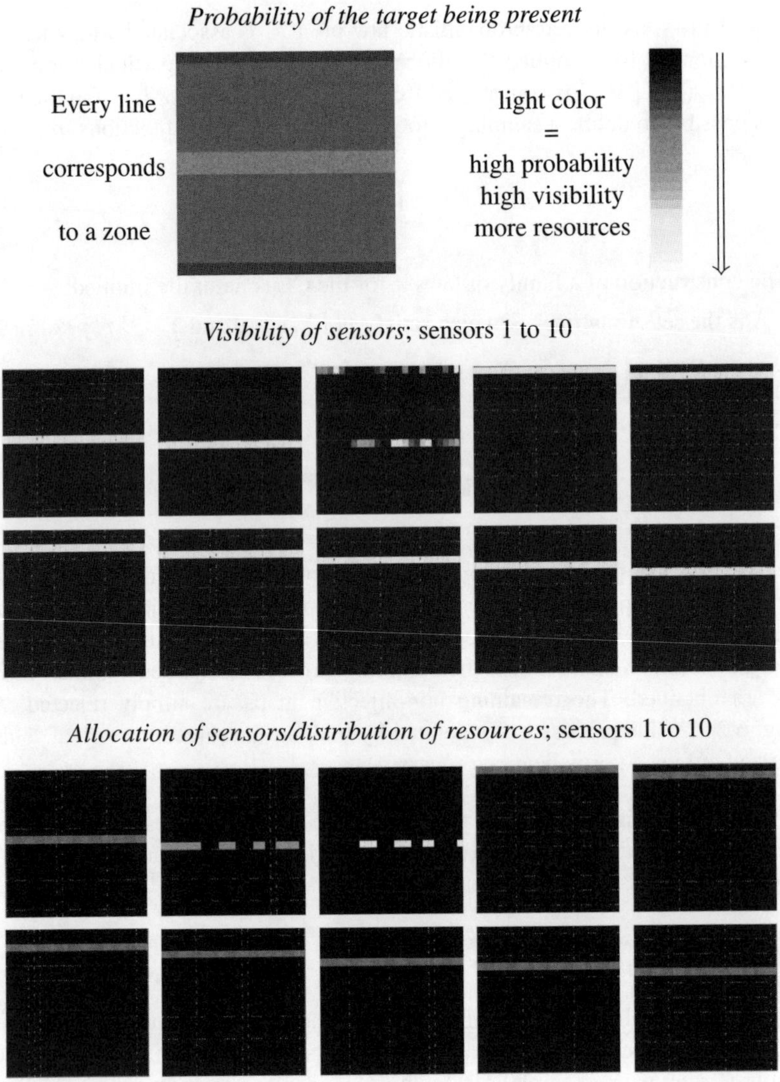

Figure 6.4. *Example*

Result and analysis. The optimal solution for this example is obtained after 30 iterations of the CE algorithm. Each sensor is specialized in the exploration of one single zone (except sensor 3). The algorithm allocates a sensor to a zone in which this particular sensor obtains an improved visibility. Sensor 3 can be allocated to two zones, which zone 1 and zone 11. The chances that the target is hidden in zone 1 are lower than in zone 11. Two sensors are therefore allocated to zone 11, sensor 2 and sensor 3. The sensors focus their research on the cells with the best visibility.

6.4. Example of implementation of CE for exploration

Throughout this section, it has been shown how CE could be applied to a path planning problem for a map-based mobile localization. It is assumed that the environment covers a territory $D \subset \mathbf{R}^2$. The objective is to choose the trajectories from an initial position $s_i \in D$ to a final position $s_f \in D$, which provide the best localizations of the mobile during the mission execution. The process of localization [DEL 99] consists of an incremental estimation of the position of the target by means of the information provided by the sensors. The estimation process is usually based on filtering methods, for instance the Kalman filter. The measurements which have been obtained are correlated to the embedded map, thus allowing for a global localization. *The posterior Cramér-Rao Bound can be used to evaluate the performance of the filter. It is a lower bound for the covariance matrix of the state estimate.*

6.4.1. *Definition of the problem*

The problem to be solved is based on two main stochastic equations. The first equation is known as the *state equation*, which provides the evolution of the mobile state between two successive time periods. The second equation is known as the *observation equation*, which links measurements to a particular state and prior knowledge of the map. The following notations are used:

– $X_k \triangleq (x_k, y_k) \in D$ the position of the mobile sensor at time k.

– $A_k \triangleq (a_x, a_y)$ the control given to the mobile sensor at time k. The controls taken into consideration are $(a_x, a_y) \in \{-d, 0, +d\} \times \{-d, 0, +d\}$ with $d \in \mathbb{R}^+$.

– Z_k is the vector of measurements at time k.

– $\mathcal{M}$ is the map in which the mobile is moving. It is built from specific features extracted from the environment and made up of N_f geometric objects. $m_i = (x^i, y^i) \in D, i \in \{1, \dots, N_f\}$ provide the positions of these objects considered as points of the map. These objects represent real elements that will be observed during the execution of the trajectory.

– $I_v(k) \subset \{1, \dots, N_f\}$ is the subset of indexes describing the objects of the map that can be seen by the mobile at time period k. Indeed, the sensor provides a local perception of the position of the mobile. Furthermore, we assume that all observations provided by the sensors originate from one and only one object of the map. There is therefore no problem in associating objects with the respective measurements.

The equation of the mobile is defined as follows:

$$X_0 \sim \pi_0 \triangleq \mathcal{N}(\overline{s_i}, P_0),$$

$$X_{k+1} = X_k + A_k + w_k, \quad w_k \sim \mathcal{N}(0, Q_k),$$

where $\{w_k\}_{k \geq 1}$ is a sequence of mutually independent Gaussian noises with average zero and respective covariance matrix Q_k.

For each visible object and each time period k, the mobile sensor has measurements describing the distance to visible objects and the angle between the axis of the sensor and the axis of the object. The horizontal axis is based on the map. The observation vector unifies all individual observations. The equation describing the observations works as follows:

$$Z_k = \{z^k(j)\}_{j \in I_v(k)},$$

with $\forall j \in I_v(k)$:

$$z^k(j) \triangleq \begin{cases} z_r^k(j) = \sqrt{(x_k - x_j)^2 + (y_k - y_j)^2} + \gamma_r^k(j), \\ z_\beta^k(j) = \arctan\left(\dfrac{y_j - y_k}{x_j - x_k}\right) + \gamma_\beta^k(j). \end{cases}$$

Variables $\{\gamma_r^k(j)\}$ and $\{\gamma_\beta^k(j)\}$ are random Gaussian noises with average zero and respective standard deviations σ_r and σ_β.

In order to address the optimization of the exploration, the map $\mathcal{M}$ is discretized to a grid of $N_x \times N_y$. cells. A state $s \in \{1, \ldots, N_s\}$ with $N_s = N_\times N_y$ is linked to each cell. Owing to the nature of the action space, there are $N_a = 8$ possible decisions in association with states s (with a notable exception at the border of the map – Figure 6.5), which control the move to eight possible neighbors. A trajectory between s_i and s_f is thus defined by the sequence of actions $V^K a = (a_1, a_2, \ldots, a_K)$ or the sequence of states $V^{K+1} s = (s_i, s_2, \ldots, s_K, s_f)$. The problem of optimal planning consists of determining the sequence $V^* a$ which minimizes the cost function of the mission. From now on, $\Im_{s_i s_f}$ denotes the trajectories which start from s_i and end at s_f.

The posterior Cramér-Rao bound as a criterion for optimization. The posterior Cramér-Rao bound is used when a random parameter is estimated from random measurements. If $\hat{X}(Z)$ provides a non-biased estimation of a random vector $X \in \mathbb{R}^r$ on the basis of random measurements Z, the posterior Cramér-Rao bound is defined as the inverse of the information matrix [VAN 68] established by Fisher J:

$$E\Big\{\big(\hat{X}(Z) - X\big)\big(\hat{X}(Z) - X\big)^T\Big\} \succeq J^{-1}(X, Z), \qquad [6.12]$$

where J is given by:

$$J_{ij} = E\Big[- \frac{\partial^2 \ln p(Z, X)}{\partial X_i \partial X_j} \Big] \quad i, j = 1, \ldots, r \qquad [6.13]$$

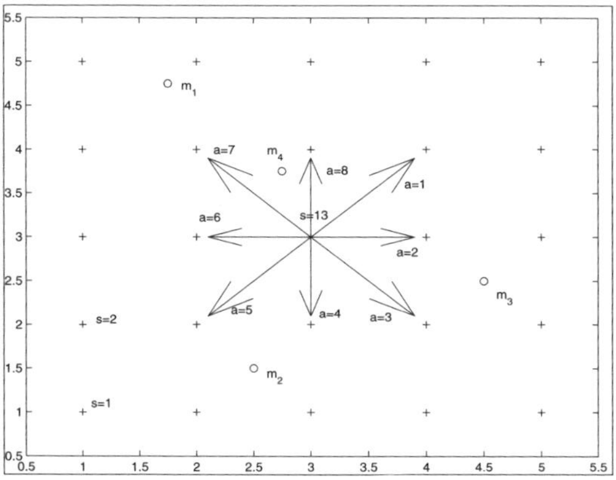

Figure 6.5. *This image shows the states s (+), the eight actions a (arrows) and the objects m_i (o)*

and $p(Z, X)$ is the joint probability for the variables X and Z. If X is of dimension r, then J has the size $r \times r$. In the context of a filtering, that is an iterative estimation of the position X_k by means of the measurements Z_k, there is a recurrent link between the matrices J_k and J_{k+1} [TIC 98]. For this application the following relationship applies:

$$J_0 = P_0^{-1}, \qquad [6.14]$$

$$J_{k+1} = \Pi_k + \left(J_k + Q_k\right)^{-1}, \qquad [6.15]$$

where Π_k represents the information obtained from the observation process. This relationship is derived by differentiating the law $p(Z_k|X_k)$ with the variable X_k. This chapter will not provide details about this calculation. However, note that a Monte Carlo simulation is used for computing the expectation because of the non-linear observation equation [PAR 02].

The optimization criterion in use is a function of matrices J_k along the trajectories. Specifically, this criterion is related to the area of the ellipsis that describes the uncertainty of the estimation. In the case of a trajectory $V_s^{K+1} = (s_i, s_2, \ldots, s_K, s_f)$, the objective function to be minimized is expressed as follows:

$$\phi\left(V_s^{K+1}\right) \propto \sum_{k=0}^{K} \alpha_k \det\left(J_k^{-1}\right), \quad \alpha_k \in \mathbb{R}^+.$$

This function depends on the determinant of J_k. It is shown that a method such as *dynamic programming* cannot solve the best trajectory for this criterion [LEC 97]. The CE method will be used in this case.

6.4.2. *Applying the CE*

In order to implement the CE algorithm, it is first necessary to define the random processes which will generate our samples. These samples are defined by a family of laws $\pi(\cdot;\lambda) \mid \lambda \in \Lambda$, which have to be chosen in accordance with the structure of the problem. Now, let τ be an element of $\Im_{s_i s_f}$, the set of valid trajectory. This trajectory is characterized by a sequence of actions which has been decided for each state within τ. The idea will be to decide for an action conditionally to a state. As a consequence, each state s is associated with a density of discrete probability fs which defines the probability of the event "choose action a":

$$f_s(i, \mathbf{p}^s) = \mathbb{P}(A = i) = p_i^s, \quad i = 1, \ldots, 8 \text{ with } \sum_{i=1}^{8} p_i^s = 1. \qquad [6.16]$$

The family adapted to this problem is thus $(f_s(\cdot, \mathbf{p}^s))_{s \leq 1 \leq N_s} \mid \mathbf{p}^s \in [0,1]^8$. Moreover, the density will be chosen uniform for the initial stage of the algorithm.

Generating trajectories. At iteration n of the CE algorithm, and assuming that the density of probability is $(f_s(\cdot, \mathbf{p}_n^s))_{s \leq 1 \leq N_s}$, the generation of a trajectory of length T in $\Im_{s_i s_f}$ is carried out according to the procedure shown in Table 6.1.

$j = 0$
while $(j < N)$ do:
    $t = 0$, $s_t = s_i$
    generate a_0^j according to $f_{s_i}(\cdot, p_n^{s_i})$
    apply action a_0^j and update s_t.
    repeat until $s_t = s_f$:
        - generate action a_t^j according to $f_{s_t}(\cdot, p_n^{s_t})$
        - apply action a_t^j and update s_t.
        - set $t = t + 1$.
    return $\tau(j) = (s_i, a_0^j, s_1^j, a_1^j, \ldots, s_{i-1}^j, a_{i-1}^j, \ldots, s_{T-2}^j, a_{T-2}^j, s_f)$
    set $j = j + 1$

Table 6.1. *Generating N trajectories at iteration* n

Updates. At iteration n of the CE algorithm, γ_{n+1} is estimated as the ρ-quantile for the sample evaluations $(\phi(\tau(j)))_{1 \leq j \leq N}$. Denote $S_{n+1} = \{j \in [\![1, N]\!] / \phi(\tau(j)) \leq \gamma_{n+1}\}$. As the density of probability is discrete $(f_s(\cdot, \mathbf{p}_n^s))_{s \leq 1 \leq N_s}$, an estimation of the parameters p_{n+1}^s is achieved using the formula [RUB 04]:

$$p_{i,n+1}^s = \frac{\sum_{j \in S_{n+1}} \delta[\{\phi(\tau(j)) \in \chi_{si}\}]}{\sum_{j \in S_{n+1}} \delta[\{\phi(\tau(j)) \in \chi_s\}]} \quad s \in 1, \ldots, N_s, \; i = 1, \ldots, 8 \qquad [6.17]$$

where $\{\phi(\tau(j)) \in \chi_{si}\}$ is the event "the trajectory $\phi(\tau(j))$ is going through state s where action i is decided" and $\{\phi(\tau(j)) \in \chi_s\}$ is the event "the trajectory $\phi(\tau(j))$ is going through state s".

6.4.3. *Analyzing a simple example*

This section presents an application of the algorithm to a simplified example. The map $\mathcal{M}$ has the size $[1, 10] \times [1, 10]$ and contains seven objects. The grid contains $N_s = 10 \times 10$ states. The initial and the final positions are situated on the grid at $(3; 2)$ and $(9; 9)$. The dynamic model is characterized by the following uncertainty:

$$P_0 = \sigma^2 \cdot \mathcal{I}_{22}, \qquad Q_k = \sigma^2 \cdot \mathcal{I}_{22}, \; \forall k \geq 1$$

where $\mathcal{I}_{22}$ is the identity matrix of size 2×2 and $\sigma^2 = 1,6 \cdot 10^{-3}$. Furthermore, the observation equation is characterized by $\sigma_a^2 = 10^{-4}$, $\sigma_\beta^2 = 10^{-8}$ for all objects. The CE algorithm has been run during 30 iterations, with 4,000 samples of trajectories generated at each iteration. The selection parameter was $\rho = 0.2$.

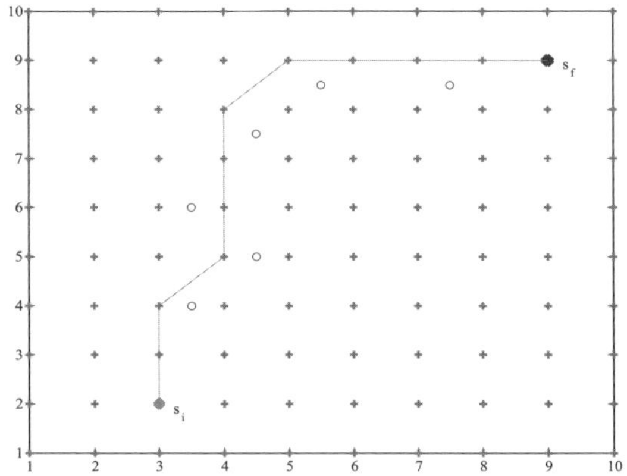

Figure 6.6. *Best possible trajectory obtained after 30 iterations of the CE algorithm*

Figure 6.6 shows the best possible trajectory which has been determined in the last iteration of the algorithm. The algorithm's behavior is coherent. The best obtained trajectory is the one that favors passages through the neighborhood of the objects of the map. Figure 6.7 presents the evolution of parameter γ and the minimal cost $\phi_{\min}$ for all generated trajectories. Figure 6.8 presents the evolution of density $f_{s_i}(., \mathbf{p}_k^{s_i})$ at the initial point. Figure 6.7 shows the decrease of parameter γ as well as the fast convergence of the algorithm. We should note the concentration of the sampling laws around the best previous samples. Furthermore, analyzing the density of probability (Figure 6.8) for the initial position s_i shows the persistence of two chosen actions. The probability is, however, higher for action 8.

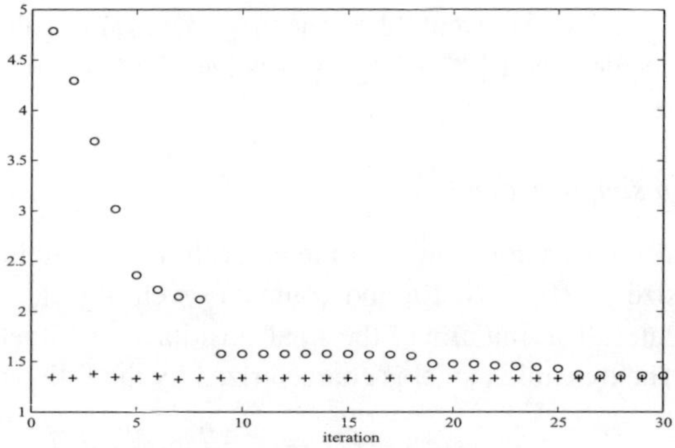

Figure 6.7. *Evolution of γ (circles) and of the minimum ϕ among the samples (crosses)*

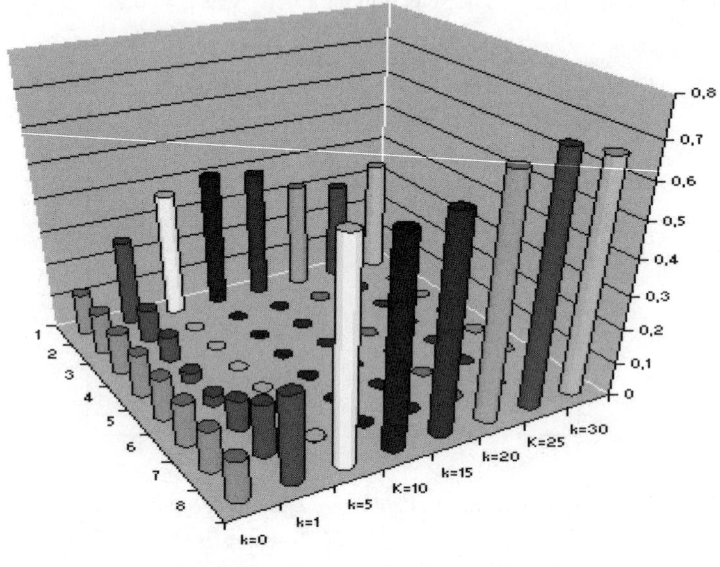

Figure 6.8. *Evolution of $f_{s_i}(\cdot, \boldsymbol{p}_k^{s_i})$*

6.5. Optimal control under partial observation

In previous sections, sensor planning has been addressed without taking into account observation feedback. Such prior planning is satisfactory when the exploration space is not excessively random. Otherwise, a dynamic planning approach should be considered. However, dynamic planning of the sensors leads to major difficulties in optimization.

Let us consider a very easy example. A moving target has to be intercepted within a delimited territory. This target is moving according to a random model and reacts by escaping the near trackers. Moving in the space is easy for the trackers but the observation of the target is difficult. Now, it is assumed to be a hill in the middle of the territory. Climbing up that hill is difficult but allows a good view from the top. Which strategy should be chosen under these conditions? Should the interception take place on the territory only, i.e. fast movement but bad observation of the target, or does it makes sense to climb to the top of the hill, which means loosing time at the beginning but allowing for a good observation and subsequently a better tracking policy. When evaluating both strategies an optimal usage of the available information has to be guaranteed. For this reason, dynamic problems are very complex from a mathematical point of view.

In the literature, some hypotheses are expressed generally. First of all, the law of evolution of the environment is assumed to be Markov. Secondly, the optimization criterion is simple enough, that is, additive over the course of time. The problem can therefore be solved theoretically [SON 71, CAS 98] by a dynamic programming approach. The solution, however, takes up too much time and space in computer storage and is therefore inaccessible, without pruning strategies. Another classical approach will approximate the decision strategy on the basis of a reinforcement learning process [BAK 04]. Even though this method is limited by the computer storage of the decision grid, there are notable progresses using hierarchical approaches.

In this section, the CE method will be applied to the learning process of optimal strategies for planning under partial observation. The aim is to describe the possible operating policies of control for each group of generic laws (typically hidden Markov models). The parameters of the law are optimized by CE. This method does not require a Markov evolution of the states nor the additive hypothesis on the objective function. The approach avoids the restrictions in term of the dimension of the observation space. On the other hand, the performance of the strategies is balanced by the complexity of the implemented models.

6.5.1. *Decision-making in partially observed environments*

This section explains a short theoretical background of control under partial observation; a simulated experiment is provided in section 6.5.3.

It is assumed that a subject has to carry out a mission in a given environment. This subject interacts with the environment. It consumes observations and produces actions. The control policy of the subject is to be optimized in order to accomplish the mission.

The environment. The environment is described by a hidden state x which evolves with time. In *the example of section 6.5.3, this state is made up of the positions* *of the target and two patrols.* The time t is discretized and flows from period 1 to the final period T. A temporal evolution of the state is represented by vector $x = x_{1:T} = x_1, \ldots, x_t, \ldots, x_T$. In the course of the mission the subject produces the decisions $d = d_{1:T}$ which has an impact on the evolution of the environment. *In the example, d is the move decision for the patrols.* The subject perceives partial and noisy observations of the environment, which are expressed as $y = y_{1:T}$. *In the example, this observation is a poor estimation of the position of the target.* The environment is then characterized by a law of the hidden state dynamic and a law of the observations, which are conditional of the decisions. This probabilistic law is subsequently denoted P:

> The hidden state x_t and the observation y_t are known on the basis of the conditional law $P(x_t, y_t \mid x_{1:t-1}, y_{1:t-1}, d_{1:t-1})$, which depends on the states, observations and past decisions. Notice that d_t itself is generated by the subject after receiving the measure y_t.

In this chapter, there is no Markov hypothesis carried out concerning P. On the other hand, it is assumed that the law $P(x_t, y_t \mid x_{1:t-1}, d_{1:t-1})$ is simulated very quickly. The law for $x, y \mid d$ is illustrated in Figure 6.9. The outgoing arrows represent the data produced by the environment, i.e. observations. The incoming arrows are the data consumed by the environment, i.e. decisions made by the subject. The variables are written in chronological order: y_t appears before d_t, since decision d_t is made after receiving y_t. The notation $P(x, y|d)$ is used to describe the law for the entire environment:

$$P(x, y \mid d) = \prod_{t=1}^{T} P(x_t, y_t \mid x_{1:t-1}, y_{1:t-1}, d_{1:t-1}).$$

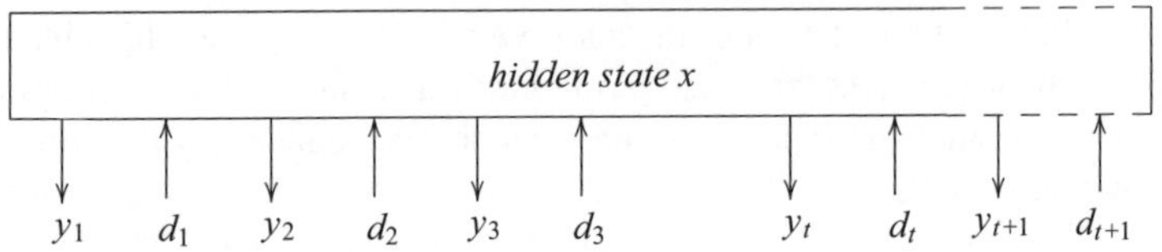

Figure 6.9. *The environment*

Evaluation and optimal planning. Evaluation and optimal planning. The mission to be accomplished will now be described and formalized. *The time which is available to accomplish the mission is limited.* T therefore represents the maximum length of time for this mission. The mission is evaluated by means of an objective function $V(d, y, x)$ which evaluates the trajectories d, y, x obtained after mission. *Typically, function V can be used to compute the time required for the accomplishment of a mission.* There is no assumption made about function V, except that the computation of $V(d, y, x)$ is fast.

Our purpose is to construct an optimal decision tree $y \mapsto (d_t(y_{1:t})|_{t=1}^{T})$, that depends on past observations, in order to maximize the average evaluation:

$$d_* \in \arg\max_d \sum_y \sum_x P\Big(x, y \mid \big(d_t(y_{1:t})\big|_{t=1}^{T}\big)\Big) V\Big(\big(d_t(y_{1:t})\big|_{t=1}^{T}\big), y, x\Big). \quad [6.18]$$

A schema for this optimization process is shown in Figure 6.10. The double arrows $\Rightarrow$ characterize the variables to be optimized. These arrows describe the flow of information between the observations and actions. The cells described as ∞ produce decisions and transmit all received and generated information to their next neighbor. This architecture illustrates a non-finite-memory problem: the decision depends on all past observations. When the evaluation V is additive and the Markovian environment applies, it is possible to build a recursive optimal solution on the principles of dynamic programming. However, note that this recursive construction involves the manipulation of the posterior probabilistic law of the hidden state as a parameter of the DP. It is difficult to handle such continuous parameters. A discrete approximation of the solution is necessary but difficult. The approach is different here: it is the probabilistic model approximating the control policy, which is optimized.

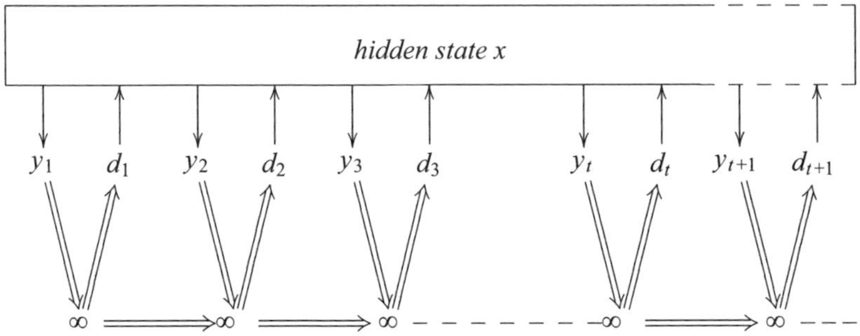

Figure 6.10. *Optimal planing*

Direct estimation of the decision tree. In equation [6.18], the value to be optimized d_* is a deterministic object. A probabilistic approach is possible. Indeed, it is an equivalent problem to find $\pi(d \mid y)$, a law of the decisions conditionally to the past observations, that maximizes the average benefit:

$$V(\pi) = \sum_d \sum_y \sum_x \prod_{t=1}^{T} \pi\big(d_t \mid d_{1:t-1}, y_{1:t}\big) P(x, y \mid d) V(d, y, x).$$

This problem is still illustrated in Figure 6.10. But now, the double arrow describes the structure of the dynamic Bayesian network for law π. Actually, there is not a true difference between the probabilistic case and the deterministic case: if the

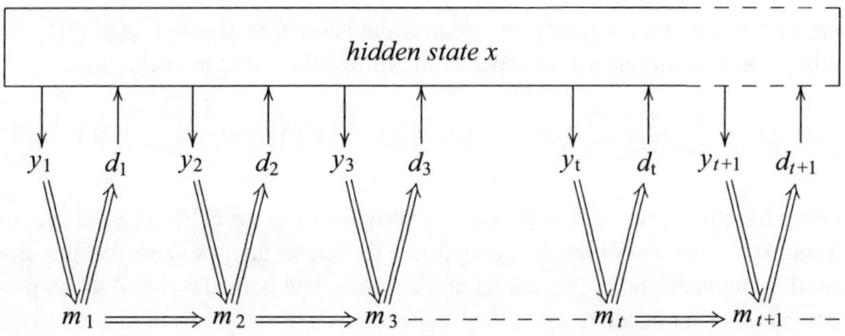

Figure 6.11. *Planning on the basis of a finite memory*

solution d_* is unique the optimal law π_* is a Dirac function around d_*. On the other hand, the probabilities become interesting in a case where π_* is just approximated, since probabilistic models are continuous, and thus better suited to approximation. Furthermore, an approximation of the optimal decision laws follows naturally, by just replacing the infinite memory ∞ with a finite memory m; see Figures 6.10 and 6.11. In fact, by limiting the hidden memory, the decision law π becomes a hidden Markov model. These models are easily optimized using CE.

Defining a family of laws for the approximation. Let M be a finite set of states that characterizes the memory of the laws. Define the variables $m_t \in M$, the memory state at time-period t. For this choice of memory M, the hidden Markov models h are defined as follows:

$$h(d \mid y) = \sum_{m \in M^T} h(d, m \mid y)$$

with

$$h(d, m \mid y) = \prod_{t=1}^{T} \big(h_d(d_t \mid m_t) h_m(m_t \mid y_t, m_{t-1})\big).$$

It is assumed that the conditional laws h_d and h_m are invariant in time. Of course, there are many possible choices for h, given M. These choices are characterized by the definition of h_d and h_m. Now, let $\mathcal{H}$ denote the set of all such laws h.

6.5.2. *Implementing CE*

Our aim is to choose MMC $h \in \mathcal{H}$ so as to attain an optimal strategy π_*:

$$\pi_* \simeq h_* \in \arg\max_{h \in \mathcal{H}} V(h).$$

From now on, $P[h]$ refers to the law of the complete system, which combines both the environment and planning components:

$$P[h](d,y,x,m) = P(x,y \mid d)h(d,m \mid y).$$

The estimation of π_* by the means of h corresponds to the following problem:

$$h_* \in \arg\max_{h\in\mathcal{H}} \sum_d \sum_y \sum_x \sum_m P[h](d,y,x,m)V(d,y,x).$$

Optimizing h_* means adjusting the parameter $h \in \mathcal{H}$ in order to narrow the probability $P[h]$ around the optimal value V. This is exactly what CE does. The original algorithm (without smoothing) has been used in this application. By choosing the selection rate $\rho \in]0,1[$ CE optimization takes the form:

1) initialize h by choosing h_d and h_m to be uniform;

2) generate N samples $\theta^n = (d^n, y^n, x^n, m^n)$ by the law $P[h]$;

3) define S as the set of the $\lfloor \rho N \rfloor$ best samples in regards to the objective $V(d,y,x)$;

4) update h by:

$$h \in \arg\max_{h\in\mathcal{H}} \sum_{n\in S} \ln P[h]\big(\theta^n\big), \qquad [6.19]$$

5) reiterate from stage 2) until convergence is reached.

Maximization [6.19] does not create any difficulties. Markov properties are particularly instrumental here: $\ln P[h]$ is divided into a sum of basic logarithms and the optimization is reduced to several small and independent sub-problems. The resolution is completed by Khun and Tucker and leads to the following update formulae:

$$\begin{cases} h_d(A \mid B) = \dfrac{\sharp\big\{(n,t) \in S \times [\![1,T]\!] / A = d_t^n \text{ and } B = m_t^n\big\}}{\sharp\big\{(n,t) \in S \times [\![1,T]\!] / B = m_t^n\big\}}, \\ h_m(A \mid B, C) = \dfrac{\sharp\big\{(n,t) \in S \times [\![1,T]\!] / A = m_t^n,\ B = y_t^n \text{ and } C = m_{t-1}^n\big\}}{\sharp\big\{(n,t) \in S \times [\![1,T]\!] / B = y_t^n \text{ and } C = m_{t-1}^n\big\}}. \end{cases}$$

6.5.3. *Example*

6.5.3.1. *Definition*

A target R moves in a territory of 20×20 cells, i.e. $[\![0,19]\!]^2$. R is tracked by two patrols B and C, which are controlled by the subject. The coordinates of R, B and C at moment t are expressed as follows: (i_R^t, j_R^t), (i_B^t, j_B^t) and (i_C^t, j_C^t). B and C receive a very limited amount of information on the position of the target and move slowly:

– moving B (or C) is done with the following orders: *turn left*, *turn right*, *go forward*, *do nothing*. there are 16 different possibilities;

– the patrols are initially positioned in the down corners of the territory, i.e. $i_B^1 = 0, j_B^1 = 19$ and $i_C^1 = 19, j_C^1 = 19$, and are directed downward;

– At each measurement scan, patrol B (respectively C) is informed whether the target is situated in their front plane or not. For example, if B is directed upward, it will detect whether $j_R < j_B$ or not. There are therefore four possible observations.

Initially the target is localized randomly in the upper half of the lattice. This positioning is based on a uniform law. The target moves in all directions (also diagonally) and no more than just one cell. This movement is probabilistic and favors escape:

$$\begin{cases} P\big(R^{t+1} \mid R^t\big) = 0 \quad \text{if } \big|i_R^{t+1} - i_R^t\big| > 1 \text{ or } \big|j_R^{t+1} - j_R^t\big| > 1, \\ P\big(R_{t+1} \mid R_t\big) \propto \big(i_R^{t+1} - i_B^t\big)^2 + \big(j_R^{t+1} - j_B^t\big)^2 \\ \qquad\qquad + \big(i_R^{t+1} - i_C^t\big)^2 + \big(j_R^{t+1} - j_C^t\big)^2 \quad \text{otherwise.} \end{cases}$$

Note that a short distance is neglected in the sum in comparison to longer distances. In other words, a patrol that is far away might hide a patrol that is very close to the target. This modeling error is deliberate, since it will be illustrative of the algorithm's capacity to generate original strategies.

The mission objective is to maintain the target in proximity of at least one patrol at each time period (i.e. a distance less than 3). The evaluation function V consists of counting the number of such hits in the course of the mission:

$$\begin{cases} V_0 = 0; \\ V_t = V_{t-1} + 1 \quad \text{if } d\big(B^t, R^t\big) \leq 3 \text{ or } d\big(C^t, R^t\big) \leq 3; \\ V_t = V_{t-1} \qquad\quad \text{otherwise.} \end{cases}$$

The entire duration of the mission is $T = 100$.

6.5.3.2. *Results*

Global behavior of the algorithm. The implemented algorithm is based on a low selection rate $\rho = 0.5$ and on a high number of samples $N = 10,000$. Note that this choice of parameters has not been optimized at all, and may be improved. Moreover, the CE smoothing was not implemented at the time of the experiment. However, the risk of degeneration has been avoided by a adding a tiny noise at the law updating stage.

Convergence. At the very initial step, the strategies are mostly irrelevant. For this particular problem the initial evaluation was 5% of the optimal value. The convergence speed was also relatively low at the beginning. This speed increases as well as the improvement of the population, until a new stable stage has been reached. At this stage about 75% of the optimal value is obtained. Swapping between stages of slow and rapid convergence has been observed several times. However, the speed usually decreases gradually.

Near-optimal policies. 69 is the best average number of hits. This number has been obtained by running the CE during a couple of days. Moreover, several instances of the run have been tested, and a large hidden Markov model $\sharp M = 256$ has been used. Two different typical behaviors have been observed within the optimal policy:

– *the patrols cooperate in order to track the target; see Figure 6.12*;

– *when the target is close to the borderline one patrol goes along the opposite borderline while the other patrol continues tracking the target*. This behavior is a consequence of the blindness of the target to a close patrol when the other one is far away.

Other tests have been carried out, for which the observations have been deleted. The average number of hits was 32. In addition, in the case where memory transitions are deleted within h, the average number of hits is 55 (usage of the previous observation only).

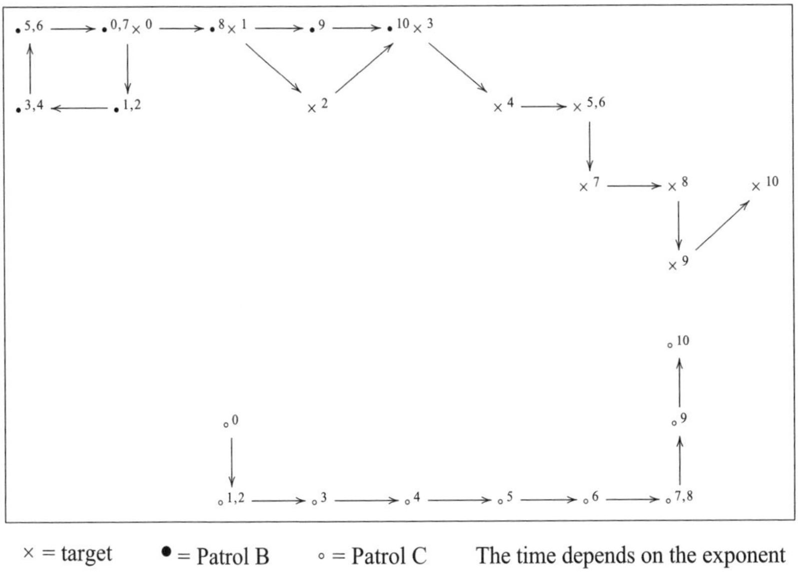

Figure 6.12. *A control sequence*

Intermediary tests. For the subsequent results in the table, the algorithm was stopped after 10,000 iterations. This fact corresponds to a strong convergence. On the currently available machines (PCs at 1 Ghz), this was several hours of processing. The table below shows the percentage that has been reached relatively to the estimated maximum, which is 69 hits. These percentages have been obtained for several memory size, #M.

$\sharp M$	16	32	64	256
Evaluation	94%	96%	97%	97%

The solutions which have been obtained are close to the best estimated solution. Moreover, it seems that a quite limited memory is sufficient for a reasonable performance.

6.6. Conclusion

The CE method for optimization is a very elegant metaheuristic which is based on the principle of importance sampling. As a simulation process, CE adapts itself to data that cannot be easily formalized. From this point of view, CE optimization is a learning approach for optimizing a strategy.

Good implementations of CE require a well adapted modeling of the sampling family in use. For dynamic problems, Markov models are favored. When observations are involved, memory-based models, like hidden Markov models, are preferred.

With CE, the problem of optimization turns into a modeling problem. Specifically, the CE method benefits from the modeling strength of probabilities. Some recent works have investigated the combination of cross-entropy with Bayesian networks. Such approaches have been used in order to reduce the combinatorics by the means of hierarchical operating policies [DAM 06].

6.7. Bibliography

[BAK 04] BAKKER B. and SCHMIDHUBER J., "Hierarchical Reinforcement Learning Based on Subgoal Discovery and Subpolicy Specialization", *8th Conference on Intelligent Autonomous Systems*, Amsterdam, The Netherlands, pp. 438–445, 2004.

[CAS 98] CASSANDRA A.R., Exact and approximate algorithms for partially observable Markov decision processes, PhD Thesis, Brown University, Rhode Island, Providence, May 1998.

[DAM 06] DAMBREVILLE F., "Cross-entropic learning of a machine for the decision in a partially observable universe", *Submitted to European Journal of Operational Research*, http://fr.arxiv.org/abs/math.OC/0605498, 2006.

[DEG 61] DE GUENIN J., "Optimum distribution of effort: an extension of the Koopman theory", *Operations Research*, 1961.

[DEB 03] DE BOER P.T., KROESE D.P., MANNOR S. and RUBINSTEIN R.Y., "A tutorial on the cross-entropy method", *Technique et Science Informatiques*, http://iew3.technion.ac.il/CE/tutor.php, 2003.

[DEL 99] DELLAERT F., FOX D., BURGARD W. and THRUN S., "Monte Carlo localization for mobile robots", *IEEE Intl. Conf. on Robotics and Automation*, 1999.

[HOM 04] HOMEM-DE-MELLO T. and RUBINSTEIN R.Y., "Rare event estimation for static models via cross-entropy and importance sampling", http://users.iems.nwu.edu/~tito/list.htm, 2004.

[KOO 57] KOOPMAN B.O., "The theory of search: III. The optimum distribution of searching effort", *Operations Research*, vol. 5, pp. 613–626, 1957.

[LEC 97] LE CADRE J.-P. and TREMOIS O., "The matrix dynamic programming property and its implications", *SIAM Journal on Matrix Analysis*, vol. 18, no. 2, pp. 818–826, 1997.

[MAR 04] MARGOLIN L., "On the convergence of the cross-entropy method", *Annals of Operations Research*, 2004.

[PAR 02] PARIS S. and LE CADRE J.-P., "Planning for terrain-aided navigation", *Conference Fusion 2002*, Annapolis, USA, pp. 1007–1014, 2002.

[RUB 04] RUBINSTEIN R. and KROESE D.P., *The Cross-Entropy method. An unified approach to Combinatorial Optimization, Monte-Carlo Simulation, and Machine Learning*, Springer, information science & statistics edition, 2004.

[SON 71] SONDIK E.J., The optimal control of partially observable Markov processes, PhD Thesis, Stanford University, Stanford, California, 1971.

[TIC 98] TICHAVSKY P., MURAVCHIK C. and NEHORAI A., "Posterior Cramer-Rao bounds for discrete-time nonlinear filtering", *IEEE Transactions on Signal Processing*, vol. 46, no. 5, pp. 1386–1396, 1998.

[VAN 68] VAN TREES H.L., *Detection, Estimation and Modulation Theory*, Wiley, New York, 1968.

Chapter 7

Optimizing Emissions for Tracking and Pursuit of Mobile Targets

7.1. Introduction

Increasingly intricate problems and a variety of available sensors based on quite different physical principles have led to the increasing necessity of optimizing the usage of resources. It is assumed here that the sensors are collocated but differ in their essential characteristics such as the type of observations, detection performance, geographical coverage and the cost of usage. This chapter only deals with tracking problems.

The problems addressed here are rather simply formalized. The system has passive as well as active measurements. Passive measurements do not include the estimation of the target's complete state and typically are limited to the estimation of the target's direction (if partially observed). Passive measurements thus imply (1D) non-linear functions with a noisy state. On the other hand, the system is able to emit (active measurements). These measurements provide a direct estimation of the distance to the target, but generate cost. This cost is limited by a global budget. The global budget covers different factors such as the surveillance capacity of the entire system, risks, etc. Several practical examples could illustrate this type of problem. For example, an aeroplane used for maritime patrol has active and passive sensors in order to accomplish its mission. Passive sensors are, for example, electronic support measurements (ESM) and an active sensor is, for example, radar. In order to maintain discretion, the emissions rely on the "less is more" concept. There are even sonar

---

Chapter written by Jean-Pierre LE CADRE.

systems which combine active and passive features of systems such as radar or infrared (IR).

Optimal measurement scheduling is thus an important tool which has been used for many different approaches. Mehra [MEH 76] has investigated different norms of the Fischer matrix. Van Keuk *et al.* [VAN 93] have also examined this problem in terms of optimizing the distribution of resources for maximizing the maintenance of existing radars tacks. Avitzour and Rogers [AVI 90] have worked on an important extension of this problem. They tried to optimize the distribution of active measures when estimating a random variable x, given that 1) the global budget has been fixed, 2) the cost for each (active) measurement is proportional to the reduction of variances for $\hat{y}$ and 3) the function of autocorrelation for the quantity to be measured $\{x(i) : i = 1, \dots, N\}$, as well as the correlations between $\{x(i) : i = 1, \dots, N\}$ and y, are known. These results were then used by Shakeri *et al.* [SHA 95] for a discrete time and vector-based system. However, note that the general framework for the analysis is linear here.

In the case analyzed in this chapter, however, the framework is non-linear. After a general presentation of the problem (section 7.2), general tools necessary for the analysis are presented in section 7.2. It is possible to examine the general formulation for the optimal distribution of active measures. This is, first of all, done for the deterministic case (section 7.3) and then for the stochastic case (section 7.4).

7.2. Elementary modeling of the problem (deterministic case)

For this deterministic case, the trajectory of the target is assumed to be deterministic. First, a measurement of the estimability of the target trajectory parameters is introduced. Then, tools based on multi-linear algebra are defined for approximating this measure. These tools are basic enough to drastically simplify the problems of optimization related to the *optimal measurement scheduling*.

7.2.1. *Estimability measurement of the problem*

Our study is first limited to rectilinear and uniform movement. The equation which describes the movement is defined below [NAR 81]:

$$\mathbf{r}(t) = \mathbf{r}(0) + t\mathbf{v}(0) - \int_0^t (t-\tau)\mathbf{a}_0(\tau)d\tau \qquad [7.1]$$

where:

– the reference time is 0;

– r and v are the (relative) vector position and vector speed of the target;

– a_0 represents the maneuver of the observer.

For passive sensors, the measurements are frequently angular bearings, defined by the following equation:

$$\beta(t) = \tan^{-1}\left[r_x(t)/r_y(t)\right]. \qquad [7.2]$$

For active measures (e.g. radar, active sonar), the distance $r(t)$ is also provided:

$$r(t) = \left[r_x^2(t) + r_y^2(t)\right]^{1/2}. \qquad [7.3]$$

The passive measurements $\hat{\beta}_k$ are available at all times (and are not cost intensive). Active measurements $\hat{r}_k$ on the other hand are only available during emission. For every active measurement there is a cost c_k, which depends on the operational context. The noises $w_{\beta,k}$ (passive measures) and $w_{r,k}$ (active measures) are modeled by a Gaussian white noise. Associated variances σ^2 depend on the position of the target relative to the system and the level of emissions.

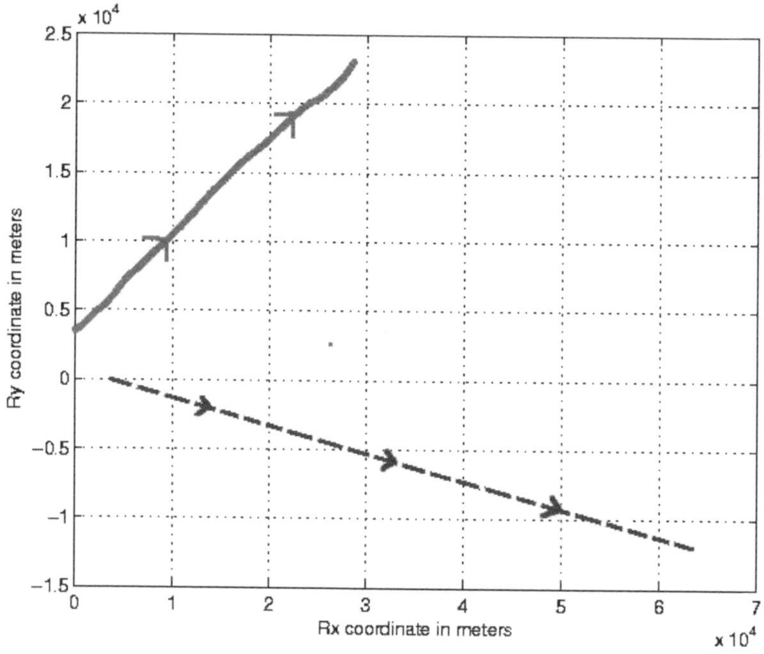

Figure 7.1. *The target-observer scenario: r_x and r_y are Cartesian coordinates. Dotted line: trajectory of the observer; straight line: trajectory of the target*

Fischer information matrix

When estimating a deterministic parameter θ, the variance of the errors is bounded by the Cramér-Rao bound:

$$\mathbb{E}(\hat{\theta}-\theta) \geq \mathbb{E}\left(\left[\frac{\partial \ln p(Z \mid \theta)}{\partial \theta}\right]2\right) = -\mathbb{E}\left(\left[\frac{\partial^2 \ln p(Z \mid \theta)}{\partial \theta^2}\right]\right). \qquad [7.4]$$

In the case of vector based parameters Θ, this inequality is replaced by a matrix-based inequality:

$$\begin{aligned}\mathbb{E}(\hat{\Theta}-\Theta) &\succeq \mathbb{E}\left(\left[\frac{\partial \ln p(Z \mid \Theta)}{\partial \Theta}\right]\left[\frac{\partial \ln p(Z \mid \Theta)}{\partial \Theta}^T\right]\right) \\ &= -\mathbb{E}\left(\left[\frac{\partial^2 \ln p(Z \mid \theta)^2}{\partial \theta_i \theta_j}^2\right]^2\right).\end{aligned} \qquad [7.5]$$

The matrix $\mathbb{E}\left(\left[\frac{\partial \ln p(Z|\Theta)}{\partial \Theta}\right]\left[\frac{\partial \ln p(Z|\Theta)}{\partial \Theta}^T\right]\right)$ is based on the dyadic product of the gradient vectors. This matrix is semi-defined and positive. It represents the average curvature of the likelihood on Θ. In the following, this matrix will be referred to as the Fischer information matrix (FIM).

The vectors $\nabla_\Theta \ln p(Z \mid \Theta)$ are calculated for Θ, which are the target's parameters (or states). Here $\Theta = (r_x, r_y, v_x, v_y)^T$ and vector Z represent the observation vectors (active and/or passive). Elementary calculations provide the vectors $\mathbf{M}_k$ or $\mathbf{N}_k$, which are instrumental for computing the FIM. These vectors are gradient vectors of observations equations: $\{\beta_k\}$ for passive and $\{r_k\}$ for active cases, relatively to the state of the target:

$$\begin{cases} \mathbf{M}_k = \dfrac{1}{r_k}\big(\cos\beta_k, -\sin\beta_k, k\cos\beta_k, -k\sin\beta_k\big)^T, \\ \mathbf{N}_k = \big(\sin\beta_k, \cos\beta_k, k\sin\beta_k, k\cos\beta_k\big)^T. \end{cases} \qquad [7.6]$$

Note that the two vectors are orthogonal. Given that angular measurements and distance are independent of one another, the calculation of the FIM is a standard operation.

$$\text{FIM} = \sum_k \left(\frac{1}{\sigma^2_{\beta,k}}\mathbf{M}_k\mathbf{M}_k^T + \frac{\delta_{r,k}}{\sigma^2_{r,k}}\mathbf{N}_k\mathbf{N}_k^T\right). \qquad [7.7]$$

In [7.7], $\delta_{r,k}$ equals 1 in case of active measurements at time k, and 0 otherwise.

In this context the "measurements" of estimability might be a function based on the FIM. The determinant which represents the volume of uncertainty associated with state $\mathbf{X}$ is a possible measurement of this estimability. The difficulty is largely based on the fact that our objective is global optimization. The sequence of controls (time steps when emission takes place) that optimizes the global cost has to be found. Unfortunately, it was proved [LEC 97] that the only functions based on the FIM, that allow a dynamic programming approach, are nearly linear functions, i.e. functions that can be expressed as linear functions by a simple, real and monotonous function.[1] However, such functions are useless, since the trace of the information matrix is trivially computed (see equation [7.7]) and not informative. As a consequence, the calculation of the function to be optimized has to undergo as much mathematical development as possible.

7.2.2. *Framework for computing exterior products*

First at all, exterior products are introduced into the vector space $\mathcal{V}$. Let $\mathcal{V}$ be an n-dimensional vector space on $\mathbb{R}$ so that $\Lambda^2\mathcal{V}$ is made of all formal summations $\sum_i \alpha_i(\mathbf{U}_i \wedge \mathbf{V}_j)$. For these summations, the symbol $\mathbf{U} \wedge \mathbf{V}$ is a bilinear and alternative function. This implies that the exterior product can only be defined by the following rules:

$$\begin{cases} (\alpha_1 \mathbf{U}_1 + \alpha_2 \mathbf{U}_2) \wedge \mathbf{V} - \alpha_1(\mathbf{U}_1 \wedge \mathbf{V}) - \alpha_2(\mathbf{U}_2 \wedge \mathbf{V}) = \mathbf{0}, \\ \mathbf{U} \wedge (\beta_1 \mathbf{V}_1 + \beta_2 \mathbf{V}_2) - \beta_1(\mathbf{U} \wedge \mathbf{V}_1) - \beta_2(\mathbf{U} \wedge \mathbf{V}_2) = \mathbf{0}, \\ \mathbf{U} \wedge \mathbf{U} = \mathbf{0} \text{ and: } \mathbf{U} \wedge \mathbf{V} + \mathbf{V} \wedge \mathbf{U} = \mathbf{0}. \end{cases} \qquad [7.8]$$

It has been proven that the space $\Lambda^2 V$, defined below, is a vector space on $\mathbb{R}$, with basis:

$$\{\mathbf{U}_i \wedge \mathbf{U}_j\} \quad 1 \leq i < j \leq n, \qquad [7.9]$$

where $\{\mathbf{U}_1, \mathbf{U}_2, \ldots, \mathbf{U}_n\}$ is a basis $\mathcal{V}$. Then, it is deduced that $\dim \Lambda^2\mathcal{V} = \frac{n(n-1)}{2} = \binom{n}{2}$. More generally, $\Lambda^p\mathcal{V}$ $(2 \leq p \leq n)$ can be created on the same principle. This space is made up out of all formal calculations (*p-vectors*):

$$\sum \alpha(\mathbf{U}_1 \wedge \mathbf{U}_2 \cdots \wedge \mathbf{U}_p), \qquad [7.10]$$

with the rules below:

$$\begin{cases} (\alpha_1 \mathbf{U}_1^1 + \alpha_2 \mathbf{U}_1^2) \wedge \mathbf{U}_2 \cdots \wedge \mathbf{U}_p = \alpha_1(\mathbf{U}_1^1 \wedge \cdots \wedge \mathbf{U}_p) + \alpha_2(\mathbf{U}_1^2 \wedge \cdots \wedge \mathbf{U}_p), \\ \mathbf{U}_1 \wedge \mathbf{U}_2 \wedge \mathbf{U}_p = \mathbf{0} \text{ if: } \exists (i,j) \text{ such that: } i \neq j \text{ and: } \mathbf{U}_i = \mathbf{U}_j, \\ \mathbf{U}_1 \wedge \cdots \wedge \mathbf{U}_p \text{ changes its sign when two } \mathbf{U}_i \text{ are interchanged.} \end{cases}$$

1. This result is important for this problem. An overview of this result is given in the Appendix (section 7.6).

In a similar way, it is easy to prove that if $\{\mathbf{U}_1, \mathbf{U}_2, \ldots, \mathbf{U}_n\}$ is a basis of $\mathcal{V}$, then $\{\mathbf{U}_{i_1} \wedge \mathbf{U}_{i_2} \wedge \cdots \wedge \mathbf{U}_{i_p}\}$ $(1 \leq i_1 < i_2 \cdots < i_p \leq n)$ is a basis for the vector space $\Lambda^p\mathcal{V}$ of dimension $\binom{n}{p}$. Note that $\dim(\Lambda^n\mathcal{V}) = 1$. The determinants are linked to the exterior product in the following way. If A is a linear application of $\mathcal{V}$, the associated mapping g_A from $\mathcal{V}^n$ to $\Lambda^n\mathcal{V}$ is defined as follows:

$$g_A(\mathbf{U}_1, \ldots, \mathbf{U}_n) \triangleq A(\mathbf{U}_1) \wedge \cdots \wedge A(\mathbf{U}_n). \qquad [7.11]$$

Furthermore, since g_A is defined by equation [7.11] and is multi-linear and alternative, there is a linear function f_A $(\Lambda^n\mathcal{V} \to \Lambda^n\mathcal{V})$ that meets the following rule:

$$g_A(\mathbf{U}_1, \ldots, \mathbf{U}_n) = f_a(\mathbf{U}_1 \wedge \cdots \mathbf{U}_n).$$

However, $\dim(\Lambda^n\mathcal{V}) = 1$, so that this linear application is just a scalar multiplication by $\det(A)$. As a consequence,

$$A(\mathbf{U}_1) \wedge \cdots A(\mathbf{U}_n) = \det(A)(\mathbf{U}_1 \wedge \cdots \mathbf{U}_n). \qquad [7.12]$$

It is therefore easy to prove that there is a natural definition of the exterior multiplication (also referred to as *wedge product*) such that $\Lambda^p\mathcal{V} \wedge \Lambda^q\mathcal{V} = \Lambda^{p+q}\mathcal{V}$. This exterior multiplication is simply defined by:

$$(\mathbf{U}_1 \wedge \cdots \wedge \mathbf{U}_p) \wedge (\mathbf{V}_1 \wedge \cdots \wedge \mathbf{V}_p) = \mathbf{U}_1 \wedge \cdots \wedge \mathbf{U}_p \wedge \mathbf{V}_1 \wedge \cdots \wedge \mathbf{V}_p. \quad [7.13]$$

The basic properties of the exterior product are as follows:

$$\begin{aligned} &(1)\ \mathbf{u} \wedge \mathbf{v} \text{ is distributive } \big(\mathbf{u} \in \Lambda^p(\mathcal{V}),\ \mathbf{v} \in \Lambda^q(\mathcal{V})\big), \\ &(2)\ \mathbf{u} \wedge (\mathbf{v} \wedge \mathbf{w}) = (\mathbf{u} \wedge \mathbf{v}) \wedge \mathbf{w} \text{ associative}, \\ &(3)\ \mathbf{v} \wedge \mathbf{u} = (-1)^{pq}\mathbf{u} \wedge \mathbf{v} \text{ alternated.} \end{aligned} \qquad [7.14]$$

All these results are very basic and are widely used in this chapter. The formalism has the advantage of making the calculations simpler. The Binet-Cauchy formula is constantly used in order to obtain explicit formulae. More precisely, let $\{\mathbf{U}_1, \ldots, \mathbf{U}_N\}$ be a set of N vectors of $\mathcal{V}$ $(N \geq n)$. Then

$$\begin{aligned} &\det\Big[(\mathbf{U}_1, \ldots, \mathbf{U}_N) \otimes (\mathbf{U}_1, \ldots, \mathbf{U}_N)^T\Big] \\ &\quad = \sum_{1 \leq i_1 < i_2 < \cdots < i_n \leq T} \big[\det(\mathbf{U}_{i_1}, \mathbf{U}_{i_2}, \ldots, \mathbf{U}_{i_n})\big]^2. \end{aligned} \qquad [7.15]$$

Consequently, it is sufficient to use an explicit description for the determinants $\det(\mathbf{U}_{i_1}, \mathbf{U}_{i_2}, \ldots, \mathbf{U}_{i_n})$ in order to obtain an explicit expression for the determinant of $\det[(\mathbf{U}_1, \ldots, \mathbf{U}_N) \otimes (\mathbf{U}_1, \ldots, \mathbf{U}_N)^T]$.

7.3. Application to the optimization of emissions (deterministic case)

Now the formalism, which has previously been established, will be used to analyze the effect of distributing active measurements in the case of the deterministic target. First of all, this is done on the basis of a uniform and rectilinear movement. The deterministic case has the advantage of being simple enough to obtain explicit and significant results. If the observer does not carry out any maneuvers, the passive problem of trajectories cannot be observed. If, however, multiple modalities are observed (e.g. active and passive measurements) it is also possible to observe problems [TRE 96]. In the case of multimodal measurements (active and passive) the scalar observation $z(t)$ is replaced by a vector-based form of observation $\mathbf{z}(t) = (z_1(t), z_2(t))^T$. The statistical structure of the problem, however, remains unchanged. Passive measurements are available at all times (at zero cost) while active measurements are relatively rare and of a high cost. Given the constraints of a budget, optimizing the distribution of active measurements makes sense. Note that the matrices $\mathcal{M}$ and $\mathcal{N}$ consist of gradient vectors associated with passive or active measurements.

$$
\begin{aligned}
\det(\text{FIM}) &= c(k) \det\left(\mathcal{M}\mathcal{M}^T + \mathcal{N}\mathcal{N}^T\right), \\
&= c(k) \det\left[(\mathcal{M}, \mathcal{N}) \begin{pmatrix} \mathcal{M}^T \\ \mathcal{N}^T \end{pmatrix}\right].
\end{aligned} \qquad [7.16]
$$

Matrices $\mathcal{M}$ and $\mathcal{N}$ are calculated on the basis of all gradient vectors (column vectors) (see equation [7.6]), i.e.

$$
\begin{cases}
\mathcal{M} = \left(\mathbf{M}_0, \ldots, \mathbf{M}_k, \ldots, \mathbf{M}_{T-1}\right), \; k = 1 \longrightarrow T-1, \\
\mathcal{N} = \left(\mathbf{N}_0, \mathbf{N}_{j_1}, \ldots, \mathbf{N}_{j_{\#a-1}}\right),
\end{cases} \qquad [7.17]
$$

where $\mathbf{M}$ and $\mathbf{N}$ are the gradients associated with passive or active measurements. This modelization corresponds to active emissions at the following points in time $\{0, j_1, \ldots, j_{\#a-1}\} \in \{0, \ldots, T-1\}$. In this context, the matrices $\mathcal{M}$ and $\mathcal{N}$ are $4 \times T$ and $4 \times (\#a)$ respectively. Despite each elementary FIM being defined in equation [7.16] being 4×4, a direct calculation of its determinant cannot be established if a direct approach is used. To avoid this problem the Binet-Cauchy formula can be used. Note that $\{i_1, \ldots, i_k\}$ is a (temporary)[2] multi-index. According to the Binet-Cauchy formula (see equation [7.15]) the following applies:

2. i_k (resp. j_l) are the indexes of measures associated with passive or active measures. It is assumed the order $k < l \Rightarrow i_k < i_l$.

$$\det(\text{FIM}) = \sum_{i_1,i_2;j_1,j_2} \left[(\mathbf{M}_{i_1} \wedge \mathbf{M}_{i_2}) \wedge (\mathbf{N}_{j_1} \wedge \mathbf{N}_{j_2})\right]^2 + \sum_{i_1;j_1,j_2,j_3} \left[(\mathbf{M}_{i_1}) \wedge (\mathbf{N}_{j_1} \wedge \mathbf{N}_{j_2} \wedge \mathbf{N}_{j_3})\right]^2, + \sum_{i_1,i_2,i_3;j_1} \left[(\mathbf{M}_{i_1} \wedge \mathbf{M}_{i_2} \wedge \mathbf{M}_{i_3}) \wedge (\mathbf{N}_{j_1})\right]^2 + \text{other terms.} \quad [7.18]$$

The "other terms" correspond to the terms: $\sum_{i_1,i_2,i_3,i_4}[\mathbf{M}_{i_1} \wedge \mathbf{M}_{i_2} \wedge \mathbf{M}_{i_3} \wedge \mathbf{M}_{i_4}]^2$ and $\sum_{i_1,i_2,i_3,i_4}[\mathbf{N}_{i_1} \wedge \mathbf{N}_{i_2} \wedge \mathbf{N}_{i_3} \wedge \mathbf{N}_{i_4}]^2$.

These two terms correspond to the zero determinants (bearings only or distances only) and associated problems cannot be observed. Calculating terms of the type $(\mathbf{M}_{i_1} \wedge \mathbf{M}_{i_2}) \wedge (\mathbf{N}_{j_1} \wedge \mathbf{N}_{j_2})$[3] can easily be carried out with the properties of exterior algebra. This is why exterior algebra is dealt with by $\Lambda^2\mathbb{R}^4$. The space of dimensions is $\binom{4}{2}$. If $\mathbf{e}_1, \ldots, \mathbf{e}_4$ is the canonical basis of $\mathbb{R}^4$ then the basis of $\Lambda^2\mathbb{R}^4$ is

$$\left\{\mathbf{e}_1 \wedge \mathbf{e}_2,\ \mathbf{e}_1 \wedge \mathbf{e}_3,\ \mathbf{e}_1 \wedge \mathbf{e}_4,\ \mathbf{e}_2 \wedge \mathbf{e}_4,\ \mathbf{e}_2 \wedge \mathbf{e}_3,\ \mathbf{e}_3 \wedge \mathbf{e}_4\right\},$$

denote by $\{\alpha_1, \ldots, \alpha_6\}$ and $\{\gamma_1, \ldots, \gamma_6\}$ the components of $\mathbf{N}_{j_1} \wedge \mathbf{N}_{j_2}$ and $\mathbf{M}_{i_1} \wedge \mathbf{M}_{i_2}$ in the canonical basis $\Lambda^2\mathbb{R}^4$. The elementary properties of multi-linearity therefore imply:

$$\begin{aligned}&\left(r\sigma_r\sigma_\beta\right)^2\left(\mathbf{N}_{j_1} \wedge \mathbf{N}_{j_2} \wedge \mathbf{M}_{i_1} \wedge \mathbf{M}_{i_2}\right)\\ &\quad= \alpha_1\gamma_6 - \alpha_2\gamma_4 + \alpha_3\gamma_5 - \alpha_4\gamma_2 + \alpha_5\gamma_3.\end{aligned} \quad [7.19]$$

Without lost of generality, $\mathbf{N}_{j_1}$ corresponds to active measurements carried out at time 0 (associated bearing: β_0) while $\mathbf{N}_{j_2}$ is associated with active measurements carried out at τ $(0 < \tau \leq (T-1))$ with the associated bearing of $\beta_0 + \delta$ $(\delta = \tau\dot{\beta})$. Furthermore, the gradient vectors $\mathbf{M}_1$ and $\mathbf{M}_2$ correspond to purely passive measurements associated in the times t and t' (associated bearing β_t and $\beta_{t'}$). This leads to:

$$\begin{cases}\sigma_r\mathbf{N}_{j_1} = \left(\sin\beta_0, \cos\beta_0, 0, 0\right)^T,\\ \sigma_r\mathbf{N}_{j_2} = \left(\sin\left(\beta_0+\delta\right), \cos\left(\beta_0+\delta\right), \tau\sin\left(\beta_0+\delta\right), \tau\cos\left(\beta_0+\delta\right)\right)^T,\\ r\sigma_\beta\mathbf{M}_{i_1} = \left(\cos\beta_t, -\sin\beta_t, t\cos\beta_t, -t\sin\beta_t\right)^T,\\ r\sigma_\beta\mathbf{M}_{i_2} = \left(\cos\beta_{t'}, -\sin\beta_{t'}, t'\cos\beta_{t'}, -t'\sin\beta_{t'}\right)^T.\end{cases} \quad [7.20]$$

3. Also equal to $\det(\mathbf{M}_{i_1}, \mathbf{M}_{i_2}, \mathbf{N}_{j_1}, \mathbf{N}_{j_2})$.

Now the components of $\mathbf{N}_{j_1} \wedge \mathbf{N}_{j_2}$ (written as: $\alpha_1, \ldots, \alpha_6$) and $\mathbf{M}_{i_1} \wedge \mathbf{M}_{i_2}$ (written as: $\gamma_1, \ldots, \gamma_6$) are computed in the canonical basis of $\Lambda^2\mathbb{R}^4$. Basic derivations lead to the following inequalities:

$$\begin{cases} \alpha_1 = -\sin\delta & \gamma_1 = -\cos\beta_t \sin\beta_{t'} + \sin\beta_t \cos\beta_{t'}, \\ \alpha_2 = \tau \sin\beta_0 \sin\left(\beta_0 + \delta\right) & \gamma_2 = (t'-t)\cos\beta_t \cos\beta_{t'}, \\ \alpha_3 = \tau \sin\beta_0 \cos\left(\beta_0 + \delta\right) & \gamma_3 = t\cos\beta_{t'} \sin\beta_t - t'\sin\beta_{t'} \cos\beta_t, \\ \alpha_4 = \tau \cos\beta_0 \cos\left(\beta_0 + \delta\right) & \gamma_4 = (t'-t)\sin\beta_t \sin\beta_{t'}, \\ \alpha_5 = \tau \cos\beta_0 \sin\left(\beta_0 + \delta\right) & \gamma_5 = t\cos\beta_t \sin\beta_{t'} - t'\sin\beta_t \cos\beta_{t'}, \\ \alpha_6 = 0 & \gamma_6 = tt'\left(\cos\beta_{t'} \sin\beta_t - \sin\beta_{t'} \cos\beta_t\right). \end{cases} \quad [7.21]$$

so that:

$$\det(\text{FIM}) = \frac{1}{r^4\sigma_r^4\sigma_\beta^4}\left[\alpha_1\gamma_6 - \alpha_2\gamma_4 + \alpha_3\gamma_5 - \alpha_4\gamma_2 + \alpha_5\gamma_3\right].$$

Until now the calculations have been exact. Adequate estimations will now be considered. Through a regressive linear approximation of the bearing (i.e. $\delta = \tau\dot{\beta}$, $\beta_t = \beta_0 + t\dot{\beta}$ and $\beta_{t'} = \beta_0 + t'\dot{\beta}$) the following estimation is deducted from $\det(\text{FIM}_{\tau,2\beta,2r})$:

PROPOSITION 7.1. *Let* $\det(\text{FIM}_{\tau,T,\beta,r})$ *be the determinant of the* FIM, *which is associated with two active measurements (separated from* τ*) and* T *passive measurements (bearing). Then:*

$$\det\left(\text{FIM}_{\tau,T}\right) \simeq \frac{\tau^2}{\left(r\sigma_r\sigma_\beta\right)^4} \sum_{0\leq t<t'\leq T} (t-t')^2\left[2 - \dot{\beta}^2(t'+t-\tau)^2\right]^2. \quad [7.22]$$

For a general case the Binet-Cauchy theorem ensures that it is sufficient to consider the following cases:

- 1 active measurement, T passive measurements;
- 2 active measurement, T passive measurements;
- 3 active measurement, T passive measurements.

For three active measurements the following result is obtained:

PROPOSITION 7.2. *Let three active measurements (at* 0, τ_1 *and* τ_2*) be given, as well as* T *passive measurements. Then:*

$$\det\left(\text{FIM}_{\tau_1,\tau_2,T}\right) \simeq \frac{T}{r^2\sigma_\beta^2\sigma_r^6} \sum_{0<\tau_1<\tau_2\leq T} \left[\tau_1\tau_2\left(\tau_1-\tau_2\right)\dot{\beta}\right]^2. \quad [7.23]$$

Proof. The active measurements are considered in $\Lambda^3\mathbb{R}^4$ (canonical basis: $\{\mathbf{e}_1 \wedge \mathbf{e}_2 \wedge \mathbf{e}_3,\ \mathbf{e}_1 \wedge \mathbf{e}_2 \wedge \mathbf{e}_4,\ \mathbf{e}_1 \wedge \mathbf{e}_3 \wedge \mathbf{e}_4,\ \mathbf{e}_2 \wedge \mathbf{e}_3 \wedge \mathbf{e}_4\}$). The gradient vectors $\mathbf{N}_0$, $\mathbf{N}_{j_1}$ and $\mathbf{N}_{j_2}$, are associated with the emissions. Denote $\{a_i\}_{i=1}^4$, $\{b_i\}_{i=1}^4$ and $\{c_i\}_{i=1}^4$ their components in this (canonical) basis. Then:

$$\begin{aligned}\sigma_r^3 \mathbf{N}_0 \wedge \mathbf{N}_{j_1} \wedge \mathbf{N}_{j_2} = {}& \alpha_1\big(\mathbf{e}_1 \wedge \mathbf{e}_2 \wedge \mathbf{e}_3\big) + \alpha_2\big(\mathbf{e}_1 \wedge \mathbf{e}_2 \wedge \mathbf{e}_4\big) \\ & + \alpha_3\big(\mathbf{e}_1 \wedge \mathbf{e}_3 \wedge \mathbf{e}_4\big) + \alpha_4\big(\mathbf{e}_2 \wedge \mathbf{e}_3 \wedge \mathbf{e}_4\big),\end{aligned} \qquad [7.24]$$

where

$$\begin{cases} \alpha_1 = a_1 b_2 c_3 - a_1 b_3 c_2 - a_2 b_1 c_3 + a_2 b_3 c_1, \\ \alpha_2 = a_1 b_2 c_4 - a_2 b_1 c_4 - a_1 b_4 c_2 + a_2 b_4 c_1, \\ \alpha_3 = a_1 b_3 c_4 - a_1 b_4 c_3, \\ \alpha_4 = a_2 b_3 c_4 - a_2 b_4 c_3. \end{cases}$$

The passive measurements are represented by vector $\mathbf{M}_t = \frac{1}{r\sigma_\beta}(\cos\beta_t \mathbf{e}_1 - \sin\beta_t \mathbf{e}_2 + t\cos\beta_t \mathbf{e}_3 - t\sin\beta_t \mathbf{e}_4)$ where

$$\begin{aligned}&\big(r\sigma_\beta\sigma_r^3\big)\big(\mathbf{N}_0 \wedge \mathbf{N}_{j_1} \wedge \mathbf{N}_{j_2} \wedge \mathbf{M}_4\big) \\ &\quad = -\alpha_1 t \sin\beta_t - \alpha_2 t \cos\beta_t - \alpha_3 \sin\beta_t - \alpha_4 \cos\beta_t.\end{aligned} \qquad [7.25]$$

The proof of equation [7.23] is achieved by using the linear approximation $\beta_t \triangleq \beta_0 + t\dot{\beta}$. □

Now, let us consider the case of a unique active measurement. The following result is obtained.

PROPOSITION 7.3. *It is assumed that there is only one active measurement and T passive measurements. Then:*

$$\det(\text{FIM})_T \simeq \frac{1}{r^6\sigma_r^2\sigma_\beta^6} \sum_{0 \le t_1 < t_2 < t_3 \le T} \big[(t_3 - t_1)(t_2 - t_1)(t_3 - t_2)\dot{\beta}\big]^2. \qquad [7.26]$$

Proof. The gradient vectors for three passive measurements are written $\mathbf{M}_{t_1}$, $\mathbf{M}_{t_2}$ and $\mathbf{M}_{t_3}$ while the only emission takes place at 0. Then $\mathbf{N}_0 = \sin(\beta_0)\mathbf{e}_1 + \cos(\beta_0)\mathbf{e}_2$, which implies that it is sufficient to compute the components (denoted α_2 and α_1) of the vector $\mathbf{M}_{t_1} \wedge \mathbf{M}_{t_2} \wedge \mathbf{M}_{t_3}$ in the basis $\mathbf{e}_1 \wedge \mathbf{e}_3 \wedge \mathbf{e}_4$ and $\mathbf{e}_2 \wedge \mathbf{e}_3 \wedge \mathbf{e}_4$. The computation leads to:

$$\begin{cases} \alpha_1 = -t_1 t_2 \sin\beta_{t_2} \cos\beta_{t_1} \cos\beta_{t_3} + t_3 t_1 \sin\beta_{t_3} \cos\beta_{t_1} \cos\beta_{t_2}, \\ \alpha_2 = t_1 t_2 \sin\beta_{t_2} \cos\beta_{t_1} \sin\beta_{t_3} - t_3 t_1 \sin\beta_{t_3} \cos\beta_{t_1} \sin\beta_{t_2}, \end{cases} \qquad [7.27]$$

so that $\det(\mathbf{N}_0, \mathbf{M}_{t_1}, \mathbf{M}_{t_2}, \mathbf{M}_{t_3}) = -\alpha_1 \cos\beta_0 + \alpha_2 \sin\beta_0$. □

Here it makes sense to recall the definition of $\dot{\beta}$, i.e. $\dot{\beta} = \left(\frac{v}{r}\right)\sin\theta$, where θ is the angle between $\mathbf{r}$ and $\mathbf{v}$, which represent the position and the speed ($\|\mathbf{v}\| = v$, $\|\mathbf{r}\| = r$). Now a summary of all results mentioned above is given.

active measures	det(FIM)	simplified form
1 mes. (à 0)	$\frac{1}{r^6\sigma_r^2\sigma_\beta^6}\sum_{0\leq t_1<t_2<t_3\leq T}\left[(t_3-t_1)(t_2-t_1)(t_3-t_2)\dot{\beta}\right]^2$	$\frac{1}{r^6\sigma_r^2\sigma_\beta^6}\dot{\beta}^2 P_1(T)$
2 mes. (à 0 and τ)	$\frac{\tau^2}{\left(r\sigma_r\sigma_\beta\right)^4}\sum_{0\leq t<t'\leq T}(t-t')^2\left[2-\dot{\beta}^2(t'+t-\tau)^2\right]^2$	
3 mes. $(0, \tau_1, \tau_2)$	$\frac{T}{r^2\sigma_\beta^2\sigma_r^6}\sum_{0<\tau_1<\tau_2\leq T}\left[\tau_1\tau_2(\tau_1-\tau_2)\dot{\beta}\right]^2$	

Any number of active measurements is available now. These measurements belong to the discrete group E, with a fixed cardinal number ($\mathrm{Card}(E) = N$) and their index is the time. Due to the Binet-Cauchy formula and previous results, it is now possible to provide a general expression of the determinants for the FIM.

PROPOSITION 7.4. *Assume active measurements belonging to the set E ($\mathrm{Card}(E) = N$) and T passive measurements. Denote FIM_E the related FIM. Then:*

$$\det(\mathrm{FIM})_E = N\cdot\det(\mathrm{FIM})_T + \sum_{\tau\in E}\det\left(\mathrm{FIM}_{\tau,T}\right) + \sum_{\tau_1,\tau_2\in E}\det\left(\mathrm{FIM}_{\tau_1,\tau_2,T}\right), \quad [7.28]$$

where $\det(\mathrm{FIM})_T$, $\det(\mathrm{FIM}_{\tau,T})$ and $\det(\mathrm{FIM}_{\tau_1,\tau_2,T})$ are determined by equations [7.26], [7.22] and [7.23].

Now, let us consider the same number of active and passive measurements. A direct calculation of $\det(\mathrm{FIM})$ is impossible due to the high number of sub-cases. However, Proposition 7.4 leads to the following result.

PROPOSITION 7.5. *The same number of active and passive measurements is available (that is T). Then:*

$$\det\left(\mathrm{FIM}_{T,T}\right) \propto T^{16}\left(\frac{1}{r^4\sigma_\beta^4\sigma_r^4} + \frac{4\dot{\beta}^2}{r^2\sigma_\beta^2\sigma_r^6} + \frac{4\dot{\beta}^2}{r^6\sigma_\beta^6\sigma_r^2}\right). \quad [7.29]$$

Previous results have been used in the analysis of information gain induced by active measurements. This gain of information is quite high. In order to verify this statement, it is sufficient to take a look at the respective vectors $\mathbf{M}_i$ (passive) and $\mathbf{N}_i$ (active). These vectors are orthogonal and $\det(\mathrm{FIM})$ is the parallelotope volume generated by these vectors. This volume is considerably increased by the inclusion of orthogonal vectors.

Optimizing the distribution of measurements requires additional considerations. The results that have been obtained above provide explicit expressions which are functions of essential parameters. Then, it is easy to prove that the term $\det(\text{FIM}_{\tau,T})$ is generally predominant in the global expression (see Binet-Cauchy) of $\det(\text{FIM})$. This is due to the following fact: the spaces generated by the gradient vectors $(\mathbf{M}_1, \mathbf{M}_2)$ and $(\mathbf{N}_1, \mathbf{N}_2)$ are (approximately) orthogonal sub-spaces of same dimension.

Distance r is assumed to be constant, as well as σ_β and σ_r. In this case, finding the best possible distribution is equivalent to the problem:

PROBLEM 7.1. *Determine τ by maximizing:*

$$\tau^2 \sum_{0 \leq t < t' \leq T} (t - t')^2 \big[1 + \dot{\beta}^2\big(- tt' + \tau(t + t')\big)\big]^2.$$

In general, $\dot{\beta}$ is an available parameter (on the basis of passive measurements). It is therefore possible to replace $\dot{\beta}$ with an estimate. Now, apart from very long scenarios, the term $\dot{\beta}^2(-t\,t' + \tau(t + t'))$ is found to be less than 1. The optimization problem is therefore simplified and can be reduced to $\tau^2 \sum_{0 \leq t < t' \leq T} (t - t')^2$. The general conclusion is that $\det(\text{FIM}_{\tau,T})$ (as well as $\det(\text{FIM})$) reaches its maximum when τ reaches its maximum. The best possible distribution of measurements comes down to concentrating the active measurements around two suprema (beginning and end). This conclusion is, of course, valid for a deterministic mobile target.

Also note that the hypotheses have been simplified (σ_β, σ_r, r are constant). However, the previous results are easily extended when the parameters (i.e. σ_β, σ_r, r) vary over the course of time.

7.3.1. *The case of a maneuvering target*

A mobile target is now considered, whose trajectory contains two legs. Then the state vector of the target, $\mathbf{X}$, is of dimension 6 ($\mathbf{X} = (r_{x,0}, r_{y,0}, v_{x,1}, v_{y,1}, v_{x,2}, v_{y,2})^T$). If the time of maneuver is known (t_m), the gradient vector takes the following form:

$$\begin{cases} r_t \sigma_b \nabla_{\mathbf{X}} \beta_t = \big(\cos\beta_t, -\sin\beta_t, t\cos\beta_t, -t\sin\beta_t, 0, 0\big)^T, & \text{for } t \leq t_m, \\ r_t \sigma_b \nabla_{\mathbf{X}} \beta_t = \big(\cos\beta_t, -\sin\beta_t, t_m\cos\beta_t, -t_m\sin\beta_t, & \\ \qquad (t - t_m)\cos\beta_t, -(t - t_m)\sin\beta_t\big)^T, & \text{for } t > t_m. \end{cases} \quad [7.30]$$

There is a similar formula for active measurements. Now, let us consider the maximization of the parallelotope hull of the ellipsoid of uncertainty. It is possible to prove that the predominant term in $\det(\text{FIM})$ is associated with exterior products $(\mathbf{M}_1 \wedge \mathbf{M}_2 \wedge \mathbf{M}_3) \wedge (\mathbf{N}_1 \wedge \mathbf{N}_2 \wedge \mathbf{N}_3)$ ($\mathbf{M}_i \rightarrow$ passive measurements $\mathbf{N}_i \rightarrow$ active measurements). Let (t_1, t_2), (t_3, t_4) and (t_5, t_6) be respectively the time associated with $(\mathbf{M}_1, \mathbf{N}_1)$, $(\mathbf{M}_2, \mathbf{N}_2)$ and $(\mathbf{M}_3, \mathbf{N}_3)$. Then:

$$\begin{aligned}\det(\text{FIM}) \approx c \sum_{t_1,\ldots,t_6} & \big[(t_1 - t_2)(t_3 - t_4)(t_5 - t_6)\big]^2 \Big\{ 2\,t_2 \cos\left(2\beta_0\right) \\ & -2\big(t_2 - 2t_m + t_2 \sin 2\beta_0\big), -2\,t_2 \dot{\beta}\big(t_2 - t_1 + (t_1 + t_2)\left(\cos 2\beta_0 + \sin 2\beta_0\right)\big), \\ & + \dot{\beta}^2\Big(\big(t_1 - t_2\big)^2 \big(t_2 - 2\,t_m\big) - t_2\big(t_1 + t_2\big)^2 \big(\cos 2\beta_0 - \sin 2\beta_0\big)\Big), \\ & - \big(t_3 + t_4 - t_5 - t_6\big)^2 \big(- t_2 + 2\,t_m + t_2\big(\cos 2\beta_0 - \sin 2\beta_0\big)\big)\Big\}^2.\end{aligned}$$

At first glance this formula seems to be complex and inapplicable. However, it is easy to prove that the term which is predominant inside the $\{\ \ \}$ in the previous equation is simply $2\,t_2 \cos(2\beta_0 - 2(t_2 - 2t_m + t_2 \sin 2\beta_0))$. Again, optimal distribution of measurements is focused on active measurements around the suprema of the legs. Let us now take a look at the general case of a trajectory which disposes of many legs (legs of identical length, expressed as j). The FIM takes the following form:

$$\begin{aligned}\text{FIM}_{1,nj} &= \text{FIM}_{1,j} + \cdots + \text{FIM}_{(n-1)j,nj}, \\ &= \sum_{m=1}^{n} \sum_{(m-1)j+1}^{mj} \big[\mathbf{d}_{m-1,n+1}(k)\, \mathbf{d}^T_{m-1,n+1}(k)\big] \otimes \mathbf{\Omega}_k,\end{aligned}$$

where

$$\mathbf{d}^T_{p,q}(k) = \big(1, j, \ldots, j, (k - pj), 0, \ldots, 0\big),$$

$$\mathbf{\Omega}_k = \begin{pmatrix} \cos^2 \beta_k & \cos \beta_k \sin \beta_k \\ -\cos \beta_k \sin \beta_k & \cos^2 \beta_k \end{pmatrix}.$$

The previous analysis thus applies to the general case.

7.4. The case of a target with a Markov trajectory

When it comes to problems in the field of tracking, a diffusion model is preferred, which extends the deterministic model. This model is defined as shown below (Cartesian coordinates):

$$
\left|
\begin{array}{l}
X_{t+1} = AX_t + HU_t + \sigma V_t, \quad \text{for which: } V_t \sim \mathcal{N}(0,Q), \\[2ex]
A = \mathrm{Id}_4 + \delta_t B \quad \text{with: } B = \begin{bmatrix} 0 & 1 \\ 0 & 0 \end{bmatrix} \otimes \mathrm{Id}_2, \\[2ex]
HU_t\text{: commands}, \\[2ex]
Q = \Sigma \otimes \mathrm{Id}_2 \quad \text{with } \Sigma = \begin{bmatrix} \alpha_3 & \alpha_2 \\ \alpha_2 & \alpha_1 \end{bmatrix}, \\[2ex]
\text{observations: } Z_t = \arctan\left(\dfrac{r_x(t)}{r_y(t)}\right) + W_t.
\end{array}
\right.
\qquad [7.31]
$$

The term of diffusion is represented by a noise (here σV_t) which expresses the uncertainty of future evolutions of the target. Even if Cartesian models plays an important role, another type of parameter is useful, i.e. the one introduced by Hammel and Aidala [HAM 85], the MPC (modified polar coordinates).

$$
\begin{bmatrix} \beta_t & \dot{\beta}_t & \dfrac{\dot{r}_t}{r_t} & \dfrac{1}{r_t} \end{bmatrix}^T. \qquad [7.32]
$$

This coordinate system is significant since it separates the observable information $(\begin{bmatrix} \beta_t & \dot{\beta}_t & \frac{\dot{r}_t}{r_t} \end{bmatrix}^T)$ from the unobservable information $(\frac{1}{r_t})$. A system with slightly modified coordinates (called LPC[4]) was introduced by [BRE 06], and is defined as follows:

$$
Y_t = \begin{bmatrix} \beta_t & \ln r_t & \dot{\beta}_t & \dfrac{\dot{r}_t}{r_t} \end{bmatrix}^T. \qquad [7.33]
$$

This system has the significant advantage that it calculates the FIM exactly. There are bijective transformations that move Cartesian coordinates to LPC and *vice versa* [BRE 06]. In the system of LPC coordinates the state equation is no longer linear and has to be deducted from equation [7.31], i.e.:

$$
Y_{t+1} = \begin{cases} f_c^{lp}\big(Af_{lp}^c(Y_t) + HU_t + \sigma W_t\big) & \text{if } r_y(t) > 0, \\ f_c^{lp}\big(-Af_{lp}^c(Y_t) + HU_t + \sigma W_t\big) & \text{if } r_y(t) < 0. \end{cases} \qquad [7.34]
$$

$$
Z_t = \beta_t + V_t.
$$

Now the notions of estimate lower bound (and FIM) are defined for a Bayesian problem. This means that there is a prior uncertainty for the state to be estimated.

---

4. Logarithmic polar coordinates.

DEFINITION 7.1 (FIM). *The problem of filtering defined previously, based on FIM (FIM), is denoted J_t and expressed by:*

$$J_t = \mathbb{E}\left\{\nabla_{Y_t} \ln p\big(Z_{1:t}, Y_{0:t}\big) \nabla_{Y_t}^T \ln p\big(Z_{1:t}, Y_{0:t}\big)\right\}, \quad [7.35]$$

where $p(Z_{1:t}, Y_{0:t})$ is the density related to the sequence of observations $Z_{1:t}$ as well as the sequence of state $Y_{0:t}$. This leads to $EQM_t \succcurlyeq J_t^{-1}$.

$$EQM_t \triangleq \mathbb{E}\big[\hat{Y}_{0:t}\big(Z_{1,t}\big) - Y_{0:t}\big]^2$$

Actually, the extension of FIM to the Bayesian case (random state) is obtained by means of an elementary lemma of matrix calculus summarized below.

LEMMA 7.1 (Cauchy-Schwarz matrix). *Let S be a block matrix that has been partitioned as follows:*

$$S = \begin{bmatrix} A & C \\ C^T & B \end{bmatrix}, \quad (A, B \succeq 0)$$

Then: $S \succeq 0$ implies $(A - CB^{-1}C^T) \succeq 0$.

Furthermore, for a vector of parameter Φ, it is obtained (integration by parts):

$$\mathbb{E}\left\{\left(\frac{\partial^2\left[\ln\left(p(Z,\Phi)\right)\right]}{\partial\theta_i \partial\theta_j}\right)_{i,j}\right\} = -\mathbb{E}\left\{\nabla_\Phi \ln p(Z,\Phi) \nabla_\Phi^T \ln p(Z,\Phi)\right\}, \quad [7.36]$$

and:

LEMMA 7.2 ($C = Id$).

FIM =

$$\begin{bmatrix} \underbrace{\mathbb{E}\left\{\left[\widehat{\Phi}(Z) - \Phi\right]\left[\widehat{\Phi}(Z) - \Phi\right]^T\right\}}_{A} & \underbrace{\mathbb{E}\left\{\left[\widehat{\Phi}(Z) - \Phi\right]\nabla_\Phi^T\left[\ln p(Z,\Phi)\right]\right\}}_{C} \\ \underbrace{\mathbb{E}\left\{\nabla_\Phi\left[\ln p(Z,\Phi)\right]\left[\widehat{\Phi}(Z) - \Phi\right]^T\right\}}_{C^T} & \underbrace{\mathbb{E}\left\{\nabla_\Phi\left[\ln p(Y,\Phi)\right]\nabla_\Phi^T\left[\ln p(Z,\Phi)\right]\right\}}_{B=\mathbf{J}(Z,\Phi)} \end{bmatrix} \quad [7.37]$$

Then, under reasonable hypothesis, $C = Id$.

Non-linear generic systems, e.g.

$$\begin{cases} X_t = F_t\big(X_{t-1}, V_t\big), \\ Z_t = H_t\big(X_t, W_t\big), \end{cases}$$

will be analyzed now. It happens that the FIM J_t is of increasing size (i.e. $4 \times t$). A direct approach is impossible as soon as t reaches reasonable values in terms of tracking. The essential idea of Tichavsky *et al.* was the usage of structures separated into blocks of matrices associated with the FIM.

LEMMA 7.3. $p(X_{0:t+1}, Z_{0:t+1}) = p_t\, p(X_{t+1} \mid X_t) p(Z_{t+1} \mid X_{t+1})$, *and therefore:*

$$\mathbf{J}(\Phi_{0:t+1}) = \begin{bmatrix} \mathbf{J}_{\Phi_{0:t-1}}^{\Phi_{0:t-1}}(p_{t+1}) & \mathbf{J}_{\Phi_{0:t-1}}^{\Phi_t}(p_{t+1}) & \mathbf{J}_{\Phi_{0:t-1}}^{\Phi_{t+1}}(p_{t+1}) \\ \mathbf{J}_{\Phi_t}^{\Phi_{0:t-1}}(p_{t+1}) & \mathbf{J}_{\Phi_t}^{\Phi_t}(p_{t+1}) & \mathbf{J}_{\Phi_t}^{\Phi_{t+1}}(p_{t+1}) \\ \mathbf{J}_{\Phi_{t+1}}^{\Phi_{0:t-1}}(p_{t+1}) & \mathbf{J}_{\Phi_{t+1}}^{\Phi_t}(p_{t+1}) & \mathbf{J}_{\Phi_{t+1}}^{\Phi_{t+1}}(p_{t+1}) \end{bmatrix} \quad [7.38]$$

The following relations between the fundamental structures of $\mathbf{J}(\Phi_{0:t+1})$ and $\mathbf{J}(\Phi_{0:t})$ are very important.

$$\mathbf{J}(\Phi_{0:t}) = \begin{bmatrix} A_t & B_t \\ B_t^T & C_t \end{bmatrix} \quad \text{and} \quad \mathbf{J}(\Phi_{0:t+1}) = \begin{bmatrix} A_t & B_t & 0 \\ B_t^T & C_t + D_t^{11} & D_t^{12} \\ 0 & D_t^{21} & D_t^{22} \end{bmatrix} \quad [7.39]$$

Now $\mathbf{J}_{\Phi_{t+1}}^{-1}$ is the lower block of $\mathbf{J}^{-1}(\Phi_{0:t+1})$. Using twice the lemma of matrix inversion on a matrix that has been subdivided into blocks, the following result is obtained:

$$\begin{aligned} \mathbf{J}_{\Phi_{t+1}}^{-1} &= D_t^{22} - \begin{bmatrix} 0 & D_t^{21} \end{bmatrix} \begin{bmatrix} A_t & B_t \\ B_t^T & C_t + D_t^{11} \end{bmatrix}^{-1} \begin{bmatrix} 0 \\ D_t^{12} \end{bmatrix} \\ &= D_t^{22} - D_t^{21} \left[C_t + D_t^{11} - B_t^T A_t^{-1} B_t \right]^{-1} D_t^{12}, \\ &= D_t^{22} - D_t^{21} \left[\mathbf{J}_{\Phi_t}^{-1} + D_t^{11} \right]^{-1} D_t^{12}. \end{aligned} \quad [7.40]$$

The following proposition is fundamental for the FIM.

PROPOSITION 7.6 (formula by Tichavský *et al.*). *For a filtering problem, the right bottom block of the FIM, denoted* J_t^{-1}, *is subject to the following recursive formula:*

$$J_{t+1}^{-1} = D_t^{22} + D_t^{33} - D_t^{21} \left(J_t^{-1} + D_t^{11} \right)^{-1} D_t^{12}, \quad [7.41]$$

where D_t^{11}, D_t^{12}, D_t^{21}, D_t^{22}, D_t^{33} *are matrices defined by:*

$$\begin{cases} D_t^{11} = \mathbb{E}\{\nabla_{Y_t} \ln p(Y_{t+1} \mid Y_t) \nabla_{Y_t}^T \ln p(Y_{t+1} \mid Y_t)\}, \\ D_t^{21} = \mathbb{E}\{\nabla_{Y_{t+1}} \ln p(Y_{t+1} \mid Y_t) \nabla_{Y_t}^T \ln p(Y_{t+1} \mid Y_t)\}, \\ D_t^{12} = \mathbb{E}\{\nabla_{Y_t} \ln p(Y_{t+1} \mid Y_t) \nabla_{Y_{t+1}}^T \ln p(Y_{t+1} \mid Y_t)\}, \\ D_t^{22} = \mathbb{E}\{\nabla_{Y_{t+1}} \ln p(Y_{t+1} \mid Y_t) \nabla_{Y_{t+1}}^T \ln p(Y_{t+1} \mid Y_t)\}, \\ D_t^{33} = \mathbb{E}\{\nabla_{Y_{t+1}} \ln p(Z_{t+1} \mid Y_{t+1}) \nabla_{Y_{t+1}}^T \ln p(Z_{t+1} \mid Y_{t+1})\}. \end{cases}$$

When it comes to applying the above recursive formula, the matrices D_t^{ij}, of course, still have to be calculated or estimated. This is usually done with the help of simulation methods. The trajectories of the target are simulated on the basis of a distribution equation as well as the corresponding observations. This simulation comes at a relatively high cost. However, it has been shown that it is possible to obtain explicit expression of matrix D_t^{ij} in accordance with the following table.

	Cartesian	**MDP**	**LPC**
D_t^{11}	yes	no	yes
D_t^{12}	yes	no	yes
D_t^{21}	yes	no	yes
D_t^{22}	yes	no	yes
D_t^{33}	no	yes	yes

Table 7.1. *Data used for explicit expression of matrix* D_t^{ij}

At this point of the chapter there will be a brief explanation of the reasons why the LPC system is a system of coordinates which expresses the matrices D_t^{ij}. All calculations are relatively long and only an overview will be provided in this chapter. Please see the original article for more detail [BRE 06].

PROPOSITION 7.7 (fundamental property [BRE 06]). *In the LPC system, the gradients* $\nabla_{Y_t} \ln p(X_{t+1} \mid X_t)$ *and* $\nabla_{Y_{t+1}} \ln p(X_{t+1} \mid X_t)$ *only rely on quadratic forms of the variables* X_t, X_{t+1}*. This leads to:*

$$\begin{cases} \nabla_{Y_t}^T \ln p(X_{t+1} \mid X_t) = (X_{t+1} - AX_t - HU_t)^T Q^{-1} A \nabla_{Y_t}\{X_t\}, \\ \nabla_{Y_{t+1}}^T \ln p(X_{t+1} \mid X_t) = (X_{t+1} - AX_t - HU_t)^T Q^{-1} \nabla_{Y_{t+1}}\{X_{t+1}\}, \end{cases} \quad [7.42]$$

and

$$\begin{cases} \nabla_{Y_t}\{X_t\} = \begin{bmatrix} r_y(t) & -r_x(t) & 0 & 0 \\ r_x(t) & r_y(t) & 0 & 0 \\ v_y(t) & -v_x(t) & r_y(t) & -r_x(t) \\ v_x(t) & v_y(t) & r_x(t) & r_y(t) \end{bmatrix}, \\ \nabla_{Y_t+1}\{X_{t+1}\} = \begin{bmatrix} r_y(t+1) & -r_x(t+1) & 0 & 0 \\ r_x(t+1) & r_y(t+1) & 0 & 0 \\ v_y(t+1) & -v_x(t+1) & r_y(t+1) & -r_x(t+1) \\ v_x(t+1) & v_y(t+1) & r_x(t+1) & r_y(t+1) \end{bmatrix}, \end{cases} \quad [7.43]$$

so that $\nabla_{Y_t}\{X_t\}$ and $\nabla_{Y_t+1}\{X_{t+1}\}$ are linear operators on X_t, X_{t+1}.

On the basis of these results, [BRE 06] has shown that it is possible to obtain an analytical formula for the terms D_t^{ij}. Detailed information on these calculations is available in [BRE 06]. First, it is necessary to introduce auxiliary matrices (denoted Γ_t). These matrices are themselves computed recursively (temporal recursion). The matrices D_t^{ij} are blocks which are extracted from Γ_t. The main advantage is that simulations are no longer needed. It is therefore possible to take optimal distribution and active measurement problems into consideration.

The following problem is now considered:

PROBLEM 7.2. *Is it possible to allocate active measurements $\mathcal{U}_{0:t}$, such as:*

$$EQM_{\ln r_l} < s \quad \forall l \in \{1, \ldots, T\}.$$

PROBLEM 7.2a. *Is it possible to allocate active measurements $\mathcal{U}_{0:t}$, such as:*

$$J^{-1}_{\ln r_l} < s \quad \forall l \in \{1, \ldots, T\}.$$

The algorithm for optimizing the distribution of measurements is presented below.

For $t = 1$ to T:

1) compute $\tilde{J}_t^{-1}$ with a passive measurement at t;

2) if $\tilde{J}^{-1}_{\ln r_l} > s$, then $\mathcal{U}_t = 1$ and re-compute J_t^{-1} if measurements are active and passive;

3) if $\tilde{J}^{-1}_{\ln r_l} < s$, then $\mathcal{U}_t = 0$ and $J_t^{-1} = \tilde{J}_t^{-1}$.

Results

Here the results for a representative scenario are presented. The parameters for the scenario are described in Table 7.2.

Parameter scenario	
duration	6000 s
$r_x^{obs}(0)$	$3,5\,\text{km}$
$r_y^{obs}(0)$	$0\,\text{km}$
$v_x^{obs}(0)$	$10\,\text{ms}^{-1}$
$v_y^{obs}(0)$	$-2\,\text{ms}^{-1}$
$r_x^{cib}(0)$	$0\,\text{km}$
$r_y^{cib}(0)$	$3,5\,\text{km}$
$v_x^{cib}(0)$	$6\,\text{ms}^{-1}$
$v_y^{cib}(0)$	$3\,\text{ms}^{-1}$
δ_t	6 s
σ	$0.05\,\text{ms}^{-1}$
σ_β	0.05 rad ($\simeq$ 3 deg.)
σ_{r_0}	2 km
σ_{v_0}	$1\,\text{ms}^{-1}$
σ_{β_0}	0.05 rad ($\simeq$ 3 deg.)

Table 7.2. *Parameters of the scenario*

Figure 7.2 shows an example of the target trajectory. Note that this trajectory is just a realization of the target diffusion law. The target is moving away from the observer quite rapidly. This leads to a uniform rectilinear movement. Active measurements are required so as to maintain the FIM criterion below a certain value. Here only the estimation of parameter $\ln(r_t)$ is required, which means that only the corresponding term $J^{-1}_{\ln r_t}$ is being analyzed.

Figure 7.3 shows the results for this scenario. These results apply for the passive mode only. PCRB depends on $\ln(r_t)$ and decreases rapidly and then stabilizes at a certain point. The horizontal line shows the values of the order and the value which cannot be exceeded.

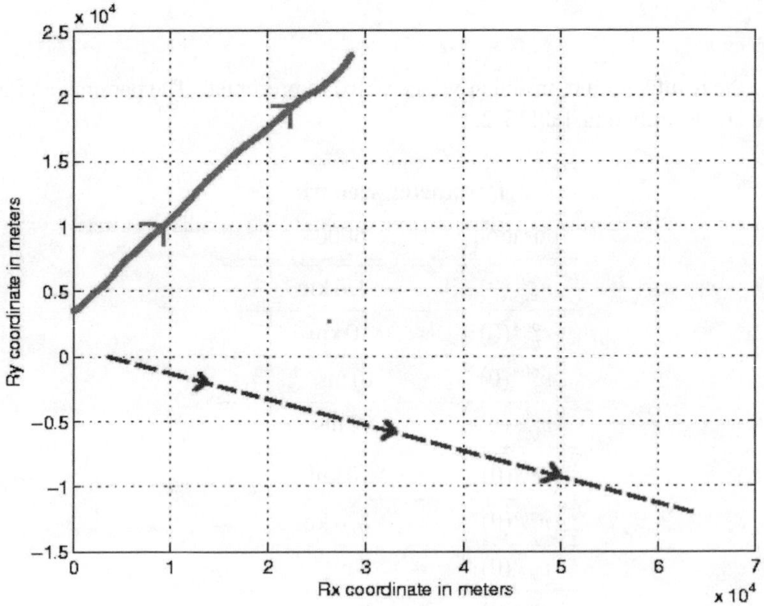

Figure 7.2. *Scenario: (a) example of the target trajectory (straight line) and the observer (dotted line) (b)*

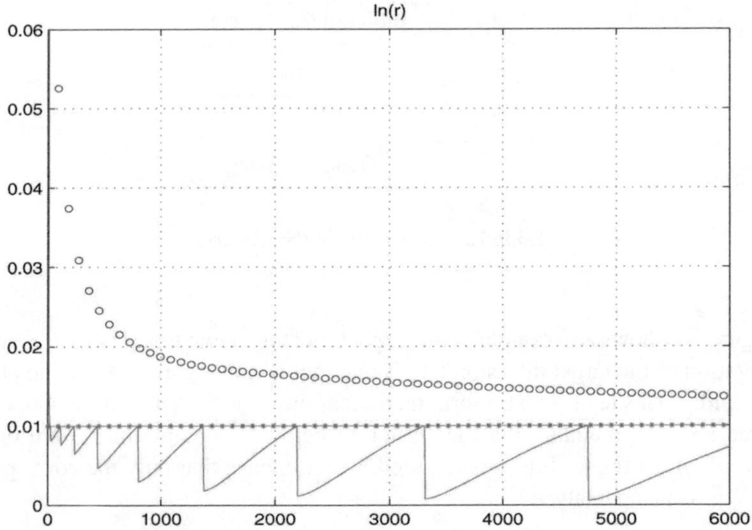

Figure 7.3. *Calculation of PCRB in relation to* $\ln r_t$ *with optimized active measurements (straight line) versus passive measurements only (dotted line)*

If the limit is infringed (without activation), an active measurement is emitted. This guideline has been respected very well (see Figure 7.3). The optimal sequencing of emissions is shown below. In Figure 7.4 the emissions are relatively frequent at the beginning of the scenario but become less frequent later on.

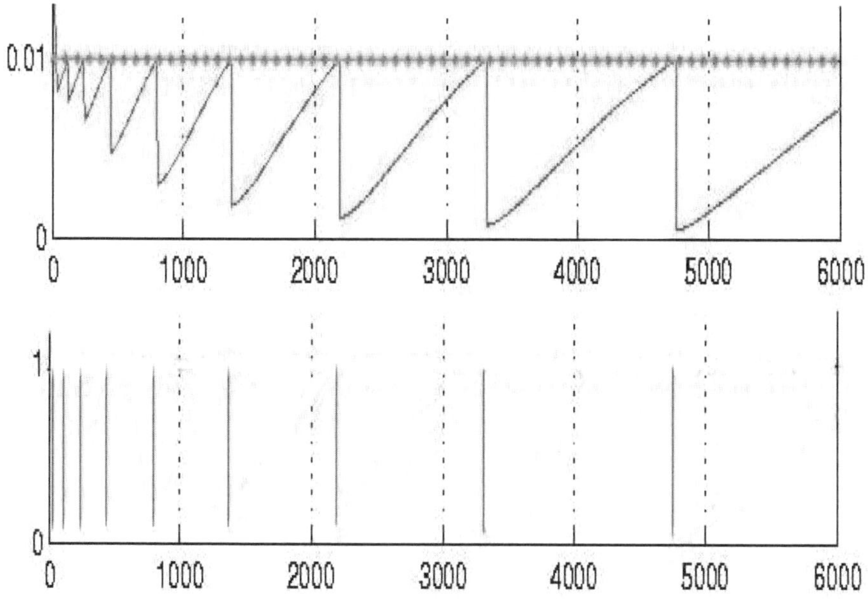

Figure 7.4. *Optimal sequencing of emissions*

7.5. Conclusion

Different problems related to the optimal sequencing of emissions have been investigated in this chapter. This chapter only deals with objective functions based on the quality of estimation (or tracking) of the target in a non-linear framework. The main difficulties of emission planning have been covered by this chapter. The explicit calculation of functional objectives has been detailed. In the case of a deterministic target trajectory, this result is applied to the evaluation of the importance of the impacting factors. This approach also reduces the problems to simple optimizations and avoids the difficulties related to the non-convexity of the functional objectives. For a Markovian target, the importance of a direct calculation of the functional objective has been shown (i.e. no simulation).

7.6. Appendix: monotonous functional matrices

Function f, mapping from $\mathcal{S}_n$ (the vector space of symmetric matrices) towards $\mathbb{R}$, satisfies the monotony property of matrices (MPM) if it meets the following condition.

DEFINITION 7.2.

– *f is of the class C^2,*

– *let A and B be two matrices of S_n, such that: $f(B) > f(A)$; then*

$$f(B+C) > f(A+C), \quad \textit{for any matrix } C \in S_n.$$

The dynamic programming recursion makes inequalities of the form:

$$\min f\left\{\sum_{j\in S}\left[C_{i,j}(d) + F_{\pi_1^*}(k+1,j)\right]p_{i,j}(d)\right\}$$

$$\leq f\left\{\sum_{j\in S}\left[C_{i,j}(d) + F_{\pi_1}(k+1,j)\right]p_{i,j}(d)\right\},$$

necessary, which have to be valid so that the optimal strategy π_1^* until time k remains optimal at time $k+1$. Now the question is to characterize functions f which meet the requirements of MPM. An answer to this question is given now.

PROPOSITION 7.8. *A matrix-valued function f, which applies for MPM, verifies:*

$$f(A) = g\big(\operatorname{tr}(AR)\big),$$

where g is any real-valued and increasing function and R is a fixed matrix.

Proof. Function f is of class C^2 and a first order approximation[5] of f around A is possible:

$$f(A+\rho C) \stackrel{1}{=} f(A) + \rho \operatorname{tr}\left[\nabla^T f(A)C\right] + 0(\rho), \quad \rho \text{ scalar.} \qquad [7.44]$$

In the previous formula, $\nabla f(A)$ describes the gradient of f on A. The notation $\operatorname{tr}[\nabla^T f(A)C]$ replaces the expression [POL 85, CAR 67] of the derivative of f, i.e. $Df_A(C)$, and corresponds to:

$$Df_A(C) = \operatorname{tr}\left[\nabla^T f(A)C\right]. \qquad [7.45]$$

Assume now that the gradients $\nabla f(A)$ are not always collinear. There are (at least) two matrices A and B for which the following applies $\nabla f(A) \neq \nabla f(B)$. Assume

5. The symbol $\stackrel{1}{=}$ means a first order approximation.

that $F^{\perp}$ denotes the orthogonal subspace of F (for the scalar product of the matrix [GRE 76]). Then:

$$(P_1) \quad \left(\nabla f(A)\right)^{\perp} \neq \left(\nabla f(B)\right)^{\perp}. \qquad [7.46]$$

At this point matrices A and B satisfy (P_1) and can be chosen mutually as close as possible (for the Frobenius norm related to the scalar product [HOR 85]). A consequence of (P_1) is the existence of a matrix C such as:

$$\operatorname{tr}\left[\nabla^T f(A)C\right] \neq 0 \quad \text{and} \quad \operatorname{tr}\left[\nabla^T f(B)C\right] = 0. \qquad [7.47]$$

Since $\operatorname{tr}[\nabla^T f(A)C] < 0$, implies $\operatorname{tr}[\nabla^T f(A)(-C)] > 0$, it is ensured without lost of generality that:

$$\operatorname{tr}\left[\nabla^T f(A)C\right] > 0 \quad \text{and} \quad \operatorname{tr}\left[\nabla^T f(B)C\right] = 0. \qquad [7.48]$$

Now, define the function $g(\rho)$ by:

$$g(\rho) \triangleq f(B + \rho C) - f(A + \rho C),$$

its first order approximation around 0 is:

$$g(\rho) \stackrel{1}{=} f(B) - f(A) - \rho \operatorname{tr}\left[\nabla^T f(A)C\right] + 0(\rho). \qquad [7.49]$$

Since function f is continuous on $\mathcal{S}_n$, it is possible to define a couple (A, B) such that:

$$f(B) - f(A) = \frac{\rho}{2} \operatorname{tr}\left[\nabla^T f(A)C\right],$$

as a consequence, the following applies:

$$\begin{aligned} &f(B + \rho C) - f(A + \rho C) \\ &\quad = -\frac{\rho}{2} \operatorname{tr}\left[\nabla^T f(A)C\right] + 0(\rho) \quad \left(\operatorname{tr}\left[\nabla^T f(A)C\right] > 0\right). \end{aligned} \qquad [7.50]$$

The equality above implies that f does not apply for the MPM. As a consequence, if f applies for the MPM, then all gradients are collinear and are therefore proportional to a single vector, denoted $\mathbf{G}$. Then:

$$\nabla f(A) = \lambda(A)\mathbf{G} \quad \forall A \in \mathcal{S}_n \quad \left(\lambda(A) \text{ scalar}\right). \qquad [7.51]$$

Owing to the intermediate value theorem and the rule of derivation of composite functions [CAR 67], the following result is obtained:

$$\nabla g\big[h(A)\big] = g'\big(h(A)\big)\nabla h(A), \quad \big(g : \mathbb{R} \longrightarrow \mathbb{R},\ h : \mathcal{S}_n \longrightarrow \mathbb{R}\big). \qquad [7.52]$$

Then f is the composition of a monotonic scalar function g with a linear form h. A linear form defined on $\mathcal{S}_n$ may always be written as $h(A) = \mathrm{tr}(AR)$, where R is any matrix (fixed and independent of A).

Conversely, it is easy to prove that $f(A) = g(\mathrm{tr}(AR))$, where g is an increasing function, verifies the MPM. □

7.7. Bibliography

[AVI 90] AVITZOUR D. and ROGERS S., "Optimal measurement scheduling for prediction and estimation", *IEEE Trans. on Acoustis, Speech and Signal Processing*, vol. 38, no. 10, pp. 1733–1739, 1990.

[BAR 88] BARAM Y. and KAILATH T., "Estimability and regulability of linear systems", *IEEE Trans. on Automatic Control*, vol. 33, no. 12, pp. 1116–1121, 1988.

[BOG 88] BOGUSLAVSKIJ I.A., *Filtering and Control*, Optimization Software, NY, 1988.

[BRE 06] BRÉHARD T. and LE CADRE J.-P., "Closed-form posterior Cramér-Rao bounds for bearings-only tracking", *IEEE Trans. on Aerospace and Electronic Systems*, vol. 42, no. 4, pp. 1198–1223, 2006.

[CAR 67] CARTAN H., *Calcul Différentiel*, Hermann, 1967.

[DAR 94] DARLING R.W.R., *Differential Forms and Connections*, Cambridge University Press, Cambridge, UK, 1994.

[GRE 76] GREUB W.H., *Linear Algebra*, 4th edition, Springer, 1976.

[HAM 85] HAMMEL S.E. and AIDALA V.J., "Observability requirements for three-dimensional tracking via angle measurements", *IEEE Trans. on Aerospace and Electronic Systems*, vol. 21, no. 2, pp. 200–207, 1985.

[HOR 85] HORN R.A. and JOHNSON C.R., *Matrix Analysis*, Cambridge University Press, 1985.

[KER 94] KERSHAW D.J. and EVANS R.J., "Optimal waveform selection for tracking systems", *IEEE Trans. on Information Theory*, vol. 40, no. 5, pp. 1536–1550, 1994.

[LEC 98] LE CADRE J.-P., "Properties of estimability criteria for target motion analysis", *IEE Proc. Radar, Sonar & Navigation*, vol. 145, no. 2, pp. 92–99, 1998.

[LEC 97] LE CADRE J.-P. and TREMOIS O., "The matrix dynamic programming property and its implications", *SIAM Journal Matrix Anal. Appl.*, vol. 18, no. 4, pp. 818–826, 1997.

[MEH 76] MEHRA K.K., "Optimization of measurement schedules and sensor design for linear dynamic systems", *IEEE Trans. on Automatic Control*, vol. 21, no. 1, pp. 55–64, 1976.

[NAR 81] NARDONE S.C. and AIDALA V.J., "Observability criteria for bearings-only target motion analysis", *IEEE Trans. on Aerospace and Electronic Systems*, vol. 17, no. 2, pp. 162–166, 1981.

[NAR 84] NARDONE S.C., LINDGREN A.G. and GONG K.F., "Fundamental properties and performance of conventional bearings-only target motion analysis", *IEEE Trans. on Automatic Control*, vol. 29, no. 9, pp. 775–787, 1984.

[POL 85] POLLOCK D.S.G., "Tensor products and matrix differential calculus", *Linear Algebra and its Applications*, vol. 67, pp. 169–193, 1985.

[POT 67] POTTER J.E. and FRASER D.C., "A formula for updating the determinant of the covariance matrix", *AIAA Journal*, vol. 5, no. 7, pp. 1352–1354, 1967.

[SHA 95] SHAKERI M., PATTIPATI K.R. and KLEINMAN D.I., "Optimal measurement scheduling for estimation", *IEEE Trans. on Aerospace and Electronic Systems*, vol. 31, no. 2, pp. 716–729, 1995.

[TRE 96] TRÉMOIS O. and LE CADRE J.-P., "Target motion analysis with multiple arrays: performance analysis", *IEEE Trans. on Aerospace and Electronic Systems*, vol. 32, no. 3, pp. 1030–1045, 1996.

[VAN 93] VAN KEUK G. and BLACKMAN S.C., "On phased array tracking and parameter control", *IEEE Trans. on Aerospace and Electronic Systems*, vol. 29, no. 1, pp. 186–194, 1993.

[YOK 92] YOKONUMA T., *Tensor Spaces and Exterior Algebra*, Transl. of Math. Monographs, vol. 108, Amer. Math. Soc., Providence, RI, 1992.

Chapter 8

Bayesian Inference and Markov Models

8.1. Introduction and application framework

Marine sciences rely on the use of different methods and different tools to be able to carry out research on what the seabed is made up of: this includes the use of geotechnical tools (isolated well-logging, on site samples) as well as side-scan sonar devices and multi-beam sounding systems which transform the characteristics of the seabed interface into images.

Investigation systems which are used and which are based on a high resolution sonar imaging system are tools that provide a solution to the problems linked to the detection of small objects that can be found on the seabed. After the emission of a sonar wave, the detection of these small objects can be carried out in one of two ways: first of all, it is possible to detect the behavior of a signal which is reflected by the object, and secondly there is a certain level of interference that is generated by the presence of this object (in the case of objects which are slightly buried in the sediment of the sea bed). It is therefore necessary that the dimensions and the location of such objects be compatible with the sonar's ability to detect them. In the case of an anechoic object, it is the formation of the acoustic shadow which makes it possible to detect and then classify the object. In this chapter we will not deal with the issues that arise during the design of a sonar system, nor will we deal with the issues which led to the development of specialized sonar devices that are used for specific tasks, such as sonar sensors which detect the presence of an object, sonar classifiers, which are used to analyze an object, but which possess a lower detection range, and acoustic cameras, which make it possible to recognize the object from a very short distance.

Chapter written by Christophe COLLET.

The differences that exist between these sonar devices can be identified through the following aspects: the frequency of emission and reception of each device (in other words the size of their antennae), the visualization mode which is used to detect and classify the object such as sectional vision, panoramic vision and lateral vision, and finally the structure of the investigation systems. The investigation systems can either be towed [THO 96], found under the body of a moving vessel [DUG 96], or on board an autonomous sensor (remote operated vehicle).

The sensors, which are used in sonar antennae, measure different pressure variations. They are thus normally used to carry out investigations on the complex amplitude module that is emitted by a signal [BUR 78] and not on the signal's intensity as is the case with radar imagery. The principle of image formation is based on dividing a region of the seabed into smaller sectors (or resolution cells). A pathway in a particular direction is formed and this makes it possible to explore the region of the seabed that is to be examined. Once the pathway has been formed, a sample of the signal which is reflected from the seabed is taken. It then becomes possible to work out the amplitude of the signal in relation to its distance from the seabed. If a large number of these paths are created electronically it therefore becomes possible to explore an entire region of the seabed that is to be analyzed, sector by sector. One solution that can be used to explore an individual sector involves creating a small number of pathways, and then moving the sonar in a direction that is perpendicular to the source of emission which comes from the signal. This solution is used in side-scan sonar devices (Figure 8.1a). The images produced by such devices are also analyzed in this chapter.

It is quite difficult to actually see the presence of any small objects or the edges of small objects in the images due to the granularity of the objects [COL 98]. This poor quality can be associated with the use of sonar systems as well as with the use of any coherent emission system. This poor quality due to the presence of speckle noise (multiplicative noise) can be modeled using either Rayleigh's [COL 98] or Weibull's [MIG 98] probability density functions (pdf). In certain cases it is possible to model this speckle noise by using more complex pdfs requiring a larger number of parameters; an example of such a probability distribution is the K law. During the modeling step and particularly when complex pdfs are used, it is vital to correctly integrate the nature of the noise, because in doing so, it means that it becomes possible to segment images that have been strongly corrupted.

8.2. Detection, segmentation and classification

The sonar images can be divided into three generic classes: shadow, reverberation and echo. The echo class is, however, not present all of the time. This idea of segmentation has been the subject of different Markov models which are used with the aim of coding a piece of *a priori* information [THO 96, MIG 98]. There is in fact no mathematical relationship between the images where the information is coded in

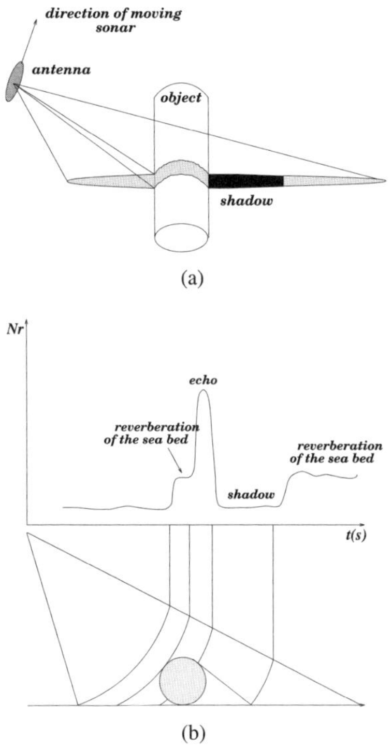

Figure 8.1. *The process of image formation using a sonar device: (a) a cylindrical object and its shadow corresponding to the region of the seabed which is acoustically hidden, (b) a diagram which shows the evolution of the signal over time. First of all we can see a weak level of volume reverberation corresponding to the microelements which are present in the column of water. We can then see the acoustic wave which is reflected by the seabed (reverberation) as well as the presence of an object which is sometimes represented by a strong reflection of the acoustic wave (the echo). Immediately after the object has been detected, a signal with a very low amplitude level is received (the shadow). The term shadow is used because it corresponds to the hidden zone of the seabed*

grayscale and the representation of the observed objects. It is possible to associate an infinite number of combinations of objects with an image which has been observed. All of these different combinations of objects are able to produce the same image. This is why it is useful to integrate *a priori* knowledge during the segmentation phase. The aim of this is to statistically reduce the number of solutions that are possible for creating the image. This *a priori* knowledge is partly made up of information which is relevant to the image which has been observed, and is also made up of more generic information. This *a priori* knowledge can be divided into both local and global knowledge. Markov modeling allows us to locally describe global properties and is useful to constrain the solution (i.e. the segmentation map) through the use of optimization algorithms (section 8.3). There are substantial benefits to be gained from using hierarchical approaches such as those mentioned in [THO 96, MIG 98] in terms of the quality of the solutions that are obtained as well as in relation to the speed at which modern algorithms can converge in comparison to the traditional algorithms that were once used.

In section 8.4 a description of the hierarchical Markov segmentation algorithm is given. The use of this algorithm makes it possible to segment the images into two different classes (shadow and reverberation). In section 8.5 we introduce another Markov algorithm allowing us to detect the third label which is the echo class. This class takes into consideration the physical formation of the sonar images. We also illustrate, using two concrete examples, the generality of these approaches as well as the variety of optimization algorithms which can be used on the segmentation map previously obtained.

The first example (section 8.6) deals with the detection and classification of cylindrical or spherical shape manufactured objects which are investigated by a high resolution sonar system. The high quality of an echo rarely leads to the classification of such manufactured objects and this is due to the stealth technology that is used to identify such objects (such as underwater mines). Nevertheless, it is possible to use the shape of the shadow produced by the object lying on the seabed thanks to signals sent by the acoustic wave. Whilst observing the shape of the shadows that are produced by such manufactured objects, we noticed that their shadows' shapes are regular and that they can be described by using simple geometric descriptors. The shapes of the shadows have an almost affine transformation and exist within a particular Markov energy function whose energy needs to be minimized (cf. section 8.6). This regularity of the shadows of such manufactured objects is the opposite of the shapes of the shadows which are produced by stones of a similar size; the shadows of these stones vary and are very irregular. These different properties will be used to classify the different objects which can be found lying on the seabed. The optimization algorithm that we have developed, and which minimizes the Markov energy function, is in this case a genetic algorithm.

The second example (section 8.7) deals with the classification of seabeds in relation to the segmentation process (shadow/reverberation). The goal consists of being able to recognize different textured shapes and this can sometimes pose a problem. The different parameters which are extracted during the segmentation process lead to a fuzzy classification of the seabeds. A Markov field model is then applied to this classification process which refines the classification of the seabeds even further because the use of this field model enforces the spatial homogenity of the classes which are associated with the segmented map (i.e. shadow, reverberation and echo labels). The optimization algorithm used in this case is the iterated conditional model (ICM) algorithm [BES 86]. This algorithm is preferred to a simulated annealing algorithm because of processing times.

8.3. General modeling

Markov fields were first used in the area of image processing in the 1980s [GEM 84]. This framework, which is used for modeling, experienced a rapid growth and success thanks to its flexibility, its ability to represent contextual information in probabilistic terms, as well as illustrating the well known Bayes theorem with which the Markov fields are associated. Within the Markov approaches that are used in image processing the observations (variables which can be observed and which rely on the analysis of the image) and the labels (information which cannot be observed and that have to be extracted from the observations) are assumed to be random field variables that we will note as Y (the observation field) and X (the label field) respectively. These fields are defined on a rectangular grid S: $Y = \{Y_s,\ s \in S\}$ and $X = \{X_s,\ s \in S\}$ where s represents a point on the grid. Let $y = \{y_s,\ s \in S\}$ and $x = \{x_s,\ s \in S\}$ be two realizations of random fields. The labels X_s take their values (either discrete or continuous) from the set of values Λ_{label}. The set of all the possible configurations of Y is defined as follows: $\Omega_{\text{obs}} = \Lambda_{\text{obs}}^{\#S}$, where $\#S$ represents the cardinal numbers of S and Λ_{obs} represents the brightness values which can be observed. The set of all the possible combinations of X is $\Omega_{\text{label}} = \Lambda_{\text{label}}^{\#S}$. This set makes up all of the label fields which are possible in the solution space. This set is huge and there is no way of exploring it in an exhaustive manner.

8.3.1. *Markov modeling*

The Markov hypothesis can be expressed by a set of solutions X which are defined on a regular grid S. Markov modeling makes it possible to specify the properties of the random X field (or label field) in relation to neighboring system $V = \{v_s,\ s \in S\}$, which is defined on a set of sites $\{s \in S\}$ (a system $V = \{V_s \subset S \mid s \in S\}$, is a neighboring system of S if, and only if, $\forall s \in S$, $s \notin V_s$ and $\forall (s, r) \in S^2$, $r \in V_s \Leftrightarrow s \in V_r$). The neighboring system, which is thus defined, leads to the generation of a set C of cliques c on S. These cliques c correspond to the set of possible configurations generated by the adopted neighboring system. The statistical relationship which links the observation field (Y) with the label field (X) in the neighboring system (V) can be explained by the equations which follow in the next few sentences. The X field is said to be Markovian and stationary in relation to a neighboring system V, so that $V = \{v_s,\ s \in S\}$. The X field is therefore a Markov field if for all x configurations of X the following is true: $P_X(x) > 0$ and $\forall s \in S$, $P_{X_s|X_r}(x_s \mid x_r,\ r \in S - \{s\}) = P_{X_s|X_r}(x_s \mid x_r,\ r \in v_s)$. The importance of Markov modeling lies in the ability to locally specify a model which is in fact a global model. The Hammersley-Clifford theorem shows the equivalence that exists between Gibbs distribution and the Markov fields [BES 74]. In this way it is possible to describe a Markov field by looking at its total probability distribution in the following way: $\forall x \in \Omega_{\text{label}}$, $P_X(x) = \frac{1}{Z}\exp\{-U_2(x)\}$ with the function $U_2(x)$ which is known as the energy function and which can be decomposed into a smaller number of local functions V_c. These local functions are known as functions of potential and are defined by the cliques of $c \in C$. These cliques are generated by the neighboring system V: $U_2(x) = \sum_{c \in C} V_c(x)$. V_c is only a function of the labels for the points $c \in C$. Z is a *normalization* constant, otherwise known as a partition function and can be defined as $Z = \sum_{x \in \Omega_{label}} \exp\{-U_2(x)\}$. The conditional laws $P_{X_s|X_t}(X_s = x_s \mid X_t = x_t,\, t \in v_s)$ can also be calculated and a link can therefore be established between the geometric shape of the neighborhood (V) and the energy term $U_2(x)$. It is also possible to simulate such Markov field in terms of $P_X(x)$ by using Gibbs sampling algorithms and Metropolis-Hastings algorithms [BES 86, DUB 89].

8.3.2. *Bayesian inference*

Bayesian inference is about modeling the labels and the observations jointly through the modeling of a coupled random field (X, Y). This field is defined by the joint distribution of $P_{XY}(x, y)$ which needs to be developed. Bayes' theorem states that $P_{XY}(x, y) = P_{Y|X}(y \mid x) \times P_X(x)$. The probability $P_{Y|X}(y \mid x)$ translates the likelihood between observations and labels. This probability is defined by a knowledge of probability laws that are linked to noise (in the case of sonar imagery this noise would be the speckle noise), whilst $P_X(x)$ makes it possible to introduce *a priori* knowledge onto the labels. If we assume the Morkovianity of the fields then the joint distribution of $P_{XY}(x, y)$ is a Gibbs distribution with an energy value of $U(x, y)$:

$$P_{XY}(x, y) = \frac{1}{Z}\exp\big\{-U(x, y)\big\} \qquad [8.1]$$

where the energy is divided into the sum of the two following terms: $U(x, y) = U_1(x, y) + U_2(x)$.

– $U_1(x, y)$, which stems from the observation likelihood, constitutes the data-driven term and characterizes the balance that exists between the labels and the observations;

– $U_2(x)$, which comes from the Markov field modeling, constitutes the *a priori* contextual term which introduces a regularizing effect into the label field.

The criterion which is often used to estimate label field is the MAP criterion (*maximum a posteriori* otherwise known as posterior mode). The search for the optimal label field $\hat{x}_{\text{opt}}$ using this criterion can also be seen as the minimization of the function $U(x, y)$. It is possible to process this optimal label field using a simulated annealing algorithm. Due to reasons associated with processing time and depending on the individual cases, we preferred using mono- or multi-scale ICM algorithms [BES 86] or genetic algorithms. Parameter estimation will be carried out using estimation algorithms such as EM [DEM 76], SEM [CEL 86] or ECI [PIE 92] depending on the model we will take into account.

8.4. Segmentation using the causal-in-scale Markov model

In this multi-scale model, the pyramid structure is made up of a hierarchy $(x^L, \ldots, x^0)$ of label fields with variable resolutions, where x^l is defined in a grid S^l which corresponds to an isotropic subsample of S of factor 2^l ($S^0 \equiv S$ and $x^0 \equiv x$). $Y = \{Y_s,\ s \in S\}$ corresponds to the observation field which can also be found on the same grid but at a much finer resolution of S. The grid is made up of N sites of s (these sites are linked to N pixels of the sonar image), and $X = \{X_s,\ s \in S\}$ represents the label field with the finest resolution where ($X^0 \equiv X$). Each label X_s is developed further in $\{e_0 = \text{shadow},\ e_1 = \text{reverberation}\}$. The segmentation of sonar images into two classes is seen as a problem of global Bayesian inference which involves the search for x^l, the label with a resolution of l so that $\hat{x}^L = \arg\max_{x^L}\{P_{X^L/Y}(x^L/y)\}$ and $\forall l = (L-1), \ldots, 0$

$$\hat{x}^l = \arg\max_{x^l} \left\{ P_{X^l/Y,X^{l+1}}(x^l/y, \hat{x}^{l+1}) \right\} \quad [8.2]$$

$$= \arg\max_{x^l} \left\{ \exp -U^l(x^l, y, \hat{x}^{l+1}) \right\} \quad [8.3]$$

$$= \arg\min_{x^l} U^l(x^l, y, \hat{x}^{l+1}). \quad [8.4]$$

Level 0 *a posteriori* distribution, which appears in equation [8.2], can be written as $P_{X/Y,X^1}(x/y, \hat{x}^1) \propto P_{X/X^1}(x/x^1) P_{Y/X}(y/x)$. In this expression $P_{X/X^1}(x/\hat{x}^1)$ models the *a priori* distribution, and $P_{Y/X}(y/x)$ under conditionally independent assumption is factorized into $\Pi_s P_{Y_s/X_s}(y_s/x_s)$, with $P_{Y_s/X_s}(y_s/x_s)$ being the probability density which is associated with each region (i.e. shadow and reverberation) of the sonar image. As far as the finest resolution is concerned the global energy function which is to be minimized can be written as follows:

$$U(x, y, \hat{x}^1) = \underbrace{-\sum_{s \in S} \ln P_{Y_s/X_s}(y_s/x_s)}_{U_1(x,y)} + \underbrace{\sum_{<s,t> \subset S} \beta_{st}\big(1 - \delta(x_s, x_t)\big)}_{U_2(x)}$$
$$+ \underbrace{\sum_{s \in S} \beta_5 \big(1 - \delta(x_s, \hat{x}^1_{\text{father}(s)})\big)}_{U_3(x, \hat{x}^1)}$$

$\delta(\cdot)$ represents the Kronecker function and $\beta_{st} = \beta_1, \beta_2, \beta_3$ or β_4 depend only on the orientation of the spatial binary clique, while β_5 is the inter-scale parameter that is associated with the causal-in-scale link. U_1 measures the balance between the observation and the label, U_2 measures the energy of the *a priori* spatial model and

U_3 measures the causal energy which symbolizes the in-scale relationship that exists with the segmentation process at the coarsest level. It is possible to successfully create a multi-grid model which is defined at every resolution level l and which can be defined as follows [HEI 94]: $U^l(x^l, y, x^{l+1}) \triangleq U(\Psi_0^l(x^l), y, \Psi_1^{l+1}(x^{l+1}))$, where Ψ_k^l is the duplication operator of S^l on S^k (for $k < l$). At the coarsest level, the multi-grid model can be defined as follows: $U^L(x^L, y) = U_1(\Psi_0^L(x^L), y) + U_2(\Psi^L(x^L))$.

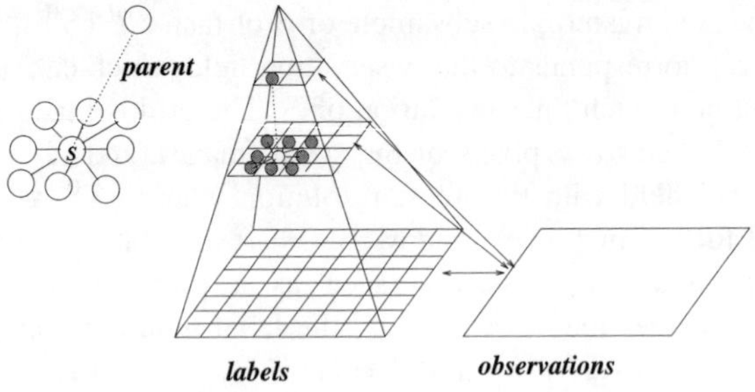

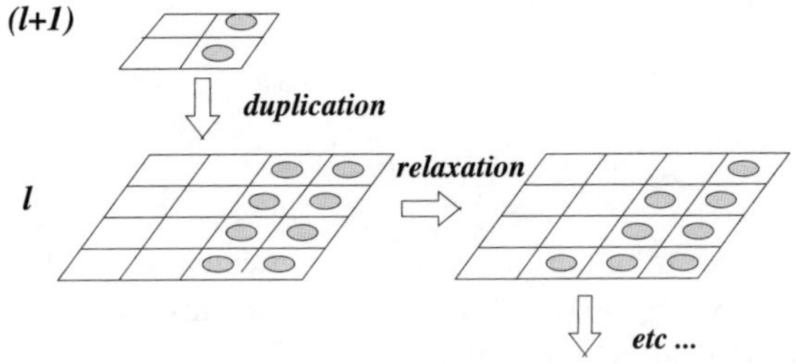

Figure 8.2. *(a) Multi-scale structure using spatial interactions of an 8-connexity neighborhood system which is causal in scale. The aim of this causal-in-scale link is to make the two segmentation classes (shadow and reverberation) much robust; (b) coarse to fine minimization strategy*

The energy of this model is described by U_2 and U_3 (named as SCM or scale causal multi-grid), and can be seen in Figure 8.2a. The amount of energy that this model possesses depends on the parameter vector $\Phi_x = (\beta_1, \ldots, \beta_5)$ while the noise U_1 is defined by a second vector Φ_y. In the case of sonar images, a Weibull density model is used [MIG 99]. It is possible to estimate all the parameters in an unsupervised way, by using the iterative conditional estimation (ICE) algorithm [MIG 99] to simultaneously estimate two parameter vectors. The ICM optimization algorithm carries out the minimization process, which can be seen in equation [8.4]. The final estimation $\hat{x}^{l+1}$ which is obtained at scale $l+1$ is interpoled by $\Psi_l^{l+1}(\hat{x}^{l+1})$ with the aim of initializing the relaxation process at the finest scale of l (Figure 8.2b).

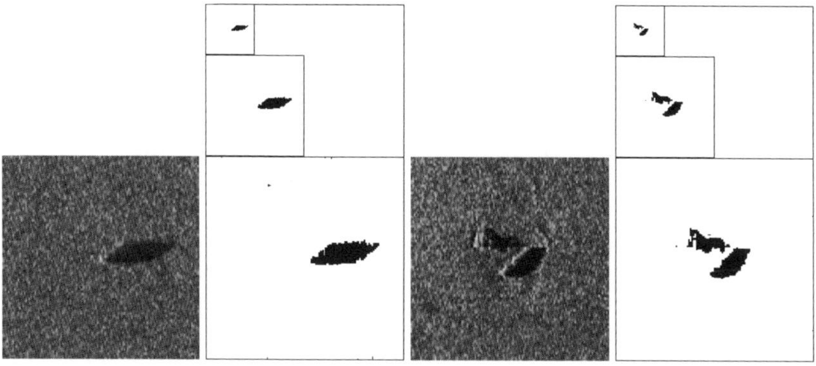

(a) (b) (c) (d)

Figure 8.3. *(a) and (c) A sonar image of a bed of sand on which there is a cylindrical shaped manufactured object (a and c) plus a stone (c); (b) and (d) the result of segmenting the image into the two classes (reverberation and shadow) using the proposed SCM algorithm*

This multi-scale minimization strategy is a rapid process which is less sensitive to local minima than standard mono-scale relaxation algorithms. The strategy is also well adapted to sonar images which have been noised with a strong specular noise and which are modeled by one of Weibull's density models [THO 96, MIG 98]. Figure 8.3 shows the results obtained for an image which has been segmented into two classes and for which the unsupervised SCM algorithm has been used.

8.5. Segmentation into three classes

Whenever an echo is present, it must be detected in spite of specular noise. One possible approach requires the use of *a priori* information we possess relating to the echo formation. In sonar imagery the object which lies on the seabed creates a possible echo which is then automatically followed by the creation of a shadow. This property can easily be integrated into a Markov model which is now going to be explained. Let $\hat{x}^{[1]}$ be the label field that is obtained after an image has been segmented into two classes by using the SCM algorithm which was mentioned earlier in this chapter. The label $\hat{x}_s^{[1]}$ belongs to $\{e_0, e_1\}$. In taking $x^{[1]}$, we now consider the subset of pixels S' ($S' \subset S$) so that $S' = \{s \in S : \hat{x}_s^{[1]} = e_1 = \text{reverberation}\}$. It is this set which then needs to be segmented into two classes in order to extract the echo from the reverberation class. Let $X^{[2]}$ be the following random binary process: $\forall s \in S', X_s^{[2]}$ takes it values from $\{e_1 = \text{reverberation},\ e_2 = \text{echo}\}$. The segmentation process will then be carried out on the restricted set of data $y^{[1]} = \{y_s, s \in S'\}$. The distribution of $(X^{[2]}/Y^{[1]} = y^{[1]},\ X^{[1]} = \hat{x}^{[1]})$ is then defined. This definition is made up of two terms: $P_{X^{[2]}/X^{[1]}}(x^{[2]}/\hat{x}^{[1]})$ which is the distribution of $X^{[2]}$ that we assume to be stationary and *de Markov*, and by the data attachment term $P_{Y_s^{[1]}/X_s^{[2]}}(y_s^{[1]}/x_s^{[2]})$

according to the label $x_s^{[2]}$. In the case of a label which is reflected from the seabed e_1:

$$P_{Y_s^{[1]}/X_s^{[2]}}(y_s^{[1]}/e_1) = \mathcal{W}_b(y_s, \alpha, \beta, \min) \quad [8.5]$$

$$= \frac{\beta}{\alpha}\left(\frac{y_s - \min}{\alpha}\right)^{\beta-1} \exp\left\{ -\left(\frac{y_s - \min}{\alpha}\right)^{\beta}\right\} \quad [8.6]$$

where α and β are the parameters of scale and shape respectively, which are positive defined according to Weibull's law (Figure 8.4(a)). The data attachment term for the echo label e_2 is written as follows:

$$P_{Y_s^{[1]}/X_s^{[2]}}(y_s^{[1]}/e_2) = \frac{2}{\gamma}\Lambda(y_s - y_{\max})\mathcal{U}(y_{\max} - y_s) \quad [8.7]$$

where $\mathcal{U}$ is the Heaviside step function and Λ is a triangle function. $y_{\max}$ corresponds to the maximum brightness of the sonar image while $\frac{2}{\gamma}$ is a normalization constant (see Figure 8.4(b)). This model is justified due to the fact that for raw data the echo corresponds to a saturation of hydrophones.

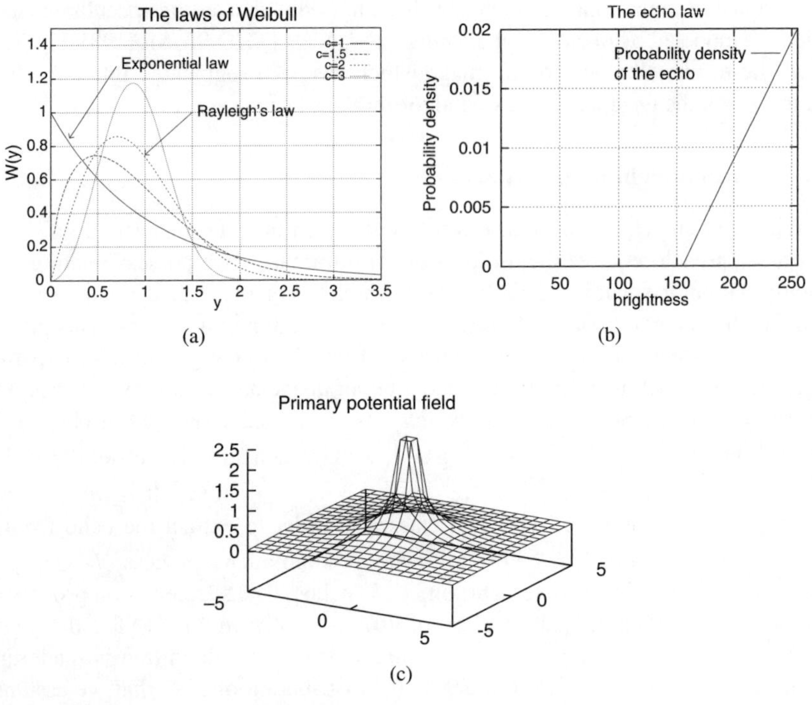

Figure 8.4. *(a) A graph showing Weibull's law for the different parameters of $\beta = c$; (b) the echo law; (c) the primary potential field which has been generated into a shadow labeled site*

We are therefore looking for $x^{[2]}$ so that [MIG 99]:

$$\hat{x}^{[2]} = \arg\max_{x^{[2]}} P_{X^{[2]}/X^{[1]},Y^{[1]}}\left(x^{[2]}/\hat{x}^{[1]}, y^{[1]}\right) \quad [8.8]$$

$$= \arg\max_{x^{[2]}} P_{Y^{[1]}/X^{[1]},X^{[2]}}\left(y^{[1]}/\hat{x}^{[1]}, x^{[2]}\right) P_{X^{[2]}/X^{[1]}}\left(x^{[2]}/\hat{x}^{[1]}\right) \quad [8.9]$$

$$= \arg\min_{x^{[2]}} \left\{U_1\left(y^{[1]}, x^{[2]}, \hat{x}^{[1]}\right) + U_2\left(x^{[2]}, \hat{x}^{[1]}\right)\right\} \quad [8.10]$$

where U_1 highlights the balance that exists between the observations and the labels $\{e_1, e_2\}$, while U_2 is the energy term which corresponds to the *a priori* model described in $P_{X^{[2]}/X^{[1]}}(x^{[2]}/\hat{x}^{[1]})$. The parameters of the energy term which determine the likelihood of U_1 can be estimated thanks to a procedure defined in [MIG 99]. U_2 expresses the constraints on the desired solution. In our case, once again we use a spatial 8-connexity neighborhood system (Figure 8.5) where β_1, β_2, β_3, β_4 represent the *a priori* parameters which are associated with horizontal, vertical, diagonal, left and right binary cliques respectively. β_0 corresponds to the singleton clique.

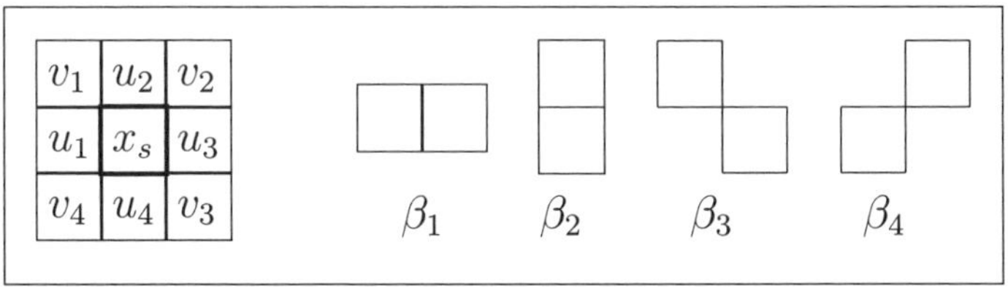

Figure 8.5. *An 8-connexity neighborhood system and the sites u_i, v_i, which are associated with that neighborhood. The four binary cliques of the 8-connexity neighborhood system and their associated β_k parameters which correspond to the Ising model*

The Potts model is used in order to favor the use of homogenous regions. The Potts model associates the binary clique $\langle s, t\rangle$ with a variable potential $\beta(1 - \delta(x_s^{[2]}, x_t^{[2]}))$, where $\beta = \beta_1 = \beta_2 = \beta_3 = \beta_4$.

The dependency of the echo class on the shadow class consists of favoring an echo label within a site s which is located not too far from a shadow region. To express such constraint, a potential associated with the singleton clique is used: $-\beta_0 \ln \Psi_{\hat{x}^{[1]}}(s)\delta(x_s^{[2]}, e_2)$. In this equation $\Psi_{\hat{x}^{[1]}}(s)$ is a field of potentials which can be defined as follows:

$$\Psi_{\hat{x}^{[1]}}(t) = \inf\left\{\underbrace{\sum_{s\in S:\hat{x}_s^{[1]}=e_0} \frac{1}{d(s,t)} \exp\left(-\frac{d(s,t)}{\sigma}\right)}_{\phi_s\left(d(s,t)\right)}, 1\right\} \quad [8.11]$$

with $d(s,t)$ being the geometric distance that exists between the pixels s and t. σ is a parameter that controls the maximum distance at which it is possible for the echo and shadow regions to interact with one another (Figure 8.4(c) represents the function $\phi_s(d(s,t))$). The total energy that needs to be minimized can be defined as follows:

$$U\big(x^{[2]}, \hat{x}^{[1]}, y^{[1]}\big) = \underbrace{-\sum_{s\in S'} \ln P_{Y_s^{[1]}/X_s^{[2]}}\big(y_s^{[1]}/x_s^{[2]}\big)}_{U_1(y^{[1]},x^{[2]},\hat{x}^{[1]})}$$
$$+ \underbrace{\sum_{<s,t>\subset S'} \beta\big(1-\delta\big(x_s^{[2]}, x_t^{[2]}\big)\big)}_{U_{21}(x^{[2]},\hat{x}^{[1]})}$$
$$+ \underbrace{\sum_{s\in S'} -\beta_0 \ln \Psi_{\hat{x}^{[1]}}(s)\cdot\delta\big(x_s^{[2]}, e_2\big)}_{U_{22}(x^{[1]},x^{[2]},\hat{x}^{[1]})}.$$

Once the actual modeling process has been defined, a traditional ICM type optimization algorithm is used [BES 86]. The initialization is made by a maximum likelihood algorithm.

In practice, on one hand due to the extremely low number of locations that are associated with an echo label, on the other hand due to the good label map initialization given by ML (maximum likelihood) algorithm, it is possible to segment the image to a single resolution level with the ICM algorithm without having to use a multi-grid strategy. Figures 8.6 and 8.7 show two examples of three-class segmentation which have been carried out on synthetic and real images. We also show (for different Markov models we defined) that it is possible for the objects to free themselves from strong specular noise and to integrate within a regularization scheme based on a Markov model, some *a priori* information related to image formation.

8.6. The classification of objects

In this section we will deal with another aspect of sonar image analysis: the classification of manufactured and natural objects which can be found lying on the seabed. This phase of low-level semantic classification (which is of great importance for the application in real life) also relies on the use of a Markov regularization approach which formalizes the *a priori* knowledge that we already possess. Contrary to the natural objects that can be found lying on the seabed, the manufactured objects are normally made up of simple and regular geometric shapes. The shadow of a cylindrical shaped object (such as a mine) is the shape of a perfect parallelogram. The *a priori* knowledge that we possess in relation to the object and its shadow can be represented by a prototype shape, which in our case is a parallelogram with its four

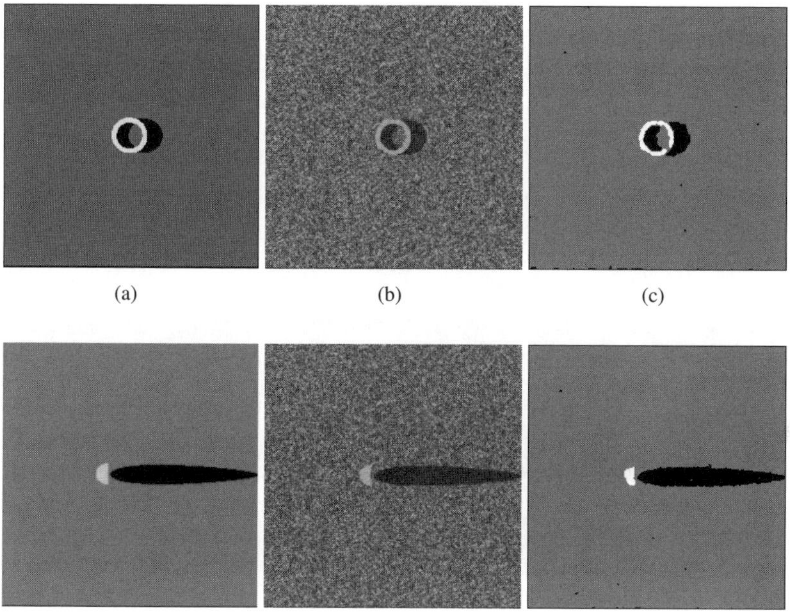

(a) (b) (c)

(d) (e) (f)

Figure 8.6. *Synthetic shapes which are similar to the shape of a circle such as a tire (a) and a sphere (d). These shapes are obtained by using a ray tracing algorithm. (b and e) are synthetic sonar images which have been artificially sounded with a speckle noise, the objects are assumed to be placed on a bed of sand. (c and f) are the resulting three-class segmentations*

sides [MIG 00a]. In other cases it might be necessary to detect ellipses corresponding to spherical shaped objects which are lying on the seabed: the shadow associated with any object can in fact be defined by m manually selected points or by m points which are evenly spread on the shadow's outline. A representation using a cubic B-spline on m control points is therefore possible. The shape of this object's shadow (or of any other object) can be created thanks to the use of a ray tracing algorithm (Figure 8.6a, d).

Let γ_0 be the prototype shape of a class of objects we are looking for. We have to define a set of linear transformations that can be carried out on γ_0 in order to consider the variability of the class of objects: different point of view, unknown *a priori* location of the object in the image, etc. Let γ_θ be a distorted version of the original γ_0 which is made possible by carrying out an affine transformation on the original and whose parameters form the vector θ. In the case of a cylindrical shape these transformations can take the form of any combination of the following functions: translation (the exact location of the object in the image is unknown), scale-factor (the exact distance of the object from the sonar device is unknown), rotation (the angle of the object from

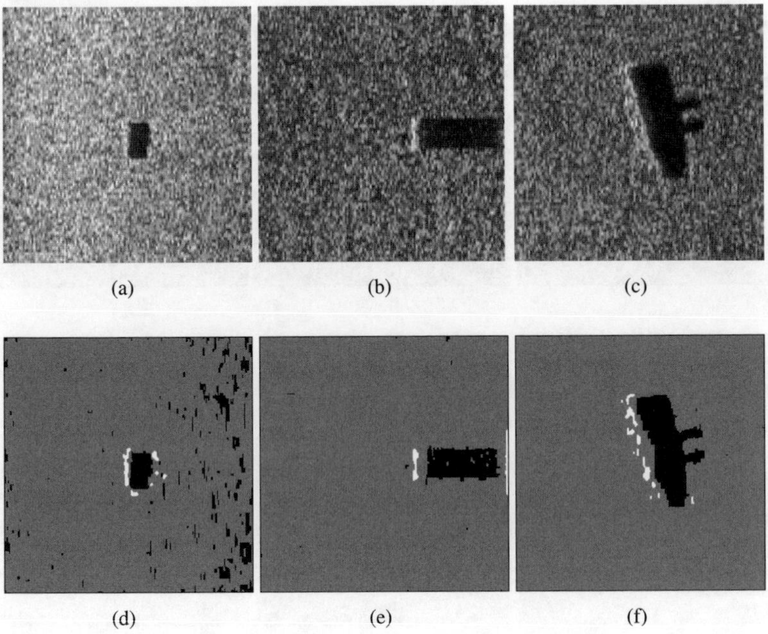

Figure 8.7. *(a) (c) (e) Real sonar images taken from manufactured objects lying on a bed of sand (two cylinders and a trolley lying on its side, it is possible to distinguish the two wheels of the trolley); (b) (d) (f) the resulting three-class segmentation*

the sonar device is unknown), slicing and stretching (with the aim of modeling the artifact during the formation of the sonar image). All of these different functions can be seen in Figure 8.8a. In the case of a spherical shaped object, the symmetry of the object in relation to the direction of the sonar beam makes it possible to limit these transformation functions to only translation, homothetic transformation and stretching (Figure 8.8b).

The detection of an object is based on an objective function ϵ which measures the balance between a prototype shape γ_θ and the two-class segmented image x. Using the Gibbs distribution, $\epsilon(\theta, x)$ statistically defines a joint model [MIG 00a]:

$$P_{\Theta,X}(\theta, x) = \frac{1}{Z} \exp\big\{ - \epsilon(\theta, x)\big\} \qquad [8.12]$$

where Θ is a random parameter vector and where Z is a normalization constant. The energy function $\epsilon(\theta, x)$ is made up of two terms:

$$\epsilon(\theta, x) = \underbrace{-\ln\left\{\frac{1}{N_{\gamma_\theta}} \sum_{s\in\gamma_\theta} \Phi_{S'}(s)\right\}}_{\epsilon_c(\theta,x)} \underbrace{-\ln\left\{\frac{1}{N_{\gamma_\theta^\bullet}} \sum_{s\in\gamma_\theta^\bullet} \delta\big(x_s, e_0\big)\right\}}_{\epsilon_r(\theta,x)} \qquad [8.13]$$

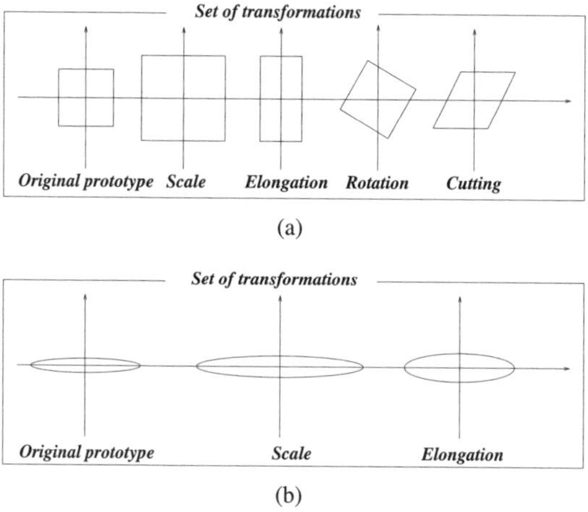

Figure 8.8. *Linear transformations which are considered in the case of a prototype shape which is associated with (a) the shadow of a manufactured cylindrical object and (b) the shadow of a spherical object*

with ϵ_c being the edge energy. ϵ_c forces the deformable prototype into positioning itself onto the edges of each object: $\Phi_{S'}$ is a field of potentials which is generated by the outlines of an object and whose equation can be seen in [8.11]. This equation is the sum of the set of edge pixels which is associated with objects that have been detected after the process of two-class segmentation (this was dealt with in section 8.4). This field of potentials makes it possible to smooth the outlines of an image so that a site that is located near such an edge can possess a value of potential which is close to one. The sum of the set of edge points is taken from all of the points of N_{γ_θ} from the distorted prototype shape γ_θ. ϵ_r is an energy term which uses a region-based approach, and which aim at placing the major part of the pixels from the deformable prototype into the shadow labeled class which is created after the image has been segmented. $\gamma_\theta^\bullet$ and $N_{\gamma_\theta^\bullet}$ represent the set of pixels and the number of pixels that can be found within a region defined by γ_θ. This detection problem has been formalized as an optimization problem that can be taken into account as a MAP (or maximum posterior mode) estimation of θ:

$$\hat{\theta}_{\text{MAP}} = \arg\max_\theta \left\{P_{\Theta/X}(\theta/x)\right\} = \arg\max_\theta \left\{\frac{1}{Z_x}\exp\left(-\epsilon(\theta,x)\right)\right\} \qquad [8.14]$$

$$= \arg\min_\theta \epsilon(\theta,x) \qquad [8.15]$$

where Z_x is the partition function which depends only on x. ϵ is the total energy function that needs to be minimized. This function is minimal whenever the deformable prototype fit exactly with the shape coming from the map created after 2-classes labeling process (section 8.4). The interior of the object exclusively contains sites which stem from the shadow class. As far as the classification phase is concerned, the value of energy which is obtained, $\epsilon(\hat{\theta}_{\text{MAP}}, x)$, is used to measure the membership function of the prototype shape that forms part of the region of the seabed that we want to investigate. $\epsilon(\hat{\theta}_{\text{MAP}}, x)$ is also used to decide whether or not the extracted object belongs to the class of objects of interest. If $\epsilon(\hat{\theta}_{\text{MAP}}, x)$ is smaller than the value of a given fixed threshold then we assume that the estimated prototype shape object fits with the object we have to detect: it is also possible to know where the object can be found in the image thanks to the translation functions contained in $\hat{\theta}_{\text{MAP}}$. If this is not the case, then it is believed that such an object is not present in the image. Due to the detection and classification strategies that are used, it is not possible to use traditional relaxation techniques (such as the gradient descent method) for minimizing the energy function. If such a method were to be used we need to make sure that initialization is properly made, close to the object we are trying to detect. However, this is not possible in practice since the aim of using such a method is to try and find the prototype shape of the shadow in the image without any previous knowledge of the prototype shape's position, angle, distance, slicing or stretching in relation to the object's shadow.

The stochastic optimization method which is based on simulated annealing algorithm [GEM 84] is the best way, nevertheless such methods require huge computing time to minimize the energy functions; even if this method does not require any specific initialization to converge toward the right solution.

These different characteristics which are mentioned above come at a high price since temperature decreasing diagrams end up with extremely long convergence times which sometimes can prevent the simulated annealing algorithm from being used for what it was intended [MIG 99, KER 96]. An alternative approach which has been adopted here consists of estimating the parameters of the energy functions through the use of a genetic algorithm [MIG 99]. This optimization procedure, which simulates the process of natural selection in the living world [GOL 89], can also be used for minimizing a continuous function of dimension L ($L = 6$ or $L = 4$ depending on the symmetry of the object that is being researched). A global minimum is obtained by carrying out successive mutations and crosses on the genes which code the different parameters of the affine transformation that is associated with each prototype of the population. $P_{\Theta,X}(\theta, x)$ from equation [8.12] corresponds to the adaptation function that we adopt to move from one generation to the next. If such an adaptation function is used, the convergence is reached 25 times faster than for other sub-optimal versions of the simulated annealing algorithm used (sub-optimal decrease of temperature law). Figures 8.9, 8.10 and 8.11 show some examples of the results of typical classifications which are obtained on our image base. It should be pointed out that a threshold of 0.2 was obtained after different tests on the different images were carried out. It should also be pointed out that the geometric shapes (i.e. manufactured objects) were correctly detected as follows: the correct part of a wreck (Figure 8.9e), the correct

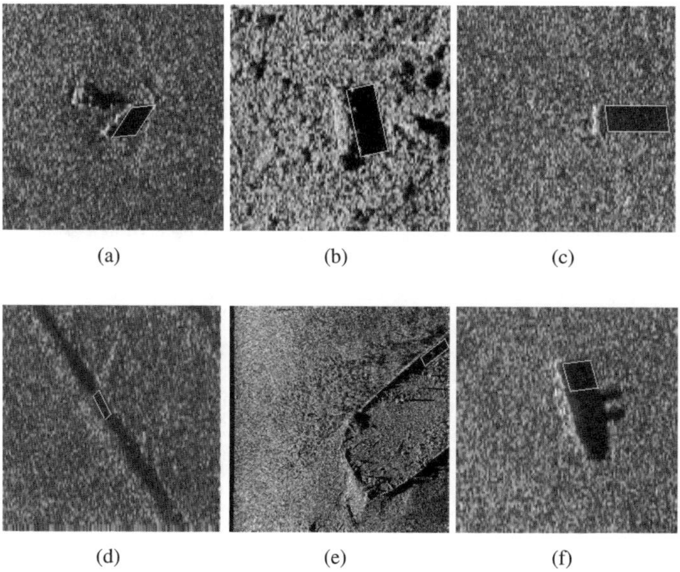

Figure 8.9. *The parallelogram prototype and the low values which are obtained thanks to the use of a genetic algorithm ($\epsilon < 0.2$) enable us to identify the shadows as being shadows which are produced by cylindrical shaped manufactured objects: (a), (b) and (c) different cylindrical objects: $\epsilon = 0.17$, $\epsilon = 0.15$, and $\epsilon = 0.14$ respectively; (d) a pipeline: $\epsilon = 0.12$; (e) a wreck: $\epsilon = 0.15$; (f) a trolley: $\epsilon = 0.2$*

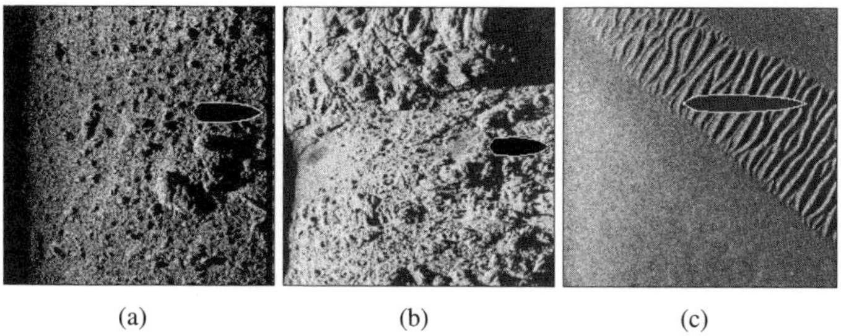

Figure 8.10. *The prototype defined by a cubic B-spline and the low values which are obtained thanks to the use of a genetic algorithm for ϵ ($\epsilon < 0.2$) make it possible to identify the presence of spherical objects: (a) $\epsilon = 0.04$; (b) $\epsilon = 0.15$; (c) $\epsilon = 0.19$. Here the shadow has been added artificially to a real image*

part of a pipe line (Figure 8.9d), a trolley (Figure 8.9f) and various other cylindrical objects (Figure 8.9a, Figure 8.9b and Figure 8.9c). The robustness toward occlusion phenomenon is illustrated for spherical objects in Figure 8.10. Figure 8.11 shows some natural objects and their corresponding values for the objective function ϵ.

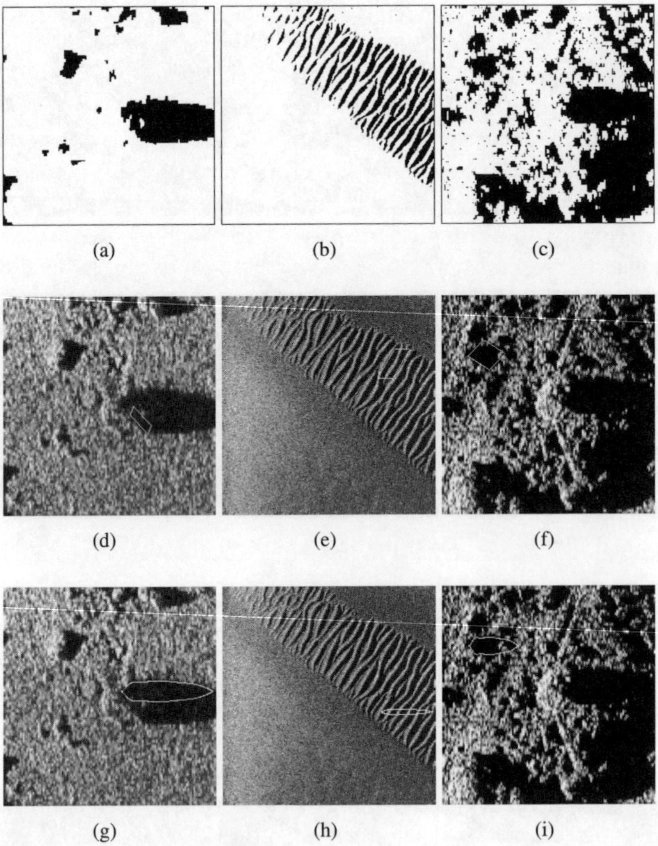

Figure 8.11. *The values ϵ (>0.2) (obtained through genetic exploration) for each prototype can reject the hypothesis of a spherical or cylindrical object being in the image: (a), (b), (c) 2 class Markov segmentation of the sonar images; (d) $\epsilon = 0.40$; (e) $\epsilon = 0.41$; (f) $\epsilon = 0.24$; (g) $\epsilon = 0.55$; (h) $\epsilon = 0.57$; (i) $\epsilon = 0.38$*

8.7. The classification of seabeds

Another application which uses the two-class segmentation maps of a sonar image (see section 8.4) is the automatic segmentation-classification process which is now proposed to segment and classify seabeds. The aim of segmenting the seabeds is to divide the acoustic images into regions according to the regions' acoustic and geological characteristics ($\Lambda = \{w_0 =$ sand, $w_1 =$ ripples, $w_2 =$ dunes, $w_3 =$ pebbles, $w_4 =$ stones$\}$).

We recommend adopting a pattern recognition-based method for such a task. Indeed, such an approach is based on the identification of the detected shadows which are produced by the different types of seabed. To this end the two classes which result from the segmentation of a sonar image are filtered with the aim of extracting the edges of an object's shadow (high-pass filter). The resulting image is divided into sub-images from which a characteristic vector is extracted. For each small image, the characteristic vector includes information relating to the density, directivity, elongation and maximum size of the shadow [MIG 00b]. Once each of these four parameters has been extracted they form a characteristic vector v, which provides information to a fuzzy classification tool which models the fuzzy *a priori* knowledge that we possess on the shadows that are produced by the different types of seabed. This model of *a priori* knowledge simply models the following facts in a simple fashion: the shadows of objects which are found on a seabed that is made up of ripples or of dunes are geometrically in place and are parallel to one another. The shadows of objects which are found on a seabed that is composed of pebbles or of stones produce shapes with a random geometry and tend to possess random orientations. This notion of fuzziness means that each parameter vector v_k that is calculated on every kth small image can associate itself with a particular degree of membership $\mu_{w_i}(v_k)$ $(0 \leq \mu_{w_i}(.) \leq 1)$ for each class of w_i. This degree of membership will then later be used to model the data-driven term; in other words it will be used to model the link that exists between the seabed class and the data which has been observed.

In order to create an accurate segmentation-classification process, the contextual information (i.e. the spatial relationship that exists between the characteristic vectors which are extracted from neighboring windows) is modeled another time with the help of a Markov field. This makes it possible to simultaneously consider the results that are obtained from the fuzzy classification process as well as considering the constraint of spatial homogenity which is enforced upon the desired solution. We consider a couple of random fields (V, W) with $V = \{V_s,\ s \in \mathcal{S}\}$ being the observation field (where v_s corresponds to the characteristic vector which is estimated on the sub-image s considered as being created from V_s), and $W = \{W_s,\ s \in \mathcal{S}\}$ is the label field which is defined on a grid $\mathcal{S}$ of N small images. Each W_s takes values in Λ. The distribution of W, $P_W(w)$ is assumed to be stationary and Markovian. In this way, the distribution of W, $P_W(w)$, leads to the issue of spatial regularization. The segmentation process leading to w is processed as a statistical segmentation problem that is resolved thanks to the use of the Bayesian formula under a MAP criterion:

$$\hat{w} = \arg\max_w P_{W/V}(w/v) = \arg\max_w \left\{P_W(w)\, P_{V/W}(v/w)\right\} \quad [8.16]$$

$$= \arg\min_w U_1(w, v) + U_2(w) \quad [8.17]$$

where $U_1(w, v)$ represents the data-driven term which measures the balance that exists between observation field and the label map, while $U_2(w)$ is an energy term which models the *a priori* information that we possess on the solution that we are looking for. The label which is used for each small image is the result of a compromise between these two equations, i.e. it is the result of the minimization of the global energy function. The neighboring model and the potential functions are the same as those mentioned in section 8.4. The characteristic-driven term $U_1(w,v)$ is defined as being: $U_1(w,v) = -\sum_{s\in\mathcal{S}} \gamma(w_s, v_s)$ with $\gamma(\cdot)$ a value that equals $\mu_{w_s}(v_s)$ if $w_s = \arg\max_{w_j}(\mu_{w_j}(\cdot))$ $(0 \leq j \leq 4)$ and zero if this is not the case [MIG 00b]. The global energy function which is to be minimized can be expressed as follows:

$$U(w,v) = \underbrace{-\sum_{s\in\mathcal{S}} \gamma(w_s, v_s)}_{U_1(w,v)} + \underbrace{\sum_{<s,t>} \beta_{st}(1-\delta(w_s, w_t))}_{U_2(w)}$$

The use of an ICM relaxation algorithm efficiently solves this optimization problem. The initialization process is carried out thanks to the classification process obtained by the fuzzy classification tool. It is not worthwhile developing a multi-grid strategy because the images which have been segmented are of a very high quality.

Figure 8.12 shows two images of seabeds and the classifications that the images have been placed into in relation to our classification system: the class is superimposed onto each image.

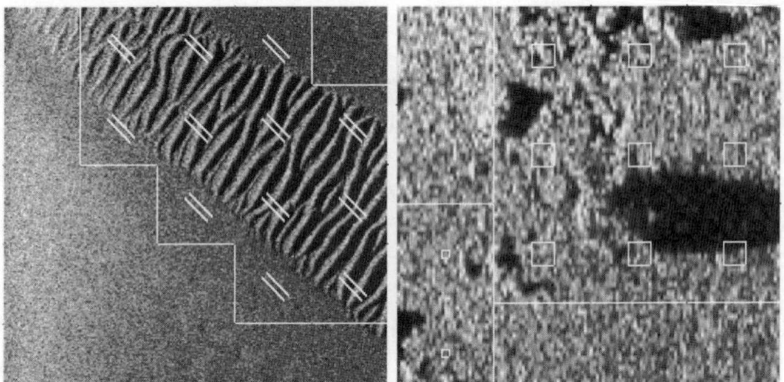

(a) (b)

Figure 8.12. *(a), (b) Images received by a towed side-scan sonar device from an area of a seabed made up of sand (by default), of sand dunes (parallel lines in image a), of pebbles (small squares at the bottom right of image b) or of stones (large squares in image b). The classification results show that the individual classification classes have been correctly detected*

8.8. Conclusion and perspectives

This chapter illustrates different methods of image processing which are made possible thanks to the use of a towed side-scan sonar device. These different methods are all based on the notion of Bayesian inference within a Markov framework, which underlines the strength and the widespread acceptance of this type of modeling. High resolution sonar imagery is an interesting branch of imagery to work in because all of the data is noised by a strong speckle effect which makes the analysis of the data a delicate process. First of all a definition of the causal-in-scale segmentation model was given (section 8.4), which is known as the scale causal multi-grid (SCM) algorithm. The SCM links a Markov field with a causal-in-scale term with the aim of segmenting an image into two classes, shadow and reverberation. *A priori* information can be expressed by using both local and global *a priori* models. The local *a priori* model can be managed at pixel level (two-class segmentation) as well as at the regional level (detection of the echo class, detection of prototype shapes with an almost linear transformation as well as the classification of different types of seabeds). The data-driven model uses either the brightness of the image by estimating the parameters of a noise model (such as those developed by Rayleigh and Weibull) or uses the two-class segmentation map (detection of the echo, classification of objects or of seabeds). The tasks of detection and classification come down to the minimization of an energy function which integrates the *a priori* model within the desired solution that we want to use in order to successfully detect and classify objects. Different optimization strategies have also been developed depending on the energy landscape in which they are used. For reasons linked to processing times, the simulated annealing algorithm is not used and has been replaced by other determinist methods such as the mono-scale ICM algorithm and the multi-scale ICM algorithm whenever access to a good initialization algorithm is possible. It is also possible to use an original stochastic optimization method based on genetic algorithms in cases where there is no access to a good initialization algorithm. It should also be pointed out that each of the four algorithms that are used is in fact an unsupervised algorithm and that they have been used and validated in a significant number of real sonar images. Their robustness and their flexibility make it possible to process large quantities of data. Recently, however, new Markov models have been developed such as the quad tree [LAF 00, PRO 03], the Markov chain and the pair-wise and triplet Markov fields [PIE 03]. These new models make it possible to generalize the analysis of multi-dimensional data such as multi-wavelengths and multi-resolutions, or to analyze the non-stationarity of images. These new models are also opening the way for a whole generation of new *a priori* type models such as the probabilistic atlas of the brain which can be obtained from segmenting magnetic resonance imaging (MRI) images [BRI 06, BRI 08, LEC 08].

8.9. Bibliography

[BES 74] BESAG J., "Spatial interaction and the statistical analysis of lattice systems", *Journal of the Royal Statistical Society*, vol. 36, pp. 192–236, 1974.

[BES 86] BESAG J., "On the statistical analysis of dirty pictures", *Journal of the Royal Statistical Society*, vol. B-48, pp. 259–302, 1986.

[BRI 06] BRICQ S., COLLET C. and ARMSPACH J.-P., "Triplet Markov chain for 3D MRI brain segmentation using a probabilistic atlas", *IEEE Int. Symposium on Biomedical Imaging: from Nano to Macro*, Virginia, USA, 6–9 April 2006.

[BRI 08] BRICQ S., COLLET C. and ARMSPACH J.-P., "Unifying framework for multimodal brain MRI segmentation based on hidden Markov chain", *Medical Image Analysis*, vol. 16, no. 6, pp. 639–652, 2008.

[BUR 78] BURCKHARDT C.B., "Speckle in ultrasound B-mode scans", *IEEE Transactions on Sonics and Ultrasonics*, vol. SU-25, no. 1, pp. 1–6, 1978.

[CEL 86] CELEUX G. and DIEBOLT J., "L'algorithme SEM: un algorithme d'apprentissage probabiliste pour la reconnaissance de mélanges de densités", *Journal of Applied Statistics*, vol. 34, no. 2, 1986.

[COL 98] COLLET C., THOUREL P., MIGNOTTE M., BOUTHEMY P. and PÉREZ P., "Une nouvelle approche en traitement d'images sonar haute résolution: la segmentation markovienne hiérarchique multimodèle", *Traitement du Signal*, vol. 15, no. 3, pp. 231–250, 1998.

[DEM 76] DEMPSTER A., LAIRD N. and RUBIN D., "Maximum likelihood from incomplete data via the EM algorithm", *Royal Statistical Society*, pp. 1–38, 1976.

[DUB 89] DUBES R.C. and JAIN A.K., "Random field models in image analysis", *Journal of Applied Statistics*, vol. 16, no. 2, pp. 131–163, 1989.

[DUG 96] DUGELAY S., GRAFFIGNE C. and AUGUSTIN J., "Deep seafloor characterization with multibeam echosounders by image segmentation using angular acoustic variations", *SPIE'96 International Symposium on Optical Science, Engineering and Instrumentation*, vol. 2847, 1996.

[GEM 84] GEMAN S. and GEMAN D., "Stochastic relaxation, Gibbs distributions and the Bayesian restoration of images", *IEEE Transactions on Pattern Analysis and Machine Intelligence*, vol. PAMI-6, no. 6, pp. 721–741, November 1984.

[GOL 89] GOLDBERG D., *Genetic Algorithm*, Addison Wesley, 1989.

[HEI 94] HEITZ F., PÉREZ P. and BOUTHEMY P., "Multiscale minimisation of global energy functions in some visual recovery problems", *CVGIP: Image Understanding*, vol. 59, no. 1, pp. 125–134, 1994.

[KER 96] KERVRANN C. and HEITZ F., "Statistical model-based segmentation of deformable motion", *Proc. International Conference on Image Processing*, Lausanne, pp. 937–940, 1996.

[LAF 00] LAFERTÉ J.-M., PÉREZ P. and HEITZ F., "Discrete Markov image modeling and inference on the quadtree", *IEEE Transactions on Image Processing*, vol. 9, no. 3, pp. 390–404, 2000.

[LEC 08] LE CAM S., SALZENSTEIN F. and COLLET C., "Fuzzy pairwise Markov chain to segment correlated noisy data", *Signal Processing*, vol. 88, no. 10, pp. 2526–2541, 2008.

[MAU 05] MAUSSANG F., Traitement d'images et fusion de données pour la détection d'objets enfouis en acoustique sous-marine, PhD Thesis, LIS – UMR CNRS 5083, Joseph Fourier University, Grenoble, November 2005.

[MIG 98] MIGNOTTE M., Segmentation d'images sonar par approche markovienne hiérarchique non supervisée et classification d'ombres portées par modèles statistiques, PhD Thesis, University of West Brittany – GTS Laboratory, July 1998.

[MIG 99] MIGNOTTE M., COLLET C., PÉREZ P. and BOUTHEMY P., "Three-class Markovian segmentation of high resolution sonar images", *Journal of Computer Vision and Image Understanding*, vol. 76, no. 3, pp. 191–204, 1999.

[MIG 00a] MIGNOTTE M., COLLET C., PÉREZ P. and BOUTHEMY P., "Hybrid genetic optimization and statistical model-based approach for the classification of shadow shapes in sonar imagery", *IEEE Trans. on Pattern Analysis and Machine Intelligence*, vol. 22, no. 2, pp. 129–141, 2000.

[MIG 00b] MIGNOTTE M., COLLET C., PÉREZ P. and BOUTHEMY P., "Markov random field and fuzzy logic modeling in sonar imagery: application to the classification of underwater-floor", *Journal of Computer Vision and Image Understanding CVIU*, vol. 79, pp. 4–24, 2000.

[MIG 00c] MIGNOTTE M., COLLET C., PÉREZ P. and BOUTHEMY P., "Sonar image segmentation using an unsupervised hierarchical MRF model", *IEEE Trans. on Image Processing*, vol. 9, no. 7, pp. 1–17, 2000.

[PIE 92] PIECZYNSKI W., "Statistical image segmentation", *Machine Graphics and Vision*, vol. 1, no. 2, pp. 261–268, 1992.

[PIE 03] PIECZYNSKI W., "Pairwise Markov chains", *IEEE Transactions on Pattern Analysis and Machine Intelligence*, vol. 25, no. 5, pp. 634–639, 2003.

[PRO 03] PROVOST J., COLLET C., ROSTAING P., PÉREZ P. and BOUTHEMY P., "Hierarchical Markovian segmentation of multispectral images for the reconstruction of water depth maps", *Computer Vision and Image Understanding*, vol. 93, no. 2, pp. 155–174, 2003.

[THO 96] THOUREL P., Segmentation d'images sonar par modélisation Markovienne hiérarchique et analyse multirésolution, PhD Thesis, University of West Brittany – GTS Laboratory, July 1996.

Chapter 9

The Use of Hidden Markov Models for Image Recognition: Learning with Artificial Ants, Genetic Algorithms and Particle Swarm Optimization

9.1. Introduction

Hidden Markov models (HMMs) are statistical tools which are used to model stochastic processes. These models are used in several different scientific domains [CAP 01] such as speech recognition, biology and bioinformatics, image recognition, document organization and indexing as well as the prediction of time series, etc. In order to use these HMMs efficiently, it is necessary to train them to be able to carry out a specific task. In this chapter we will show how this problem of training an HMM to carry out a specific task can be resolved with the help of several different population-based metaheuristics.

In order to explain what HMMs are, we will introduce the principles, notations and main algorithms which make up the theory of hidden Markov models. We will then continue this chapter by introducing the different metaheuristics which have been considered to train HMMs: an evolutionary method, an artificial ant algorithm and a particle swarm technique. We will finish the chapter by analyzing and evaluating six different adaptations of the above metaheuristics that enable us to learn HMMs from data which comes from the images.

Chapter written by Sébastien AUPETIT, Nicolas MONMARCHÉ and Mohamed SLIMANE.

9.2. Hidden Markov models (HMMs)

HMMs have existed for a long time. They were defined in 1913 when A. A. Markov first designed what we know as Markov chains [MAR 13]. The first efficient HMM algorithms and the key principles of HMMs only appeared during the 1960s and 1970s [BAU 67, VIT 67, BAU 72, FOR 73]. Since then, several variants of the original HMMs have been created and a number of HMM applications have been a success. In order to give precise definitions, we need to define what we consider to be a discrete hidden Markov model throughout this chapter.

9.2.1. *Definition*

A discrete hidden Markov model corresponds to the modeling of two stochastic processes. The first process corresponds to a hidden process which is modeled by a discrete Markov chain and the second corresponds to an observed process which depends on the states of the hidden process.

DEFINITION 9.1. *Let* $\mathbb{S}=\{s_1,\ldots,s_N\}$ *be the set of* N *hidden states of the system and let* $S=(S_1,\ldots,S_T)$ *be a tuple of* T *random variables defined on* $\mathbb{S}$. *Let* $\mathbb{V}=\{v_1,\ldots,v_M\}$ *be the set of the* M *symbols which can be emitted by the system and let* $V=(V_1,\ldots,V_T)$ *be a tuple of* T *random variables defined on* $\mathbb{V}$. *It is therefore possible to define a first order discrete hidden Markov model by using the following probabilities:*

– the initialization probability of hidden states: $P(S_1=s_i)$,

– the transition probability between hidden states: $P(S_t=s_j \mid S_{t-1}=s_i)$,

– the emission probability of the symbols for each hidden state: $P(V_t=v_j \mid S_t=s_i)$.

If the hidden Markov model is stationary then the transition probability between hidden states and the emission probability of the symbols for each hidden state are independent of the time $t>1$. For every value of $t>1$, we can define $A=(a_{i,j})_{1\leq i,j\leq N}$ with $a_{i,j}=P(S_t=s_j \mid S_{t-1}=s_i)$, $B=(b_i(j))_{1\leq i\leq N,\,1\leq j\leq M}$ with $b_i(j)=P(V_t=v_j \mid S_t=s_i)$ and $\Pi=(\pi_1,\ldots,\pi_N)'$ with $\pi_i=P(S_1=s_i)$. A first order stationary hidden Markov model denoted by λ is therefore defined by the triplet (A,B,Π). Throughout this chapter we will continue to use the notation λ and we will use the term hidden Markov models (HMM) for the first order stationary HMMs. Let $Q=(q_1,\ldots,q_T)\in\mathbb{S}^T$ be a sequence of hidden states and let $O=(o_1,\ldots,o_T)\in\mathbb{V}^T$ be a sequence of observed symbols. The probability of producing a sequence of hidden states Q and of producing a sequence of observations O according to the HMM λ is written as follows:[1]

$$P\big(V=O,\ S=Q \mid A,B,\Pi\big)=P\big(V=O,\ S=Q \mid \lambda\big)$$

1. Strictly, we should consider the hidden Markov model λ as the realization of a random variable and note $P(V=O,\ S=Q \mid l=\lambda)$. However, to simplify the notations, we have chosen to leave out the random variable which refers to the model.

The conditional probabilities lead to:

$$P(V = O,\ S = Q \mid \lambda) = P(V = O \mid S = Q, \lambda)P(S = Q \mid \lambda)$$
$$= \left(\prod_{t=1}^{T} b_{q_t}(o_t)\right) \cdot \left(\pi_{q_1} \prod_{t=1}^{T-1} a_{q_t, q_{t+1}}\right)$$

From an HMM λ, a sequence of hidden states Q and a sequence of observations O, it is possible to calculate the balance between the model λ and the sequences Q and O. In order to be able to do this, we only need to calculate the probability $P(V = O, S = Q \mid \lambda)$. This probability corresponds to the probability of a sequence of observations O being produced by the model λ after a sequence of hidden states Q has been produced.

Whenever the sequence of states is unknown, it is possible to evaluate the likelihood of a sequence of observations O according to the model λ. The likelihood of a sequence of observations corresponds to the probability $P(V = O|\lambda)$ that the sequence of observations has been produced by the model when considering all the possible sequences of hidden states. The following formula is therefore validated:

$$P(V = O \mid \lambda) = \sum_{Q \in \mathbb{S}^T} P(V = O,\ S = Q \mid \lambda)$$

Whenever HMMs are used, three fundamental issues need to be resolved (for the model λ):

– evaluating the likelihood of $P(V = O|\lambda)$ for a sequence of observations O according to the model λ. This probability is calculated using the forward or backward algorithm with a complexity level of $O(N^2T)$ [RAB 89];

– determining the sequence of hidden states Q^* which has most likely been monitored with the aim of producing the sequence of observations O: the sequence of hidden states Q^*, which is defined by the equation below, is determined by Viterbi's algorithm with a complexity level of $O(N^2T)$ [VIT 67];

$$Q^* = \arg\max_{Q \in \mathbb{S}^T} P(V = O,\ S = Q \mid \lambda)$$

– training/learning one or several HMMs from one or several sequences of observations whenever the exact number of hidden states of the HMMs is known. Training the HMMs can be seen as a maximization problem of a criterion with the fact that the matrices A, B and Π are stochastics. A number of different criteria exist which can be used.

9.2.2. *The criteria used in programming hidden Markov models*

In the following, an optimal model for a given criterion is noted as λ^* and is used to represent the set of HMMs that exist for a given number of hidden states N and a given number of symbols M. O is the sequence of observations that is to be learned. During our research, we found that five different types of training criteria exist:

– the first type of criteria is the maximization of the likelihood. The aim of these criteria is to find the HMM λ^* which validates the equation:

$$\lambda^* = \arg\max_{\lambda \in \Lambda} P(V = O|\lambda)$$

However, no specific or general methods exist for this type of criteria. Nevertheless, the Baum-Welch (BW) algorithm [BAU 67] and the gradient descent algorithm [KAP 98, GAN 99] make it possible, from an initial HMM, to improve it according to the criterion. These algorithms converge and lead to the creation of local optima of the criterion;

– the second type of criteria is the maximization of *a posteriori* probabilities. The aim of these criteria is to determine the HMM λ^* which validates the equation:

$$\lambda^* = \arg\max_{\lambda \in \Lambda} P(\lambda|V = O)$$

This type of criteria, which is associated with the Bayesian decision theory [BER 98], can be divided into a number of smaller criteria depending on the aspects of the decision theory that are being investigated. The simplest forms of this criterion lead to maximum likelihood problems. However, the more complex forms of this criterion do not allow for this possibility. For the more complex forms it is possible to use a gradient descent algorithm in order to investigate such issues. In all cases, the resulting models are local optima of the criteria;

– The third type of criteria is the maximization of mutual information. The aim of these criteria is to simultaneously optimize several HMMs with the aim of maximizing the discrimination power of each HMM. This type of criteria has been used on several occasions in several different forms [RAB 89, SCH 97, GIU 02, VER 04]. One of the solutions that is available and which can be used to maximize these criteria is the use of a gradient descent algorithm. Once again the results which are obtained are local optima of the criteria.

– The fourth type of criteria is the minimization of classification errors of the observation sequences. Several forms of these criteria have been used [LJO 90, GAN 99, SAU 00]. In the three previous types of criteria, the criteria are non-derivable and they are also not continuous. In order to maximize the three previous types of criteria, the criteria are usually approximated with derivable functions. Gradient descent techniques are then used.

– The fifth and final type of criteria is segmental k-means criteria. The aim of these criteria is to determine the HMM λ^* which validates the equation:

$$\lambda^* = \arg\max_{\lambda \in \Lambda} P\big(V = O,\ S = Q^*_\lambda \mid \lambda\big)$$

Q^*_λ is the sequence of hidden states which is obtained after the Viterbi algorithm [VIT 67] has been applied to the HMM λ. As far as segmental k-means is concerned, the aim is to find the HMM which possesses the sequence of hidden states which has been monitored the most. One of the main characteristics of this criterion is that it is neither derivable nor continuous. One solution consists of approximating the criterion by applying a derivable or at least a continuous function to it; however developing such a solution for the moment is a difficult task. It is possible, however, to use the segmental k-means algorithm [JUA 90] to partially maximize this criterion. The way in which this algorithm functions is similar to the way in which the Baum-Welch algorithm functions for maximum likelihood. The segmental k-means algorithm iteratively improves an original HMM and also makes it possible to find a local optimum of the criterion.

We have noticed that numerous criteria can be considered when it comes to training HMM. The criteria which have been described in this section are not the only criteria that can be used but they are the ones that are used the most often. For the majority of these criteria there is at least one corresponding algorithm that exists and which can be used in order to maximize the criteria. These algorithms, which are applied to an initial model, lead to the creation of a new model which in turn improves the value of the said criterion. This approach makes it possible to calculate a local optimum of the criterion in question, but not a global optimum. In certain cases, a local optimum is sufficient, but this is not always the case. It is therefore necessary to be able to find optimal models, or at least find models which resemble, from the point of view of the criterion t an optimal solution of a criterion. One possibility consists of using metaheuristics [DRE 03] with the aim of investigating and exploring HMMs.

9.3. Using metaheuristics to learn HMMs

The aim of this section is to describe the principal metaheuristics that are used to train HMMs. Before we introduce the different metaheuristics, we believe that it is a good idea to introduce the three different types of solution spaces which have been used up until now.

9.3.1. *The different types of solution spaces used for the training of HMMs*

It is possible to consider three types of solution spaces which can be used to train HMMs [AUP 05a]: Λ, $\mathbb{S}^T$ and Ω. In order to describe the different solution spaces, we consider HMMs possessing N hidden states and M symbols, and sequences of observations O with a length of T.

9.3.1.1. *Λ solution space*

The Λ solution space is the solution space which is the most commonly used for training HMMs. This type of solution space corresponds to the set of stochastic matrix triplets (A, B, Π) which define a HMM. Λ is therefore isomorphic to the Cartesian product $\mathbb{G}_N \times (\mathbb{G}_N)^N \times (\mathbb{G}_M)^N$ where $\mathbb{G}_K$ is the set of all stochastic vectors with K dimensions. $\mathbb{G}_K$ possesses the fundamental property of being convex which implies that Λ is also convex.

9.3.1.2. $\mathbb{S}^T$ *solution space*

The $\mathbb{S}^T$ solution space corresponds to the set of hidden state sequences where each sequence is of length T. The fundamental properties of this type of solution space are that it is discrete, finite and is composed of N^T elements. An HMM is defined by its triplet of stochastic matrices, then any given solution of $\mathbb{S}^T$ cannot be directly used as an HMM. Instead, a labeled learning algorithm is used in order to transform the hidden state sequence into an HMM. The labeled learning algorithm estimates probabilities by computing frequencies. $\gamma(Q)$ is the model which is created after the labeled learning algorithm has been applied to the sequence of observations O and to the sequence of states $Q \in \mathbb{S}^T$. The set $\gamma(\mathbb{S}^T) = \{\gamma(Q) \mid Q \in \mathbb{S}^T\}$ is a finite subset of Λ. The function γ is neither injective, nor surjective and as a consequence it is not possible to use an algorithm such as the Baum-Welch algorithm with this type of solution.

9.3.1.3. Ω *solution space*

The Ω solution space was initially defined with the aim of providing a vector space structure for the training of HMMs [AUP 05a]. With this in mind, we consider the set $\mathbb{G}_K$ of stochastic vectors with K dimensions. We define $\mathbb{G}_K^*$ as a subset of stochastic vectors with dimension K. None of the components of this subset have a value of zero, in other words $\mathbb{G}_K^* = \{x \in \mathbb{G}_K \mid \forall i = 1 \dots K,\ x_i > 0\}$. Therefore $r_K : \mathbb{R}^K \mapsto \mathbb{R}^K$ is a regularization function which is defined on $\mathbb{R}^K$ by the following equation:

$$r_K(x)_i = x_i - \max_{j=1\dots K} x_j$$

where $r_K(x)_i$ is the ith component of the vector $r_K(x)$.

We define $\Omega_K = r_K(\mathbb{R}^K) = \{x \in \mathbb{R}^K \mid r_K(x) = x\}$ and three symmetric operators $\oplus_K : \Omega_K \times \Omega_K \mapsto \Omega_K$, $\odot_K : \mathbb{R} \times \Omega_K \mapsto \Omega_K$ and $\ominus_K : \Omega_K \times \Omega_K \mapsto \Omega_K$ so that for all values $(x, y) \in (\Omega_K)^2$ and $c \in \mathbb{R}$:

$$x \oplus_K y = y \oplus_K x = r_K(x + y)$$

$$c \odot_K x = x \odot_K c = r_K(c \cdot x)$$

$$x \ominus_K y = y \ominus_K x = r_K(x - y) = x \oplus \big(-1 \odot_K y \big).$$

Then $(\Omega_K, \oplus_K, \odot_K)$ is a vector space. Let $\psi_K : \mathbb{G}_K^* \mapsto \Omega_K$ and $\phi_K : \Omega_K \mapsto \mathbb{G}_K^*$ two operators. These operators can transform the elements of $\mathbb{G}_K^*$ into elements of Ω_K, and vice-versa with the help of the equations below (for all values of $x \in \mathbb{G}_K^*$ and for all values $y \in \Omega_K$):

$$\psi_K(x)_i = \ln x_i - \max_{j=1...K} \ln x_j,$$

$$\phi_K(y)_i = \frac{\exp y_i}{\sum_{j=1}^{K} \exp y_j}.$$

We define Ω as $\Omega = \Omega_N \times (\Omega_N)^N \times (\Omega_M)^N$ and Λ^* as $\mathbb{G}_N^* \times (\mathbb{G}_N^*)^N \times (\mathbb{G}_M^*)^N$. By generalizing the operators $\oplus_K$, $\odot_K$, $\ominus_K$, ψ_K and ϕ_K to the Cartesian products Ω and Λ^*, and by removing the K index, we can prove that $(\Omega, \oplus, \odot)$ is a vector space, and that $\psi(\Lambda^*) = \Omega$ and $\phi(\Omega) = \Lambda^*$. It is also important to note that $\Lambda^* \subset \Lambda$.

9.3.2. *The metaheuristics used for the training of the HMMs*

The six main types of metaheuristics that have been adapted to the HMM training problem are: simulated annealing [KIR 83], Tabu search [GLO 86, GLO 89a, GLO 89b, HER 92, GLO 97], genetic algorithms [HOL 75, GOL 89], population-based incremental learning [BUL 94, BUL 95], ant-colony-based optimization algorithms such as the API algorithm (API) [MON 00a, MON 00b] and particle swarm optimization (PSO) [KEN 95, CLE 05].

As mentioned earlier in this chapter numerous metaheuristics have been used to train HMMs. When some of the metaheuristics were applied to the HMMs it led to the creation of several adaptations of these metaheuristics (see Table 9.1). These adaptations were not all created with the aim of maximizing the same criterion[2] and they do not explore nor investigate the same solution space (see Table 9.1).

Algorithm/solution	Λ	$\mathbb{S}^T$	$(\Omega, \oplus, \odot)$
Simulated annealing	[PAU 85]	[HAM 96]	
Tabu search	[CHE 04]		
Genetic algorithm	[SLI 99, THO 02]	[AUP 05a]	
Incremental population-based learning	[MAX 99]		
API	[MON 00a]	[AUP 05a]	[AUP 05c]
Particle swarm optimization	[RAS 03]		[AUP 05b]

Table 9.1. *Adaptations of the metaheuristics in relation to the three different solution spaces used for training HMM*

2. The criterion of maximum likelihood is, nevertheless, the criterion that is used the most.

The use of these metaheuristics raises the issue of size: we do not know which metaheuristic is the most efficient. The metaheuristics have not been compared and no normalized data set has been created which would make it possible to compare the results of performance studies. Even if the data sets were normalized we would still have to deal with the issue of understanding exactly how the algorithms are parameterized. The parameters of the algorithms often correspond to a default choice (or to an empirical choice) rather than corresponding to the best parameters that could make up the algorithm, or at least to the parameters which guarantee a high quality solution. In the next section we will go into more detail about the six adaptations of the generic metaheuristics as well as setting them and finally comparing them with the aim of overcoming some of the problems mentioned in this section. In order to make this possible, we will take the criterion of maximum likelihood as well as the data which comes from the images into consideration.

9.4. Description, parameter setting and evaluation of the six approaches that are used to train HMMs

The algorithms which have been used in this study all consider the maximum likelihood criterion. The six different approaches are the result of the adaptation to the problem of the genetic algorithms, of the API algorithm and of the particle swarm optimization algorithm.

9.4.1. *Genetic algorithms*

The genetic algorithm [HOL 75, GOL 89] which we have considered in this study is the one that has also been described in [AUP 05a] and can also be seen in Figure 9.1. This algorithm is the result of adding an additional mutation and/or optimization phase of the parents to the GHOSP algorithm which is described in [SLI 99]. For the purpose of our study we have taken two different adaptations of the genetic algorithm into consideration.

The first adaptation of the genetic algorithm uses solutions in Λ. We have called this adaptation of the genetic algorithm AG-A. The chromosome which corresponds to the $\lambda = (A, B, \Pi)$ HMM is the matrix $(\Pi\ A\ B)$ which is obtained by the concatenation of the three stochastic matrices that describe the HMM. The selection operator is an elitist selection operator. The crossover operator is a classical one point (1X) crossover. It randomly chooses a horizontal breaking point within the model (see Figure 9.2). The mutation operator is applied to each coefficient with a probability of p_{mut}. Whenever the mutation operator is applied to a coefficient h, the mutation operator replaces the coefficient h with the value $(1 - \theta)h$ and also adds the quantity θh to another coefficient. This other coefficient is chosen at random and can be found on the same line in the same stochastic matrix. For each mutation the coefficient θ is a number which is uniformly chosen from the interval $[0; 1]$. The optimization operator then takes $\mathcal{N}_{\text{BW}}$ iterations from the Baum-Welch algorithm and applies them to the individual (person or child) that needs to be optimized.

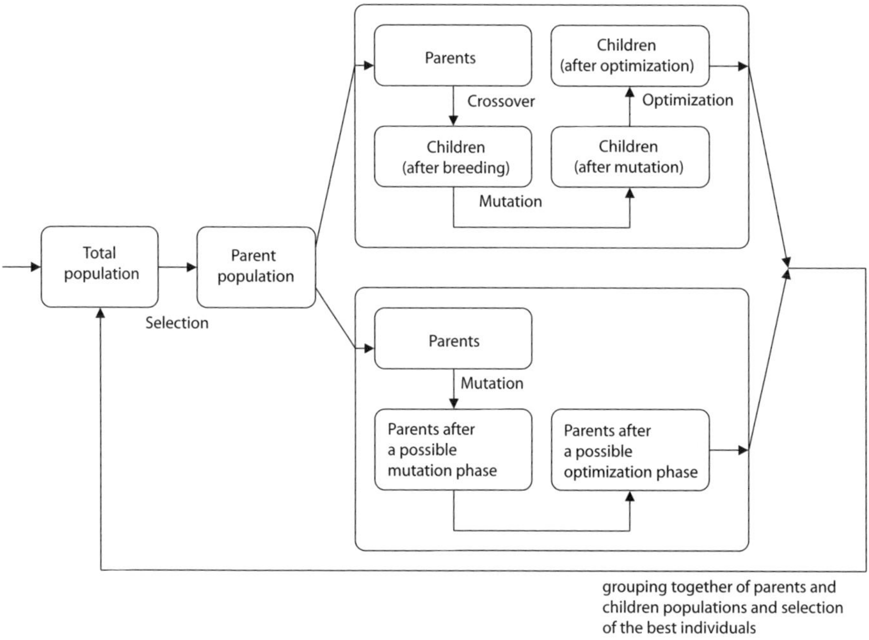

Figure 9.1. *The principle of the genetic algorithm used in programming the HMMs*

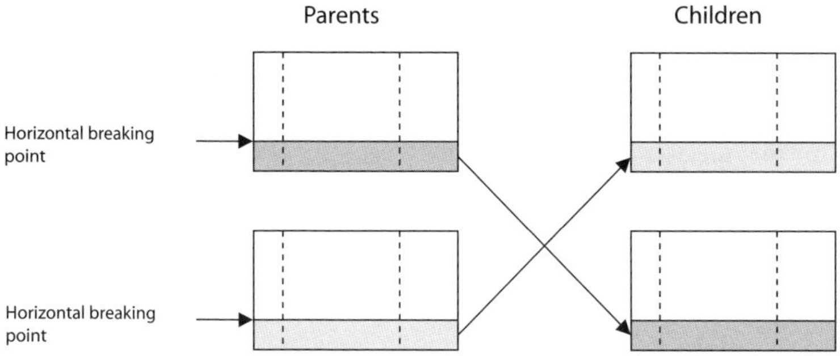

Figure 9.2. *The principle of the crossover operator (1X) which can be found in the AG-A adaptation of the genetic algorithm*

The second adaptation of the genetic algorithm uses the $\mathbb{S}^T$ solution space in order to find an HMM. We have called this adaptation of the genetic algorithm AG-B. The individuals are represented by a sequence of hidden states which label the sequence of observations that need to be learned. The corresponding HMM is found thanks to the use of a labeled learning algorithm (for the statistic estimation of the probabilities). The selection operator which is used is once again the elitist selection operator. The crossover operator corresponds to the classical 1X operator of genetic algorithms that use binary representation. The mutation operator modifies each element from the sequence of states with a probability of p_{mut}. The modification of an element from the sequence involves replacing a state in the solution. The process of optimizing the parents and the children is not carried out because of the non-bijectivity of the labeled learning operator.

The parameters of the two algorithms are as follows: $\mathcal{N}$ (the size of the population), p_{mut} (the probability of mutation), *MutateParents* (do we apply the mutation operator to the parent population?), *OptimizeParents* (do we apply the optimization operator to the parent population?), and $\mathcal{N}_{\text{BW}}$ (the number of iterations which is taken from the Baum-Welch algorithm).

9.4.2. *The API algorithm*

The API algorithm [MON 00a, MON 00b] is an optimization metaheuristic which was inspired by the way in which a population of primitive ants (*Pachycondyla apicalis*) feed. Whenever the ants' prey is discovered, this species of ant has the ability to memorize the site at which the prey has been found (the hunting site). The next time the ants leave the nest, they go back to the last site they visited as this is the last one they remember. If, after a number of several different attempts (named as local patience), the ants are unable to find any new prey on the site they give up on the site and do not use it in the future. From time to time the nest is moved to a hunting site. The result of adapting these principles to the global optimization problem has been called the API algorithm (see Algorithm 9.1).

The experiments described in [AUP 05a] show that factors such as the size of the ant colony and the memory capacity of the ants are anti-correlated. As a result of this, we assume that the size of the ant's memory capacity is one.

In this chapter we deal with three adaptations of the API algorithm. In order to specify the three different adaptations of the algorithm a definition of the initialization operator which refers to the initial position of the nest has been given and we also define the exploration operator.

The first adaptation [MON 00a] of the API algorithm is known as API-A, and the aim of this adaptation of the algorithm is to explore the Λ solution space. The choice of the initial position of the nest is obtained by randomly choosing an HMM within the set Λ. The exploration operator from an existing solution depends on one particular parameter: the amplitude. If we take $\mathcal{A}$ as the amplitude, then the exploration operator applies the function $AM_{\mathcal{A}}$ to the coefficients of the model:

$$AM_{\mathcal{A}}(x) = \begin{cases} -v & \text{if } v < 0 \\ 2 - v & \text{if } v > 1 \\ v & \text{if not} \end{cases} \quad \text{and } v = x + \mathcal{A} \cdot \left(2\mathcal{U}\left([0, 1[\right) - 1\right)$$

```
Randomly choose the initial position of the nest
The memory of the ants is assumed as being empty
While there are still iterations that need to be executed do
   For each of the ants do
      If the ant has not chosen all of its hunting sites Then
         The ant creates a new hunting site
      If not
         If the previous solution is unsuccessful Then
            Randomly choose a new site to explore
         If not
            Choose the last site that was explored
         End if
         Find a solution around a new site to explore
         If the new solution is better than the hunting site Then
            Replace the hunting site with this solution
         If not
            If too many unsuccessful attempts Then
               Forget about the hunting site
            End if
         End if
      End if
   End for
   If it is time that the nest should be moved Then
      Move the next to the best solution that has been found
      Empty the memory of the ants
   End if
End while
```

Algorithm 9.1. *The API algorithm*

$\mathcal{U}(X)$ is a random number which is chosen in the set X. $\mathcal{N}_{\text{BW}}$ is the number of iterations taken from the Baum-Welch algorithm. It is possible to apply these iterations to the HMMs which are discovered by the exploration operator.

The second adaptation [AUP 05a] of the API algorithm is known as API-B, and the aim is to explore the $\mathbb{S}^T$ solution space. The selection operator, which chooses the initial position of the nest, uniformly chooses each of the T states of the sequence that exist in $\mathbb{S}$. The local exploration operator which is associated with a solution x, and which has an amplitude of $A \in [0;1]$ modifies the number of states L that exist in the sequence x. The number of states L is calculated in the following way:

$L = \min\{A \cdot T \cdot \mathcal{U}([0;1]), 1\}$. The states to modify are uniformly chosen in the sequence and their values are generated in $\mathbb{S}$. The different positions (a.k.a. models) explored are not optimized due to the non-bijectivity of the labeled learning operator.

The third adaptation [AUP 05a] of the API algorithm is called API-C and the aim of this adaptation of the algorithm is to explore the Ω solution space. The generation operator chooses the nest's initial position uniformly in Λ^*. The exploration operator for a solution $x \in \Omega$, and for an amplitude A chooses a solution with $y = \psi(\mathcal{U}(\Lambda^*))$ by computing:

$$x \oplus \left(\frac{-\ln \mathcal{U}(]A;1[)}{\|y\|_{\max}} \odot y \right)$$

Note that $\|\cdot\|_{\max}$ is the traditional maximum distance ($\max_i |x_i|$).

The parameters of the algorithms are as follows: $\mathcal{N}$ (the size of the ant colony), $\mathcal{A}^i_{\text{site}}$ (the amplitude of exploration around the nest in order to choose a hunting site), $\mathcal{A}^i_{\text{local}}$ (the amplitude of exploration around a hunting site), $\mathcal{T}_{\text{Movement}}$ (the number of iterations that exists between two movements of the nest), $e_{\max}$ (the number of consecutive, unsuccessful attempts to find a hunting site before giving up) and $\mathcal{N}_{\text{BW}}$ (the number of iterations taken from the Baum-Welch algorithm). Two different types of amplitude parameters are also taken into consideration: a set of parameters which is common to all of the ants (known as a homogenous parameters), and a set of parameters which is specific to each ant (known as a heterogenous parameters). As far as the heterogenous parameters are concerned, the following equations are used for the ants $i = 1 \ldots \mathcal{N}$:

$$\mathcal{A}^i_{\text{site}} = 0.01 \left(\frac{1}{0.01}^{\frac{i}{\mathcal{N}}} \right)$$

$$\mathcal{A}^i_{\text{local}} = \mathcal{A}^i_{\text{site}}/10.$$

9.4.3. *Particle swarm optimization*

Particle swarm optimization (PSO) [KEN 95, CLE 05] is a technique that is used to make $\mathcal{N}$ particles move towards a particular set of solutions. At each instant of t, each particle possesses an exact location which is noted as $x_i(t)$, as well as possessing an exact speed of movement which is noted as $v_i(t)$. $x_i^+(t)$ is used to refer to the best position that is ever found by the particle i for a given instant of t. $V_i(t)$ is used to refer to the set of particles which can be found in the neighborhood of the particle i. $\hat{x}_i(t)$ is used to refer to the position of the best particle of $V_i(t)$, in other words:

$$\hat{x}_i(t) = \arg \max_{x_j \in V_i(t)} f\big(\hat{x}_j(t-1)\big).$$

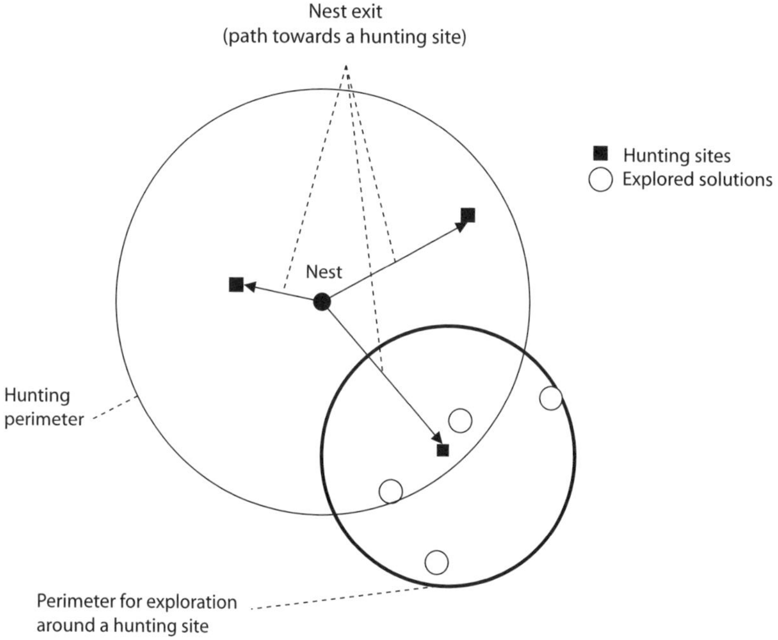

Figure 9.3. *The principle of the exploration techniques of an ant used in the API algorithm*

This equation refers to the maximization of the criterion f. Traditional PSO is controlled by three different coefficients which are ω, c_1 and c_2. These different coefficients all play a role in the equations which refer to the movement of the particles. ω controls the movement's inertia, c_1 controls the cognitive component of the equations and c_2 controls the social component of the equations.

The adaptation of PSO algorithm which we have used in our study aims at finding a HMM model within the Ω solution space. The corresponding algorithm is shown in Algorithm 9.2.

The parameters of the algorithms are as follows: $\mathcal{N}$ (number of particles), ω (parameter of inertia), c_1 (cognitive parameter), c_2 (social parameter), V (size of the particles' neighborhood) and $\mathcal{N}_{\text{BW}}$ (number of iterations taken from the Baum-Welch algorithm). The neighborhood of a particular particle can be compared to a social neighborhood that is divided into rings. Whenever the size of the neighborhood is V for a particular instant of t, the ith particle's neighborhood $V_i(t)$ is said to be constant and equal to V_i. Whenever the particles form a circular shape then V_i is made up of the $V/2$ particles which precede the particle itself and of the $V/2$ particles which come after the particle. Figure 9.4 shows what a size two neighborhood looks like for particles one and five.

For all particles of i **do**
$\mathbf{x}_i(0) = \mathcal{U}(\psi(\Lambda))$
$\mathbf{v}_i(0) = \mathcal{U}(\psi(\Lambda))$
$\mathbf{x}_i^+(0) = \mathbf{x}_i(0)$
End for
While there remains any iterations that need to be carried out **do**
// Movement of particles
For each particle **do**
$\mathbf{x}' = \mathbf{x}_i(t-1) \oplus \mathbf{v}_i(t-1)$ *// movement*
$\mathbf{x}_i(t) = \psi(BW(\phi(\mathbf{x}')))$ *// optimization*
$\mathbf{v}_i(t) = \mathbf{x}_i(t) \ominus \mathbf{x}_i(t-1)$ *// effective movement calculation*
If $P\big(V = O \mid \phi(\mathbf{x}_i(t))\big) > P\big(V = O \mid \phi\big(\mathbf{x}_i^+(t-1)\big)\big)$ **Then**
$\mathbf{x}_i^+(t) = \mathbf{x}_i(t)$
If not
$\mathbf{x}_i^+(t) = \mathbf{x}_i^+(t-1)$
End if
End for
// update speeds
For each particle **do**
Calculate neighborhood $\mathcal{V}_i(t)$ with time t
$\hat{\mathbf{x}}_i(t) = \mathbf{x}_j(t)^+$ with $j = \arg\max_{k \in \mathcal{V}_i(t)} P(V = O|\phi(\mathbf{x}_k^+(t)))$
$\mathbf{v}_i(t) = \omega \odot \mathbf{v}_i(t-1)$
$\oplus\big[c_1 \cdot \mathcal{U}([0,1])\big] \odot \big(\mathbf{x}_i^+(t) \ominus \mathbf{x}_i(t)\big)$
$\oplus\big[c_2 \cdot \mathcal{U}([0,1])\big] \odot \big(\hat{\mathbf{x}}_i(t) \ominus \mathbf{x}_i(t)\big)$
End for
End while

Algorithm 9.2. *Adaptation of the PSO algorithm*

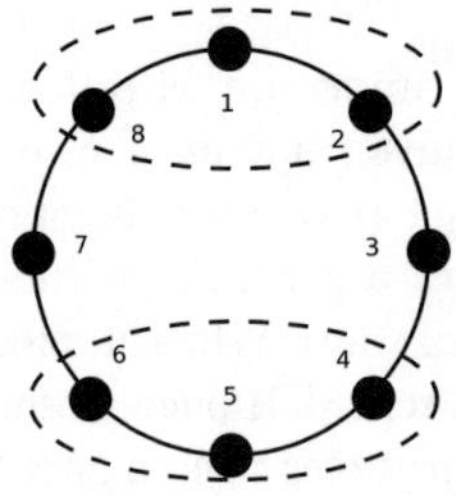

Figure 9.4. *An example of a circular neighborhood used in PSO*

9.4.4. *A behavioral comparison of the metaheuristics*

The algorithms that have been mentioned in this section are based on metaheuristics which use other techniques that are used to explore different solution space. These metaheuristics are made up of a population of agents which interact with one another with the aim of being able to find the best solution possible. These agents interact with one another using different methods and different individuals and they do this at different times. The interactions which take place within a genetic algorithm are carried out by crossing the best quality solutions with one another and this is made possible by exchanging and sharing the genes of the solutions at each iteration of the algorithm. With this in mind, and if the codes of the adapted genes are present, it becomes possible for each iteration that occurs within the algorithm to transfer the best characteristics from the parent solutions to the child. In PSO the interaction of the different agents takes place at each iteration of the algorithm thanks to the updates of the velocity vectors. However, it is not possible to transfer the best characteristics of the particles such as the position of the particle. Instead of this, PSO looks for the best particles that exist within a particle's neighborhood. The API algorithm uses an extremely different approach: the agents only interact with each other whenever the nest is moved. This type of interaction can be referred to as a direct interaction in the sense that the nest is a solution which is made available to each of the ants. Each of these approaches has its advantages and disadvantages. Interacting with each of the iterations, as is the case for the genetic algorithm and for the PSO algorithm, makes it possible to find the correct solutions in the quickest time possible. However, this frequency of interaction runs the risk of all of the agents moving in the same direction, which in turn reduces the diversity of the exploration process. In addition, if the agents find several similar solutions as far as the criterion of optimization is concerned, it becomes highly possible that poles of attraction will be created. This in turn means that the agents will only move between these poles and that any explorations that were carried out before the creation of these poles will be lost. However, it is sometimes necessary to concentrate the exploration efforts on several poles in order to guarantee the creation of an optimal or almost optimal solution. On the other hand, rare interactions can also occur and this is the case with the API algorithm. Such interactions can reduce the effectiveness of the search for an optimal solution since the research that is carried out is blind research, with no knowledge of the remaining solutions that are available. However, this rarity of interactions can also turn out to be advantageous because these rare interactions reduce the number of iterations that the agents need to consider before making a decision as to what pole they will move towards. The decision is made as soon as the nest is moved. This type of quick decision making can also turn out to be damaging in certain cases. In all of the examples that have been mentioned in this section, the moment when a decision is made (the interaction) can be an advantage or a disadvantage depending on the characteristics of the type of solutions involved. Hesitating between several solutions is better when we want to find the maximum optimal solution but this idea of hesitating between several solutions can reduce the speed at which the algorithm converges. The search for an optimal solution is carried out for each of the algorithms by a process of guided random research. In the case of the genetic algorithm, this search for the optimal solution is carried out by the mutation operator and in the case of the API algorithm it is carried out by the local exploration operator. These two different operators play similar roles. In the PSO algorithm the search for an optimal solution is carried out thanks to random contributions which come from the individual components of the algorithm whenever the velocity vectors are updated. In the genetic algorithm the search for an optimum solution is carried out by statistically reinforcing the zones that we are interested in exploring thanks to the use of the elitist selection operator. Each individual solution is part of the sample which is taken around the zone that we are interested in exploring. In the PSO algorithm the solutions are exploited by analyzing the spatial density of the particles which exist around the correct solutions. The exploitation of the solutions in the API algorithm is carried out by using two pieces of information: the position of the nest and the way in which the ants forage for food around a hunting site. These two mechanisms can be seen as a hierarchical reinforcement around the zones which we are interested in exploring. The nest determines the focal point around the zone, and around which the ants establish their hunting sites. The hunting site of an ant determines the zone in which it will search for its prey. The API algorithm has the capacity to do something which the other two algorithms are incapable of doing, which is that the API algorithm is able to forget about the exploration sites which proved to be unsuccessful. This is possible thanks to the API's patience on the exploration sites and to its patience on waiting for the nest to be moved. The ability to do this is a huge advantage for the API algorithm since it can abandon a non-profitable zone and concentrate on another zone with the aim of not wasting any effort.

9.4.5. *Parameter setting of the algorithms*

The explanation of the experimental study that was carried out in relation to how these six approaches can be used to train a HMM is divided into two phases. The first phase involves carrying out a search for parameters on a reduced set of data. After this is complete, we will use the configurations of these parameters to evaluate the performance of the algorithms when they are applied to a bigger set of data.

DEFINITION 9.2. *Let $f_{A,X}$ be the probability distribution of the random variable which measures the performance of the algorithm A for a fixed configuration of X parameters. Let be $\Delta_{v=x} = (*, \dots, *, v = x, *, \dots, *)$ the set of configurations of parameters for which the parameter v has a value x. The probability distribution $f_{A,\Delta_{v=x}}$ is defined as:*

$$f_{A,\Delta_{v=x}} = \frac{1}{|\Delta_{v=x}|} \sum_{X \in \Delta_{v=x}} f_{A,X}$$

A configuration $X = (x_1, \dots, x_K)$ of parameters for a stochastic algorithm A is said to be strong, if for every value x_i, the probability distribution $f_{A,\Delta_{v_i=x_i}}$ has a high mean and a reduced standard deviation.

Let $EG = \{(e_1, g_1), \dots, (e_L, g_L)\}$ be a set of samples so that e_i is a configuration of $\Delta_{v=x}$ parameters and g_i is the measurement of the configuration's performance. The probability distribution $f_{A,\Delta_{v=x}}$ can be approximated by the following probability distribution:

$$\frac{1}{|EG|} \sum_{(e,g)\in EG} \mathcal{N}(g, \sigma)$$

where $\mathcal{N}(m, s)$ is the normal distribution of mean m and of standard-deviation s and considering the following equation:

$$\sigma = 0.1 \cdot \left(\max_{(e,g)\in EG} g - \min_{(e,g)\in EG} g \right)$$

Figure 9.5 shows the probability distributions which are the result of five different mutation rates which have been applied to the first image which can be seen in Figure 9.7 by using the AG-B algorithm.

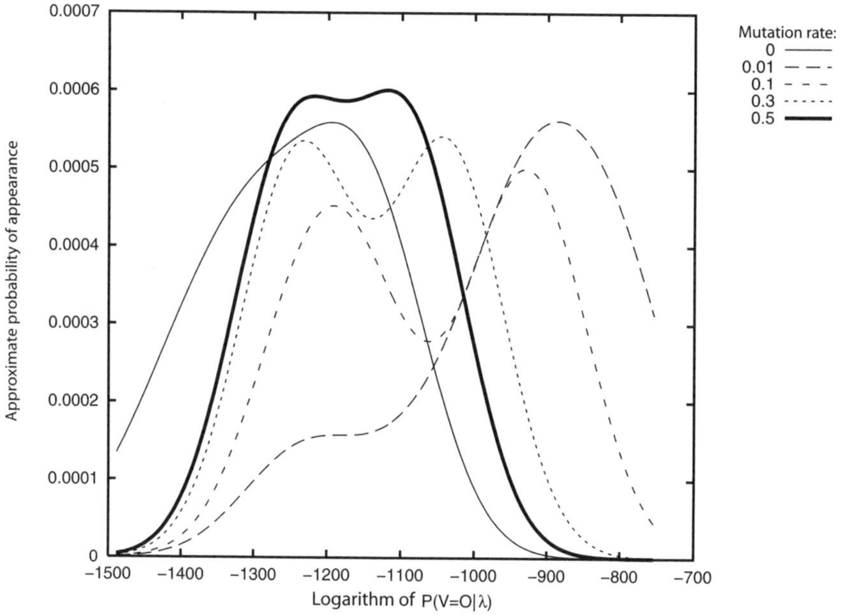

Figure 9.5. *Approximate distributions showing the probabilities of certain performances appearing, taking into consideration five different mutation rates for the AG-B algorithm*

In order to determine the configurations of the strong parameters, we only need to compare the probability distributions. With the aim of carrying out a fair comparison of the algorithms, we have considered the configurations of parameters which provide the same opportunities to all the algorithms. Whenever the Baum-Welch algorithm is not used, the algorithms are able to evaluate approximately 30,000 HMMs (i.e. 30,000 uses of the forward algorithm). Whenever the Baum-Welch algorithm is used, the algorithms are able to carry out approximately 1,000 iterations of the Baum-Welch algorithm (2 or 5 iterations are carried out for each explored solution). The data that we have used for these experiments was obtained from images of faces taken from the ORL database [SAM 94]. The four faces which can be seen in Figure 9.7 are the faces that were used. These images are in 256 gray-level. We have recoded them into 32 gray-level and they have been linearized into gray-level sequences (see Figure 9.6).

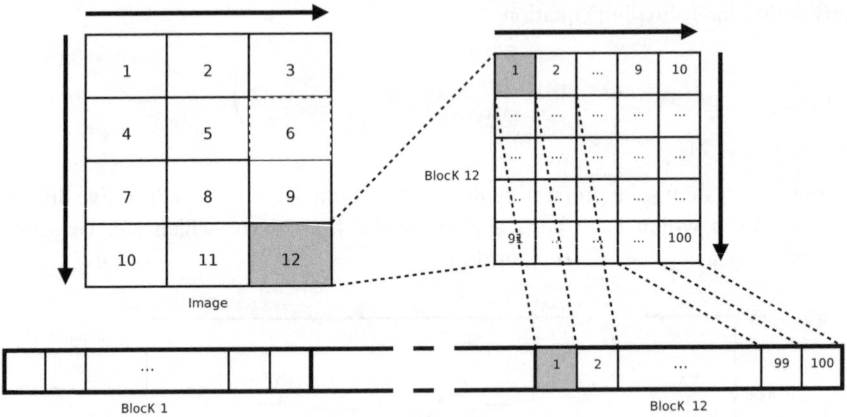

Figure 9.6. *The principle of coding an image into a sequence of observations with blocks of 10×10 pixels*

Figure 9.7. *The first faces of the first four people from the ORL image database [SAM 94]*

Several different parameter settings have been taken into consideration for all of the conditions that have been mentioned above. The parameter settings that we obtained from the four images in Figure 9.7 are given in Table 9.2.

Genetic Algorithms					
	$\mathcal{N}_{\text{BW}}$	$\mathcal{N}$	p_{mut}	*MutateParents*	*OptimizeParents*
AG-A1	0	5	0.01	No	No
AG-A2	2	5	0.3	Yes	Yes
AG-A3	2	20	0.3	No	Yes
AG-A4	5	20	0.5	Yes	Yes
AG-A5	5	5	0.5	No	Yes
AG-B	N/A	5	0.01	No	N/A

API Algorithm							
	$\mathcal{N}_{\text{BW}}$	$\mathcal{N}$	$\mathcal{A}^i_{\text{local}}$	$\mathcal{A}^i_{\text{site}}$	e_{max}	$\mathcal{T}_{\text{movement}}$	Type of parameter
API-A1	0	50	0.1	0.1	1	15	Homogenous
API-A2	2	5	0.1	0.1	4	15	Homogenous
API-A3	5	20	0.1	0.1	4	15	Homogenous
API-A4	0	2	N/A	N/A	5	10	Heterogenous
API-A5	2	5	N/A	N/A	4	15	Heterogenous
API-A6	5	5	N/A	N/A	3	5	Heterogenous
API-B1	0	2	0.2	0.1	1	5	Homogenous
API-B2	0	2	N/A	N/A	1	5	Heterogenous
API-C1	0	2	0.9	0.8	1	5	Homogenous
API-C2	2	20	0.1	0.2	4	5	Homogenous
API-C3	5	50	0.1	0.2	4	15	Homogenous
API-C4	0	2	N/A	N/A	5	4	Heterogenous
API-C5	2	20	N/A	N/A	2	5	Heterogenous
API-C6	5	50	N/A	N/A	1	20	Heterogenous

PSO						
	$\mathcal{N}_{\text{BW}}$	$\mathcal{N}$	ω	c_1	c_2	V
OEP1	0	20	0.8	1.5	2.5	9
OEP2	2	20	0.8	1.5	0	N/C
OEP3	5	50	0.4	0	0.5	6

N/A means non-applicable. N/C means uncharacteristic.

Table 9.2. *The configurations of the parameters*

9.4.6. *Comparing the algorithms' performances*

In order to compare the algorithms with the configurations of the parameters, we decided to use four sets of images with each set of images being made up of between four and ten images. The characteristics of the sets of the images that were used can be seen in Table 9.3.

Database	Content	Dimension (in pixels)	Grey level
[TCD 06]	Flowers	110×125	32
[TEX 06]	Patterns	100×100	2
[AGA 04]	Countryside	100×40	8
[AGA 04]	Cars	$[120; 266] \times [73; 176]$	8

Table 9.3. *Characteristics of the images used for programming the HMMs*

In order to compare the effectiveness of the metaheuristics with a traditional approach we used algorithms Random0, Random2 and Random5. 30,000 HMMs were randomly chosen in the Λ solution space for algorithm Random0. The result of this algorithm is the best model that has been explored. For algorithms Random2 and Random5 we explored between 500 and 200 random HMMs. We then applied two or five iterations from the Baum-Welch algorithm to each random HMMs. We considered HMMs which had 10 and 40 hidden states.

Each image is transformed into a sequence by following the same principles that are used for setting the algorithms. The images are learned several times and the mean likelihood logarithm is computed. $m(I, A)$ is the mean for an image I and for an algorithm A. If we note $\mathbb{A}$ as the set of algorithms that must be compared, then $e(I, A)$ can be defined as follows:

$$e(I, A) = \frac{m(I, A) - \min_{X \in \mathbb{A}} m(I, X)}{\max_{X \in \mathbb{A}} m(I, X) - \min_{X \in \mathbb{A}} m(I, X)}$$

The number $e(I, A)$ measures the efficiency of algorithm A on a scale from zero to one for the image I. The most effective algorithm leads to $e(I, A) = 1$ and the least effective algorithm leads to $e(I, A) = 0$. In order to compare all of the algorithms with one another we suggest that the measurement $\bar{e}(A)$ is used. Let $\mathbb{I}$ be the set of images that are used. Then, we define:

$$\bar{e}(A) = \frac{1}{|\mathbb{I}|} \sum_{I \in \mathbb{I}} e(I, A).$$

Table 9.4 shows the results of the experiment. As we can see, the algorithms are divided into three groups. The first group is made up mainly of those algorithms which do not use the Baum-Welch algorithm. These algorithms tend not to perform as well as those algorithms which do use the Baum-Welch algorithm. We can also note that algorithm API-A1 provides worse statistics than if we were to carry out purely random research. It would seem that there is less chance of finding a good model by carrying out a poorly guided exploration of the different types of solutions than if a completely random exploration of the solutions were to be carried out. This example shows that it is necessary to carry out a comparative study of the metaheuristics if they are to be used to train the HMMs. The second group of algorithms contains those algorithms which have an average performance level (from 44.31% to 85.94%). The final group of algorithms contains the highest performing and the most effective algorithms. As we can see, these algorithms use two or five iterations from the Baum-Welch algorithm (but not all in the same way), and they search for HMMs in the Λ and Ω solution spaces. We can also see that as the performance of the algorithms increases and the lower the variability: their standard deviation decreases.

algorithm A	$\bar{e}(A)$ average	Standard deviation	$\mathcal{N}_{\text{BW}}$	Solution space
API-A1	0.13%	0.53%	0	Λ
Random0	10.86%	5.07%	0	Λ
API-B1	13.07%	9.77%	0	$\mathbb{S}^T$
AG-B	16.48%	12.16%	0	$\mathbb{S}^T$
API-B2	19.05%	12.43%	0	$\mathbb{S}^T$
API-A4	20.38%	7.97%	0	Λ
OEP1	44.31%	12.65%	0	Ω
Random2	66.60%	7.07%	2	Λ
API-C4	78.39%	3.72%	0	Ω
API-A2	82.17%	9.95%	2	Λ
Random5	82.29%	7.03%	5	Λ
API-C1	85.68%	5.60%	0	Ω
AG-A1	89.80%	6.84%	0	Λ
API-A3	90.13%	9.21%	5	Λ
AG-A2	93.84%	5.11%	2	Λ
AG-A4	95.94%	3.77%	5	Λ
AG-A5	97.66%	2.11%	5	Λ
API-C6	98.32%	1.37%	5	Ω
API-C3	98.38%	1.62%	5	Ω
API-C5	98.50%	1.48%	2	Ω
OEP3	98.71%	1.70%	5	Ω
AG-A3	98.72%	1.45%	2	Λ
API-C2	98.88%	1.14%	2	Ω
API-A5	98.89%	1.02%	2	Λ
OEP2	99.34%	0.87%	2	Ω
API-A6	99.74%	0.32%	5	Λ

Table 9.4. *A measurement of the algorithms' efficiency*

9.5. Conclusion

In this chapter we have discussed the problem of learning HMMs using metaheuristics. First of all, we introduced the HMMs and the traditional criteria that were used to train them. We also introduced the different types of solution spaces which are available nowadays, and which can be used to train the HMMs. The fact that there are different types of solution spaces available is a fact that is too often forgotten. The three types of search spaces that we investigated are complementary representations and include: a discrete representation ($\mathbb{S}^T$), a continuous representation (Λ) and a vector-space representation (Ω).

Our experiments have shown that there is less chance of finding a good HMM in the $\mathbb{S}^T$ solution space in comparison to the two other types of solution spaces, Λ and Ω; or at least this is the case with the search methods we used.

The use of metaheuristics for training HMMs has already been investigated and many encouraging results were found. In this chapter we introduced a comparative study which was inspired by three biologically-inspired approaches. Each approach focuses on a particular population of solutions. We have concentrated our comparison on one critical point, a point which can easily be analyzed and critiqued, i.e. the choice of parameters. We have tried to configure our algorithms in as fair a way as possible. However, we must bear in mind that the results and conclusions that we have obtained depend on the particular area of study that we chose to investigate, in other words, the training of sequences created from images whilst considering the maximum likelihood as the training criterion.

Our findings have also shown that a moderate use of a local optimization method (which in our case is the Baum-Welch algorithm) is in fact able to improve the performance of the metaheuristics that we chose to investigate.

The conclusions which can be drawn from this experimental study have led to new questions that need to be answered: is it possible to apply the results that we have found to other domains of study? Is it possible to apply the results to other criteria and to other metaheuristics? We strongly believe that it is best to act carefully when trying to answer such questions and to avoid too much generalization. We hope that this study can act as a framework for any future and more rigorous comparative studies.

9.6. Bibliography

[AGA 04] AGARWAL S., AWAN A. and ROTH D., "Learning to detect objects in images via a sparse, part-based representation", *IEEE Transactions on Pattern Analysis and Machine Intelligence*, vol. 26, no. 11, pp. 1475–1490, 2004.

[AUP 05a] AUPETIT S., Contributions aux modèles de Markov cachés: métaheuristiques d'apprentissage, nouveaux modèles et visualisation de dissimilarité, PhD Thesis, Department of Information Technology, University of Tours, Tours, France, 30th November 2005.

[AUP 05b] AUPETIT S., MONMARCHÉ N. and SLIMANE M., "Apprentissage de modèles de Markov cachés par essaim particulaire", BILLAUT J.-C. and ESSWEIN C. (Eds.), *ROADEF'05: 6th Conference of the French Society of Operational Research and Decision Theory*, vol. 1, pp. 375–391, François Rabelais University Press, Tours, France, 2005.

[AUP 05c] AUPETIT S., MONMARCHÉ N., SLIMANE M. and LIARDET S., "An exponential representation in the API algorithm for hidden Markov models training", *Proceedings of the 7th International Conference on Artificial Evolution (EA'05)*, Lille, France, October 2005, CD-Rom.

[BAU 67] BAUM L.E. and EAGON J.A., "An inequality with applications to statistical estimation for probabilistic functions of Markov processes to a model for ecology", *Bull American Mathematical Society*, vol. 73, pp. 360–363, 1967.

[BAU 72] BAUM L.E., "An inequality and associated maximisation technique in statistical estimation for probabilistic functions of Markov processes", *Inequalities*, vol. 3, pp. 1–8, 1972.

[BER 98] BERTHOLD M. and HAND D.J. (Eds.), *Intelligent Data Analysis: An Introduction*, Springer-Verlag, 1998.

[BUL 94] BULAJA S., Population-based incremental learning: a method for integrating genetic search based function optimization and competitive learning, Report no. CMU-CS-94-163, Carnegie Mellon University, 1994.

[BUL 95] BULAJA S. and CARUANA R., "Removing the genetics from the standard genetic algorithm", in PRIEDITIS A. and RUSSEL S. (Eds.), *The International Conference on Machine Learning (ML'95)*, San Mateo, CA, Morgan Kaufman Publishers, pp. 38–46, 1995.

[CAP 01] CAPPÉ O., "Ten years of HMMs", http://www.tsi.enst.fr/cappe/docs/hmmbib.html, March 2001.

[CHE 04] CHEN T.-Y., MEI X.-D., PAN J.-S. and SUN S.-H., "Optimization of HMM by the Tabu search algorithm", *Journal Information Science and Engineering*, vol. 20, no. 5, pp. 949–957, 2004.

[CLE 05] CLERC M., *L'optimisation par Essaims Particulaires: Versions Paramétriques et Adaptatives*, Hermes Science, Paris, 2005.

[DRE 03] DREO J., PETROWSKI A., SIARRY P. and TAILLARD E., *Métaheuristiques pour l'Optimisation Difficile*, Eyrolles, Paris, 2003.

[DUG 96] DUGAD R. and DESAI U.B., A tutorial on hidden Markov models, Report no. SPANN-96.1, Indian Institute of Technology, Bombay, India, May 1996.

[FOR 73] FORNEY JR. G.D., "The Viterbi algorithm", *Proceedings of IEEE*, vol. 61, pp. 268–278, March 1973.

[GAN 99] GANAPATHIRAJU A., "Discriminative techniques in hidden Markov models", Course paper, 1999.

[GIU 02] GIURGIU M., "Maximization of mutual information for training hidden Markov models in speech recognition", *3rd COST #276 Workshop*, Budapest, Hungary, pp. 96–101, October 2002.

[GLO 86] GLOVER F., "Future paths for integer programming and links to artificial intelligence", *Computers and Operations Research*, vol. 13, pp. 533–549, 1986.

[GLO 89a] GLOVER F., "Tabu search – part I", *ORSA Journal on Computing*, vol. 1, no. 3, pp. 190–206, 1989.

[GLO 89b] GLOVER F., "Tabu search – part II", *ORSA Journal on Computing*, vol. 2, no. 1, pp. 4–32, 1989.

[GLO 97] GLOVER F. and LAGUNA M., *Tabu Search*, Kluwer Academic Publishers, 1997.

[GOL 89] GOLDBERG D.E., *Genetic Algorithms in Search, Optimization and Machine Learning*, Addison-Wesley, 1989.

[HAM 96] HAMAM Y. and AL ANI T., "Simulated annealing approach for Hidden Markov Models", *4th WG-7.6 Working Conference on Optimization-Based Computer-Aided Modeling and Design, ESIEE, France*, May 1996.

[HER 92] HERTZ A., TAILLARD E. and DE WERRA D., "A Tutorial on tabu search", *Proceedings of Giornate di Lavoro AIRO'95 (Enterprise Systems: Management of Technological and Organizational Changes)*, pp. 13–24, 1992.

[HOL 75] HOLLAND J.H., *Adaptation in Natural and Artificial Systems*, University of Michigan Press: Ann Arbor, MI, 1975.

[JUA 90] JUANG B.-H. and RABINER L.R., "The segmental k-means algorithm for estimating parameters of hidden Markov models", *IEEE Transactions on Acoustics, Speech and Signal Processing*, vol. 38, no. 9, pp. 1639–1641, 1990.

[KAP 98] KAPADIA S., Discriminative training of hidden Markov models, PhD Thesis, Downing College, University of Cambridge, 18 March 1998.

[KEN 95] KENNEDY J. and EBERHART R., "Particle swarm optimization", *Proceedings of the IEEE International Joint Conference on Neural Networks*, vol. 4, IEEE, pp. 1942–1948, 1995.

[KIR 83] KIRKPATRICK S., GELATT C.D. and VECCHI M.P., "Optimizing by simulated annealing", *Science*, vol. 220, no. 4598, pp. 671–680, 13 May 1983.

[LJO 90] LJOLJE A., EPHRAIM Y. and RABINER L.R., "Estimation of hidden Markov model parameters by minimizing empirical error rate", *IEEE International Conference on Acoustic, Speech, Signal Processing*, Albuquerque, pp. 709–712, April 1990.

[MAR 13] MARKOV A.A., "An example of statistical investigation in the text of "Eugene Onyegin" illustrating coupling of "tests" in chains", *Proceedings of Academic Scienctific St. Petersburg*, VI, pp. 153–162, 1913.

[MAX 99] MAXWELL B. and ANDERSON S., "Training hidden Markov models using population-based learning", BANZHAF W., DAIDA J., EIBEN A.E., GARZON M.H., HONAVAR V., JAKIELA M. and SMITH R.E. (Eds.), *Proceedings of the Genetic and Evolutionary Computation Conference (GECCO'99)*, vol. 1, Orlando, Florida, USA, Morgan Kaufmann, p. 944, 1999.

[MON 00a] MONMARCHÉ N., Algorithmes de fourmis artificielles: applications à la classification et à l'optimisation, PhD Thesis, Department of Information Technology, University of Tours, 20th December 2000.

[MON 00b] MONMARCHÉ N., VENTURINI G. and SLIMANE M., "On how *Pachycondyla apicalis* ants suggest a new search algorithm", *Future Generation Computer Systems*, vol. 16, no. 8, pp. 937–946, 2000.

[PAU 85] PAUL D.B., "Training of HMM recognizers by simulated annealing", *Proceedings of IEEE International Conference on Acoustics, Speech and Signal Processing*, pp. 13–16, 1985.

[RAB 89] RABINER L.R., "A tutorial on hidden Markov models and selected applications in speech recognition", *Proceedings of the IEEE*, vol. 77, no. 2, pp. 257–286, 1989.

[RAS 03] RASMUSSEN T.K. and KRINK T., "Improved hidden Markov model training for multiple sequence alignment by a particle swarm optimization – evolutionary algorithm hybrid", *BioSystems*, vol. 72, pp. 5–17, 2003.

[SAM 94] SAMARIA F. and HARTER A., "Parameterisation of a stochastic model for human face identification", *IEEE Workshop on Applications of Computer Vision*, Florida, December 1994.

[SAU 00] SAUL L. and RAHIM M., "Maximum likelihood and minimum classification error factor analysis for automatic speech recognition", *IEEE Transactions on Speech and Audio Processing*, vol. 8, no. 2, pp. 115–125, 2000.

[SCH 97] SCHLUTER R., MACHEREY W., KANTHAK S., NEY H. and WELLING L., "Comparison of optimization methods for discriminative training criteria", *EUROSPEECH '97, 5th European Conference on Speech Communication and Technology*, Rhodes, Greece, pp. 15–18, September 1997.

[SLI 99] SLIMANE M., BROUARD T., VENTURINI G. and ASSELIN DE BEAUVILLE J.-P., "Apprentissage non-supervisé d'images par hybridation génétique d'une chaîne de Markov cachée", *Signal Processing*, vol. 16, no. 6, pp. 461–475, 1999.

[TCD 06] "T.C. Design, Free Background Textures, Flowers", http://www.tcdesign.net/free_textures_flowers.htm, January 2006.

[TEX 06] "Textures Unlimited: Black & white textures", http://www.geocities.com/texturesunlimited/blackwhite.html, January 2006.

[THO 02] THOMSEN R., "Evolving the topology of hidden Markov models using evolutionary algorithms", *Proceedings of Parallel Problem Solving from Nature VII (PPSN-2002)*, pp. 861–870, 2002.

[VER 04] VERTANEN K., An overview of discriminative training for speech recognition, Report, University of Cambridge, 2004.

[VIT 67] VITERBI A.J., "Error bounds for convolutionnal codes and asymptotically optimum decoding algorithm", *IEEE Transactions on Information Theory*, vol. 13, pp. 260–269, 1967.

Chapter 10

Biological Metaheuristics for Road Sign Detection

10.1. Introduction

The automatic detection and recognition of traffic signs in road scenes has received particular attention for the past two decades, especially in the context of driver assistance systems design. Aimed at performing in real time, the methods which have been developed until now generally combine a segmentation phase based on color or shape and a classification phase. Despite the excellent performances asserted by certain authors, these algorithms remain sensitive to occultation by other components of the road scene. Moreover, color information is highly dependent on the wear and tear of traffic signs, as well as on the illumination conditions. With this in mind, it is difficult to ensure that the results of the segmentation phase are completely reliable [GAV 99]. Characteristics linked to the gradient of the images are also sensitive to image perturbations. Any error at this stage is indeed comparable to a partial occultation of the sought object: the characteristics of the segmented objects are distorted and the classification process becomes much more difficult.

In this chapter, we consider a different application, namely traffic sign inventory from large road scene databases. Such image databases are created by inspection vehicles driving around the French road network. The images are taken by standard numerical cameras under natural illumination conditions. The main area of interest of these inventories is the French secondary road network, the longest network. On the one hand, difficult situations such as illumination variations, backlights, worn-out

---

Chapter written by Guillaume DUTILLEUX and Pierre CHARBONNIER.

or occulted signs occur more frequently in this context than on the French national road network or on French motorways. On the other hand, the images may be processed off-line, hence the real-time constraint can be relaxed and a more robust technique, belonging to the family of deformable models [GRE 89, AMI 91] can be used. The principle of the method consists of using a mathematical model, which is a prototype of the sought object, and distorting and translating it until it fits the sought object in the analyzed image. The quality of adjustment and the acceptability of the necessary deformation are measured, in the Bayesian framework, by a likelihood and a prior value, respectively. Localizing a particular object of interest then becomes an optimization problem in the maximum *a posteriori* (MAP) framework. The residual value of the optimized function gives an indication of whether the object is present in the image which has been analyzed, or not.

In practice it is quite difficult to deal with such an optimization problem which is characterized by the presence of numerous local minima. Therefore, the optimization should not be dealt with by using traditional numerical methods, such as the conjugate gradient or the simplex method. In this chapter we propose to use biological metaheuristics, i.e. methods inspired from biological mechanisms.

The following sections deal first with the state of the art of automatic road sign detection. The components of the functional to be minimized are defined in the following section. The third section gives a brief introduction to algorithms which have not been dealt with elsewhere in this book. The final section of the chapter proposes a systematic evaluation of the method over a representative image sequence. The individual performances of the different algorithms are also compared.

10.2. Relationship to existing works

In this section, we propose a brief overview of vertical road sign detection algorithms. Generally speaking, it is possible to distinguish between three different categories of methods. The first uses the appearance of the road sign, i.e. the values of the pixels, without any initial segmentation. For example, in [BET 95] a transformed image is obtained by applying an anisotropic scaling and a rotation to a gray-level prototype image, and by translating it onto the analyzed image. A simulated annealing maximizes the normalized correlation between the deformed prototype and the analyzed image, with respect to the five parameters of the geometric transformation. Since the research undertaken by Moghaddam and Pentland [MOG 97], active appearance models have turned out to be quite successful when it comes to detecting and recognizing faces or objects. The reasons for their success are that they make it possible to represent classes of objects and that they also allow us to take into account different variations of the shape, orientation or illumination. In addition, the similarity measures involved in appearance-based methods can be made robust to partial occultation [DAH 04]. However, such methods require a relatively significant amount of processing time.

The second type of methods, possibly the most widespread one, combines a segmentation phase with a classification phase, according to a traditional framework in pattern recognition. A great variety of methods have been proposed. For further information about these methods, the reader should see [LAL 95] and more recent states of the art in e.g. [ESC 04] or [LOY 04]. It is possible to distinguish between two sub-categories of approaches. Approaches which are based on color use, most of the time, color representations that do not depend on the luminance to tackle the problem of illumination variation. Approaches which are based on contour information use either a traditional detection-closing-analysis scheme, or Hough's transform (or one of its variants). Other voting techniques such as Loy's Fast Radial Symmetry Transform [LOY 02] and its extension to the analysis of polygonal objects [LOY 04] have started to be used more recently. A major disadvantage of all of these methods which depend heavily on the quality of segmentation, is that they are very sensitive to occultation as well as to appearance variations.

The third type of methods is the deformable model method, which in some way is an intermediate category between the above-mentioned two families. The principle of deformable models it to adjust a prototype onto the image, as is the case with the first type of methods mentioned above. The object representation involved in deformable models however is simpler; most of the time, only the outline of the object is used. As is the case with the second type of methods, the image which is analyzed is a simplified version of the original image obtained by a rough segmentation process, also called *pre-segmentation*, whose purpose is to extract characteristics that allow shape adjustment. These characteristics may be contaminated by noise or can be incomplete, which is counterbalanced by the use of a model. In [GAV 99] a distance transform [BOR 88] is calculated from the contours extracted from the image. This distance map is then matched with the same representation calculated beforehand within a hierarchy of shape prototypes. The algorithm applies a multi-scale strategy with the aim of reducing the processing time of the segmentation. In [AOY 96] the pre-segmentation phase leads to the creation of an edge map by Laplacian filtering and thresholding. A genetic algorithm is then used to detect speed limit road signs by adjusting a circular prototype. Deformable models were also used to classify silhouettes of vehicles [JOL 96]. The pre-segmentation process consists of computing image gradients and a rough motion analysis from successive images. A functional is defined in a Bayesian framework, which combines an *a priori* on admissible deformations and a likelihood term computed from the pre-segmentation maps. The optimization is carried out using a global algorithm, namely, a simulated annealing. In a similar philosophy, a hybrid method combining a genetic algorithm with a local gradient descent was used to successfully detect shadows in sonar images [MIG 00].

Our research has been inspired from the work produced by [JOL 96, MIG 00] as well as the contemporary work carried out on the recognition of road signs led by De La Escalera *et al.*, whose work is also based on the use of genetic algorithms or simulated annealing [ESC 03, ESC 04].

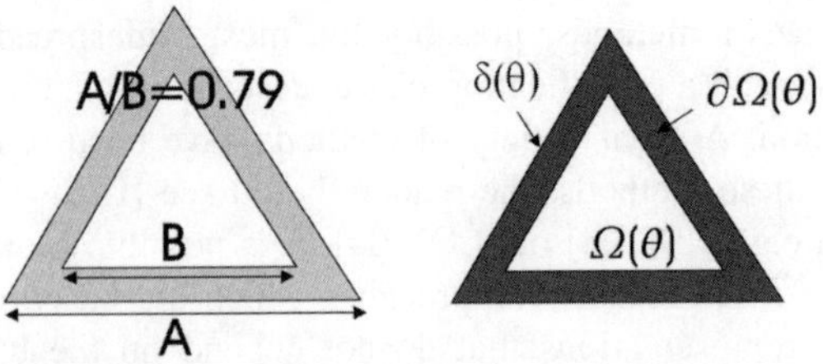

Figure 10.1. *Template and integration domains*

10.3. Template and deformations

The template is tailored to the end application. As far as the detection of danger road signs is concerned, the prototype is created from two nested triangles (see Figure 10.1). The prototype can be deformed and placed onto the image by using linear transformations like in [MIG 00]:

$$\begin{pmatrix} x' \\ y' \end{pmatrix} = \begin{pmatrix} \cos\alpha & -\sin\alpha \\ \sin\alpha & \cos\alpha \end{pmatrix} \begin{pmatrix} 1 & h \\ 0 & 1 \end{pmatrix} \begin{pmatrix} s_1 & 0 \\ 0 & s_2 \end{pmatrix} \begin{pmatrix} x \\ y \end{pmatrix} + \begin{pmatrix} t_x \\ t_y \end{pmatrix} \qquad [10.1]$$

The transformation parameters are grouped together in a vector and from here on will be referred to as Θ. This model can easily be adapted to other polygonal or elliptic shapes so that it is possible to segment the different types of road signs.

10.4. Estimation problem

Following [JOL 96], we express the road sign segmentation task from a given image I in the Bayesian framework as an estimation problem in the sense of maximum *a posteriori* (MAP):

$$\hat{\Theta} = \arg\max_{\Theta} P(\Theta \mid I). \qquad [10.2]$$

Using the Bayes theorem, $P(\Theta \mid I)$ is proportional to the product of an *a priori* probability of the parameter values $P_p(\Theta)$, and a likelihood level $P_v(I, \Theta)$, which measures the adequacy of the prototype to the image. These two different distributions are both exponential distributions. Hence, $P(\Theta \mid I)$ can be expressed as a Gibbs distribution with energy $E(I, \Theta) = U_v(I, \Theta) + U_p(\Theta)$. A more detailed description of each of these two terms will now be given.

10.4.1. *A priori energy*

The *a priori* energy on shape makes it possible to define a relevance level of a configuration which is created from a particular instance of the vector Θ. Measuring such energy is based on the following three criteria:

– the usual position of the road signs in the image (normally on the right-hand side of the road scene),

– the different sizes that are possible for a particular road sign,

– the usual orientation of the road sign.

Each criterion of parameter θ is supposed to be located in a relevance interval $[\theta_{\min}, \theta_{\max}]$ and is represented by a V-shaped penalty function:

$$\begin{aligned} \theta < \theta_{\min}, \quad & g(\theta) = a\big(\theta_{\min} - \theta\big) \\ \theta_{\min} \le \theta \le \theta_{\max}, \quad & g(\theta) = 0 \\ \theta > \theta_{\max}, \quad & g(\theta) = b\big(\theta - \theta_{\max}\big). \end{aligned} \qquad [10.3]$$

The limits of the relevance interval can be considered as data derived from a "ground truth", i.e. a manual analysis that was carried out on a representative set of image sequences. The slopes a and b are modeled empirically. The global *a priori* term $U_p(\Theta)$ is the sum of the different penalty terms.

10.4.2. *Image energy*

It is possible to distinguish vertical road signs from their environment by analyzing their color and shape. Therefore, the image energy function $U_v(I, \Theta)$ is made up of two elements: an edge-oriented term $U_c(I, \Theta)$, which is based on gradient information, and a region-oriented term $U_r(I, \Theta)$, which is based on color information.

10.4.2.1. *Using gradient information*

As is the case with a lot of manufactured objects, vertical road signs possess straight and aligned edges (see Figure 10.2). The edge-oriented element [JOL 96] simultaneously uses the direction of the gradient and its magnitude. To be more specific, on every edge of the polygon the edge-oriented element integrates the product of the magnitude of the gradient $\|\nabla I\|$ with a function of the scalar product of the direction vector of the segment $\overrightarrow{S}$ and the local direction of the gradient of the image $\overrightarrow{\nabla I}$:

$$U_c(I, \Theta) = 1 - \frac{\int_{\delta(\Theta)} \|\nabla I\| \times h(\overrightarrow{S} \cdot \overrightarrow{\nabla I})}{\sqrt{\int_{\delta(\Theta)} \|\nabla I\|^2} \sqrt{\int_{\delta(\Theta)} h^2(\overrightarrow{s} \cdot \overrightarrow{\nabla I})}} \qquad [10.4]$$

where $h(x) = \max(0, 1 - |x|)$. The value of the criterion which is defined is close to zero when the model is aligned with the edges of a particular object of interest, and is close to one in the opposite case. This criterion is evaluated along the exterior triangle of the prototype.

Figure 10.2. *Image, magnitude (reversed grayscales) and angle of the gradient*

With the aim of making the likelihood term more "attractive" it is possible to alter its value in the following way. An edge map is calculated by extracting the local maxima of the gradient which exceed a certain threshold level (see left image in Figure 10.3). With the help of the fast marching algorithm [SET 96] it is possible to determine the distance $d(x, y)$ that exists between each point (x, y) of the image grid and the closest edge point. In practice, however, this distance is propagated only up to a certain maximum threshold level $d_{\max}$. The map $d_{\max} - d(x, y)$ (see right image in Figure 10.3) is introduced into the edge energy function U_c in place of the magnitude of the gradient.

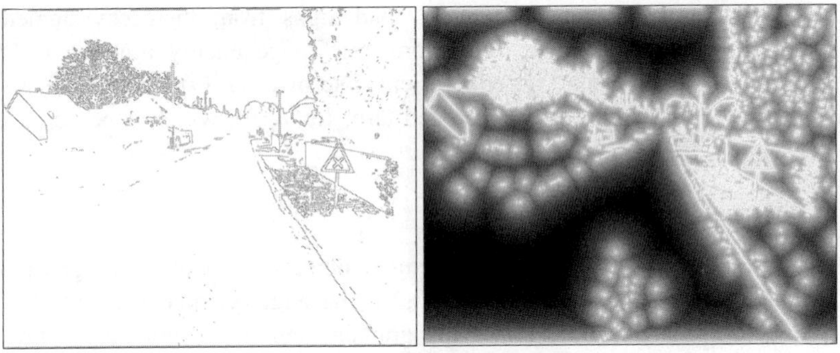

Figure 10.3. *Edge map and associated distance transform,* $d_{\max} - d$ *(histogram drawn to scale in order to improve the view of the image)*

10.4.2.2. *Exploiting color information*

Warning road signs are characterized by their red color. In order to identify the pixels which exhibit a predominant red color rapidly, the red green blue (RGB) color model is used. The pre-detector assumes a pixel to be red if its components respect the following two conditions:

$$\begin{aligned} R &> \alpha(G + B) \\ R - \max(G, B) &> \beta\big[\max(G, B) - \min(G, B)\big]. \end{aligned} \qquad [10.5]$$

The first criterion involves placing a constraint on the normalized red component $R/(R+G+B)$, which must be dominant. The normalization of the red component makes the pre-detector robust to illumination variations. The second criterion checks that the pixel does not tend towards yellow nor towards magenta. In practice $\beta = 2\alpha$ is suitable in most cases, which then only leaves parameter α to be tuned. The same value of a parameter can be used for an entire image sequence. Usual values range from 0.5 to 0.75. The pre-detector introduced here performs as well as the one proposed by De La Escalera *et al.* [ESC 04], and this is without having to change the color model, which, in turn, reduces the computational cost of the pre-detection phase.

An incorrect white balance is, unfortunately, quite a common occurrence in such image sequences that are to be processed. Since the images to be analyzed always contain a part of the road which can act as a reference zone, a type of "gray-world" balancing [BUC 80] can be automatically carried out in most cases. Figure 10.4 shows an example of a red dominance map which illustrates the robustness of the method against luminance variations.

Figure 10.4. *Pre-detection of the color red on an RGB image with backlight*

For a shape Θ the region-oriented term is calculated from the red dominance map. More specifically, it is made up of two terms which are linked to two zones of the chosen model: the external crown $\partial\Omega$ and the interior of the small triangle Ω (see Figure 10.1):

$$U_e(I,\Theta) = \frac{\iint_{\partial\Omega(\Theta)} I(x)dx}{\iint_{\partial\Omega(\Theta)} dx} \quad \text{and} \quad U_i(I,\Theta) = \frac{\iint_{\Omega(\Theta)} I(x)dx}{\iint_{\Omega(\Theta)} dx} \qquad [10.6]$$

The global region-oriented element is finally written as follows:

$$U_r(I,\Theta) = \min\big(1, 1 + U_e(I,\Theta) - U_i(I,\Theta)\big). \qquad [10.7]$$

When Θ is perfectly adjusted, the value of U_r is zero. On the other hand, $U_r(I, \Theta)$ has a value of one whenever Θ is in a zone of the image which is either completely "red" or "not red".

We have written the problem of road sign detection as the minimization of a cost function $E(I, \Theta)$ whose different terms have just been described. The question is now to find an algorithm which is capable of efficiently minimizing E. This will be dealt with in the next section.

10.5. Three biological metaheuristics

The non-convex nature of E over the set of possible transformations Θ, means that an adapted algorithm must be used for the recognition of road signs. Due to the fact that the value of E does not possess any interesting mathematical properties it is difficult to choose *a priori* an algorithm that can be used. This means that an experimental comparison of certain algorithms must be carried out. We decided to compare three different algorithms inspired from the world of biology. Each of these algorithms has very different characteristics. One of the algorithms is evolution strategy (ES), another is particle swarm optimization (PSO) and the final one is clonal selection (CS). The three algorithms have been implemented by the authors in Scilab language.

10.5.1. *Evolution strategies*

ES will be considered as a reference algorithm in this experimental comparison because it has been used by the authors since the beginning of the development of a road sign detection tool. If we consider a particular optimization problem and link it to a particular species in its own environment, it is possible to create an analogy between the population of organisms of this particular species and the set of solutions of the optimization problem. It is also possible to create an analogy between the performance of these organisms and the cost-function value. The operators of crossover and mutation which are used to produce new candidate solutions are inspired by the main genetic mechanisms involved in the evolution of a species with sexual reproduction. This analogy has led to the creation of algorithms usually classified as evolutionary algorithms. A member of the evolutionary algorithms class conforms to the diagram which can be seen in Figure 10.5. Genetic algorithms are the most famous member of this class. ESs and evolution programs [BÄC 96] are also found in this class. It must be pointed out that evolution programs are seldom used. Since we address a continuous variable optimization problem, we have decided to focus only on ES, which will be introduced in the next section.

10.5.1.1. *Algorithm*

10.5.1.1.1. Gaussian representation

ES, which was designed for numerical optimization, does not go into as much depth as a genetic algorithm when it comes to the analogy with living beings, since it

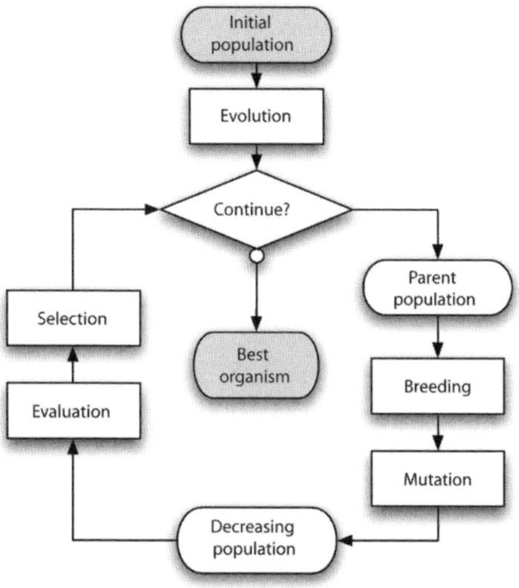

Figure 10.5. *Flow-chart of an evolutionary algorithm*

operates only on the phenotype[1] of the living object. It is therefore not necessary to encode the variables that need to be optimized. In ESs the individuals are represented by Gaussian random vectors, which are noted as $\vec{N}(\Theta, \mathbf{C})$. The average vector is the vector Θ which contains n variables of the optimization problem. Vector $\vec{N}$ also contains the covariance matrix $\mathbf{C}$ which is divided into two vectors $\vec{\sigma}$ and $\vec{\alpha}$. Vector $\vec{\sigma}$ contains the variances, and $\vec{\alpha}$ the out-of-diagonal terms of C, or rotations. Even though it is possible to introduce dependencies between different variables, this is in fact very seldom carried out. Therefore, $\mathbf{C}$ is usually diagonal.

10.5.1.1.2. Initialization

The initial population of the average vector Θ can come from *a priori* information on the expected position of the minimum. It can also be randomly sampled from a uniform distribution, properly scaled and centered. The values of the covariance matrix are fixed non-zero arbitrary values.

10.5.1.1.3. Operators

In order to make the population evolve, ES uses traditional mutation, crossover and selection operators in succession. The first two operators are applied to all of the

1. An organism's full hereditary information, even though it may not be seen, is known as the genotype, while the phenotype refers to only the information which can be observed.

components of the organism, i.e. average, as well as variances and strict covariances. In the next part of this chapter we assume that the problem is made up of n variables. $N(a,b)$ refers to the selection of a Gaussian random variable with an average a, and a standard deviation b.

The crossover operator relies on the phenotype of two organisms S and T, in the same way as the breeding operator relies on the genotype for GA:

$$(\Theta', \vec{\sigma}', \vec{\alpha}') = \mathbf{R}\big((\Theta_S, \vec{\sigma}_S, \vec{\alpha}_S), (\Theta_T, \vec{\sigma}_T, \vec{\alpha}_T)\big). \quad [10.8]$$

Bäck describes six detailed formulae, which deal with the recombination of the initial vectors' components [BÄC 96]. Referring to the literature, he points out that it is better to use two different formulae depending on whether we consider the variables of the optimization problem or the covariances. For the variables, he recommends using the process of discrete recombination; for the covariances, he recommends using a process of recombination known as pan-mictic recombination:

$$\begin{aligned}
\Theta'_i &= \Theta_{S,i} \mid \Theta_{T,i}, & \forall i \in \{1,\ldots,n\} \\
\sigma'_i &= \sigma_{S,i} + \frac{1}{2}\big(\sigma_{T,i} - \sigma_{S,i}\big), & \forall i \in \{1,\ldots,n\} \\
\alpha'_i &= \alpha_{S,i} + \frac{1}{2}\big(\alpha_{T,i} - \alpha_{S,i}\big), & \forall i \in \big\{1,\ldots,n(n-1)/2\big\}.
\end{aligned} \quad [10.9]$$

where $\mid$ refers to the random choice operation between two values. The mutation operator is applied to the descendant, which results from randomly choosing one of the two values. The mutation operator is also applied to the variances and rotations, and then to the variables as follows:

$$\begin{aligned}
\sigma'_i &= \sigma_i e^{\tau' N(0,1) + \tau N_i(0,1)} & \forall i \in \{1,n\} \\
\alpha'_i &= \alpha_i + \beta e^{N_i(0,1)} & \forall i \in \big\{1, n(n-1)/2\big\} \\
\Theta' &= \Theta + \vec{N}(\vec{0}, \vec{\sigma}', \vec{\alpha}')
\end{aligned} \quad [10.10]$$

With these equations in mind we recognize a process which is similar to the cooling schedule of the simulated annealing. The internal parameters of ES evolve through the iterations. The factors τ, τ' and β have been established as follows in literature:

$$\begin{aligned}
\tau &\propto \left(\sqrt{2\sqrt{n}}\right)^{-1} \\
\tau' &\propto \left(\sqrt{2n}\right)^{-1} \\
\beta &\approx 0.0873
\end{aligned} \quad [10.11]$$

The cost-function value is calculated for each of the λ organisms of the offspring. The organisms are ranked and then a number of μ organisms are chosen. The notation which is used for this demographic scheme is (μ, λ)-ES. The other form of notation which can be used is $(\mu + \lambda)$-ES. This latter form corresponds to the selection among parents and offspring organisms. This scheme is said to be elitist. Unlike genetic algorithms, the selection of organisms is carried out strictly according to the ranking of the organism. The lowest ranked organisms have no chance of survival.

10.5.1.2. *Theory and convergence*

The probabilistic framework in which the ESs have been defined has made it possible to obtain theoretical results somewhat stronger than for the other algorithms presented in this chapter. For example, an analytical expression of the probability of the speed of convergence has been established for (μ, λ)-ES. Nevertheless, it is still quite difficult to use this expression for parent populations whose size is larger than one [BÄC 96, Chapter 2.1.7]. In addition, a certain number of results are available for academic test functions, such as the sphere or the corridor. The rule for adjusting the standard deviation of the model which has been presented above comes from theoretical investigations. Bäck compares the ESs and the genetic algorithms for five different academic functions, exhibiting different topologies. ES turns out to be the best algorithm to use, by far [BÄC 96].

10.5.1.3. *Practice*

The theory behind ESs does not mention anything about the size of the parent population to be dealt with. It seems that the parent population can be somewhat smaller than is the case for GA, without having a negative effect on the ES. A size of parent population that is equal to the size of the number of variables can typically be used. For example, a parent population with less than four individuals strongly limits which organisms the recombination operator can choose from. If there happens to be a lot of variables, it is possible to deal with a much smaller number of organisms than the total number of variables. The rule is to produce seven times as many descendants as there are parents. Elitist ESs are not recommended, thus a (μ, λ)-ES is preferable.

10.5.1.4. *Applying the ES to the issue of detection*

After carrying out a parametric study on μ, a $(4, 28)$-ES was chosen. The components of $\vec{\sigma}$ were tuned to a size of 1/100 of the range of each parameter. Since no information on dependencies between the parameters is available, we take $\vec{\alpha} = \vec{0}$.

10.5.2. *Clonal selection*

Clonal selection (CS) was chosen after a series of tests using several different algorithms was carried out on a set of classical test functions by one of the authors. It turns out that CS was the best performing algorithm of all of the algorithms that were tested and that its results on several optimizations showed a very low standard-deviation.

10.5.2.1. *Some information about the immune system*

In this section we provide a brief description of the immune system of vertebrates which acts as the basis for CS. For more detailed information on how the immune system functions, see more specialized literature on the topic, such as [ALB 02] or [JAN 04]. A perfectly functioning vertebrate immune system ensures the protection and specific defense of the organism against pathogens, which continually try to attack the organism. Specific defense refers to the immune system fighting against a particular type of pathogen. These pathogens, which we will refer to as antigens, can be present in many different shapes or forms such as heavy metals, synthetic organic compounds, bacteria etc.

The immune system secretes a particular type of protein known as an antibody which in turn neutralizes and destroys the antigens. The production of an antibody is partly coded by the immunity genes. This notion of coding leads to the creation of a certain number of basic protein building blocks, which work together in order to create an antibody. The process which leads to the creation of functional antibodies by producing and arranging these building blocks into a sequence is known as somatic recombination. The diversity of molecules which is produced by the process of somatic recombination is staggering. It is estimated that there are 10^{11} different antibodies which are present within the human body at any given moment. This is one of the amazing aspects of the immune system.

When the immune system intercepts an antigen, it is assumed that there is an antibody that possesses a certain affinity, or complementarity with the antigen. A part of the immune system's response is to modulate the multiplication of the cells which produce the antibodies in relation to their affinity with the antigen. These cells which produce antibodies are known as B lymphocyte cells. This process is illustrated in Figure 10.6. A B lymphocyte cell only produces one type of antibody. The multiplication of the B lymphocyte cells is carried out by a process of cloning. The clones which are produced have two functions. They either produce new antibodies with the aim of neutralizing other occurrences of the antigen, or they act as a memory in case of a further attack carried out by the antigen. In the case of any further attack carried out by the antigen, the immune system's response becomes more intense, in other words more antibodies are produced and they are produced at a quicker rate. The lifespan of a memory B lymphocyte ranges from several months to several years.

This very brief and crude introduction results in the creation of an analogy between the affinity of the antigen/antibody, and the optimization process. As far as the optimization process is concerned, the immune system of vertebrates possesses two valuable characteristics.

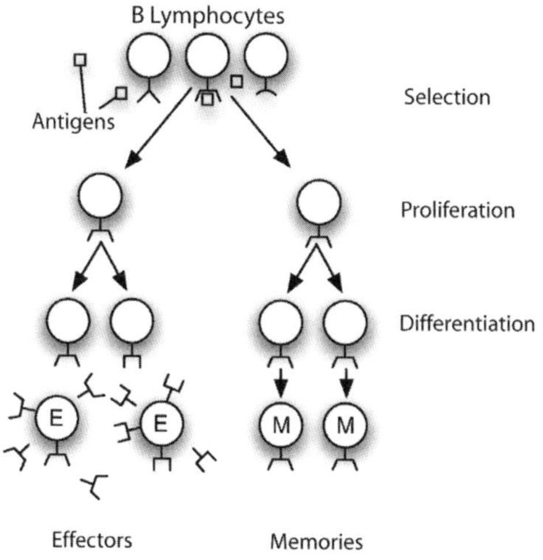

Figure 10.6. *The principle of clonal selection*

On one hand, the immune system is capable of producing a large range of possible solutions to help with the issue of optimization. On the other hand, it possesses a learning mechanism. De Castro and von Zuben developed CS [CAS 02] after studying the controlled multiplication of B lymphocytes. This algorithm is introduced in the next part of this section. Other authors have used the immune network as their basis for CS. By relying on the immune network it is possible to provide an explanation of the long-term memory of past immune responses, with a moderate cost for the living host organism. These aspects of the immune system which have just been mentioned are not the only aspects which can be used in analyzing engineering issues, and the optimization process is not the only process which can be used as well. Another possibly more straightforward application is pattern recognition. For an overview of all of the different applications of the artificial immune system that are available, see [DAS 98].

10.5.2.2. *An introduction to CS*

The flow chart of clonal selection adapted to optimization can be seen in Figure 10.7.

The algorithm manipulates a population of antibodies. Each antibody contains a set of data for the parameters which are to be optimized. It is possible to create the initial population of N antibodies by random selection. For the next iterations, this population of antibodies is made up of $N - d$ antibodies resulting from previous iterations and d new randomly generated antibodies. The population is then evaluated with respect to the cost function. The population is then ranked according to merit

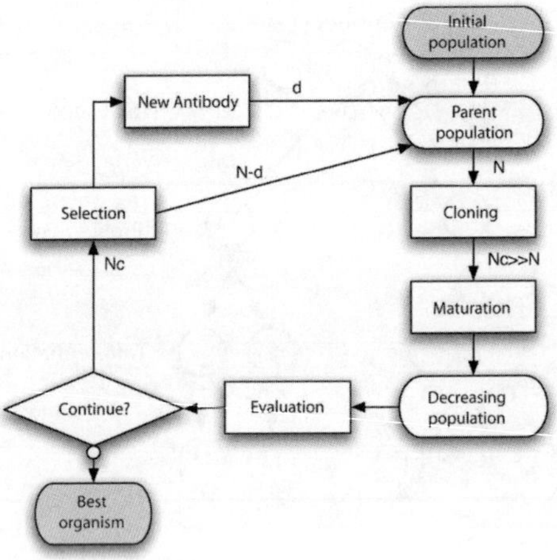

Figure 10.7. *Flow-chart of the clonal selection*

in descending order. Ranking the antibodies like this makes it possible to choose the $n \leq N$ best individuals, which will then undergo a cloning phase. The total number of clones N_c can be written as follows:

$$N_c = \sum_{i=1}^{N} \text{arrondi}\left(\frac{\beta N}{g(i)}\right). \qquad [10.12]$$

where g is a function which can either be dependent on the affinity between the antibody and the antigen, or can be a constant value and β is a multiplicative factor. According to de Castro and von Zuben the first rule is suitable whenever it is a global minimum that is being looked for – global form of the algorithm. The second rule is used whenever several good quality minima are being looked for – local form of the algorithm. The aim of the maturation phase is to introduce variations into the clones that have been created. [CAS 02] does not go into a lot of details on the maturation phase but does state that the mutation rate m should decrease when the affinity of the organism f increases, in our case, when the value of the cost function E decreases. De Castro and von Zuben suggest defining m by $m = \exp(-\rho f)$, where ρ is a control parameter. After the maturation phase, the N_c clones are evaluated and the best $N - d$ clones are retained before the entire cycle is restarted up until the stopping condition is met.

10.5.2.3. *Performance and convergence*

Clonal selection is a recent algorithm and the underlying theory is less developed than is the case for the other algorithms which have been introduced in this chapter.

There is no information about the algorithm's probability of convergence or the algorithm's speed of convergence. The effects of the internal parameters were only studied in an empirical way for multimodal functions. Remarkably, the complexity of the CS is $O(N_c)$, where N_c is the number of clones, while the complexity of an evolutionary algorithm is $O(N^2)$, where N refers to the size of the population. An empirical evaluation over a large number of test functions which was carried out by the Regional Ponts-et-Chausseées Laboratory in Strasbourg, France, highlights the excellent performances of this algorithm. It was indeed the best algorithm over (all but one) studied functions. Another point which needs to be mentioned is the low scattering of results obtained for the CS, which is not the case for the other algorithms. For the choice of parameters, $N = 10$, $\beta = 1$ are general purpose values. The optimal value of ρ may vary. It is best not to have a value of d larger than two, for fear of seeing the optimization process transform into Brownian motion. The optimization of these last two parameters does not lead to the same values, depending on whether the local or the global form of the algorithm is used.

10.5.2.4. *Applying the CS to the optimization problem*

We placed the value $g(i) = 1/i$ into equation [10.12], $N = n = 10$, $\beta = 1$, and $\rho = 2$. In the absence of a more precise definition of the mutation operator in [CAS 02], we used a similar operator to the operator that is used in the canonic genetic algorithm.

10.5.3. *Particle swarm optimization*

Particle swarm optimization (PSO) which was developed by Kennedy and Eberhart [KEN 95, KEN 01] has already been described in Chapter 9. Due to this fact it is not relevant to describe this algorithm again in this chapter. PSO combines a very simple formulation with a high level of efficiency. The PSO which has been implemented here is the "variant 1 with constriction" PSO proposed in [CLE 02] by Clerc and Kennedy. The size of the population is 20 particles. According to Kennedy and Clerc, the value of χ is calculated by having $\kappa = 0.8$ and having $\Phi_{\max} = 4.5$.

10.6. Experimental results

10.6.1. *Preliminaries*

10.6.1.1. *Additional information on the road sign detectors*

Even if the algorithms can achieve global optimization, it is best to provide them with an initial population as close as possible to the regions of interest. The first population uses, when possible, connected components extracted from the red dominance maps supplied by the pre-segmentation stage (see section 10.4.2.2). The bounding box of each sufficiently large connected component is calculated. Then, the Θ transformation that yields the largest triangle inside the bounding box is calculated.

The size of the resulting population is, in general, too small and it is necessary to generate more individuals. This is done by randomly sampling over a uniform distribution, properly scaled and centered (the resulting Θ transformations must lead to the creation of patterns which respect the *a priori* criterion that has been described in section 10.4.1).

On the other hand, local optimization is not completely missing in the detector. It can happen that the initialization from the connected components gives an individual that almost fits the road sign. Therefore, before carrying out the optimization process, whilst using one of the three metaheuristics that were mentioned in section 10.5, a gradient descent is systematically applied to all individuals which have been created from the connected components. In this way, a certain number of the road signs can be localized at a low computational cost. Moreover, if the algorithm is stopped because the likelihood energy becomes very small, applying a gradient descent makes it possible to refine the convergence without extending the overall computation time.

10.6.1.2. *Establishing the ground-truth*

It is necessary to have a reference sequence, whose properties are known, in order to assess the performance of the detector. In this sequence, all the images with danger road signs are identified and the coordinates of the vertices of these signs are stored. The result of this survey will be referred to as the "ground truth" in the remainder of this chapter. It can only be made by a human operator. For this purpose, a graphical interface that makes it possible to navigate along image sequences was designed at the Regional Ponts-et-Chaussées Laboratory in Strasbourg, France. Of course, the ground truth depends on the human operator who made it. In order to avoid any possible bias in our experiment, the ground truth was made by six different people from the chosen image sequence. So, the real ground truth on which the different detectors are compared in the following is an average ground truth i.e. the arithmetic mean of coordinates. The ground truth was prepared for a sequence of images called CD3. The images were taken along the French secondary road called RN74, in the department of the Vosges (88). With CD3, the maximum standard deviation on each coordinate of the center-of-mass in a given image is 1.85 pixels. The sampling rate of the image sequence is 5 m. Thus, it is not uncommon to see a road sign appear several times within consecutive images, with different sizes. Since the aim was to identify all of the danger road signs along a given route and not to identify all of the appearances of danger road signs in the image sequence, a chaining process was carried by one of the operators. The idea of this chaining process is to group all of the images that contain a view of the same road sign.

10.6.1.3. *Definition of receiver operating characteristic (ROC) curves*

With the aim of evaluating the different road sign detectors, we plotted a ROC curve for each of them, in relation to the average ground truth. Such curves represent the correct detection rate vs. the false positive rate, as a function of a threshold on the value of the cost function. In the present case, there is no absolute definition of a ROC curve. Hence, we chose to measure the distance (in terms of center-of-mass) between the average ground truth and the result of the optimization algorithm. We consider that a maximum distance of 4 pixels between these two points is acceptable. The expressions of the false positive rate and the correct detection rate are as follows:

$$t_{FA}(c, D) = \frac{n_{FA}(c, D)}{N_{FA}(D)} \qquad [10.13]$$

$$t_{BD}(c, D) = \frac{n_{BD}(c, D)}{N_{BD}(D)} \qquad [10.14]$$

where FA means False Alarm, BD is the (French) acronym for "correct detection", c is the value of the threshold which is set for the cost-function. In other words, only shapes whose cost is less than c are taken into account in the calculation of t_{FA} and t_{BD}. In [10.13] and [10.14], D refers to the distance threshold. For the purpose of our experiment we decided to use Euclidean distance, but it is possible to use any other type of distance. A correct detection occurs if a shape satisfies the condition on c and if its center-of-mass is located less than 4 pixels away from the average ground truth. If only the first of these conditions is met then this is a false alarm. The other possibility for a false alarm is when the condition on c is met, while there is no road sign in the image. This second possibility never occurs in our case because the optimization process is applied to images that belong to the ground truth, i.e. which contain at least one road sign, by definition. N_{FA} is the maximum possible number of false detections. For a fixed image size, this number only depends on the distance threshold. More specifically, $N_{FA} = lh - S$, where l (respectively, h) is the width (respectively the height) of the image in pixels, and S is the number of pixels in the detection zone. N_{BD} is the number of road signs to be found. The calculation of t_{BD} is carried out by only using the correct detections of distinct road signs.

In addition to the ROC curves, we also drew histograms of the distance (in terms of center-of-mass) between the optimal solution and the average ground-truth. These statistics are only calculated for road signs which meet the cost-function criterion.

10.6.2. *Evaluation on the CD3 sequence*

This section focuses on the results that were obtained from the CD3 sequence by the six different detectors that can be devised, since there are two possible contour-based terms and three optimization algorithms.

With an edge-oriented term based on the magnitude of the gradient (see equation [10.4]), ES and PSO detect the presence of all of the road signs. CS, however, is not perfect as can be seen in Figure 10.8.

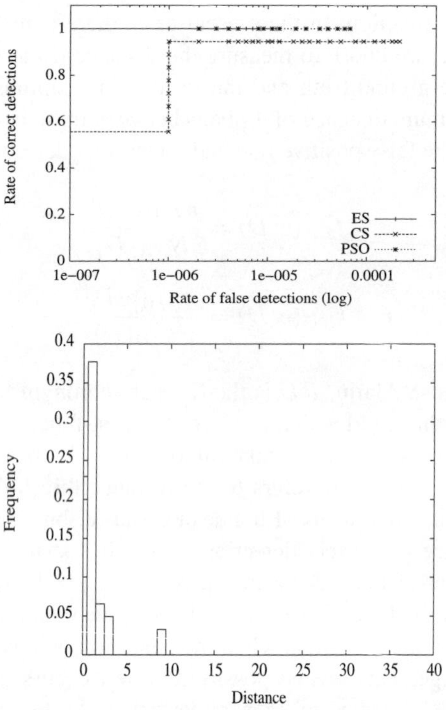

Figure 10.8. *ROC curves and a typical distance histogram of a detector that uses an edge-oriented term based on the magnitude of the gradient*

The residual likelihood value after optimization is never equal to zero, which explains why the ROC curve does not begin at zero. If the threshold value c is too small then none of the images will be considered. If this threshold is increased then all of the correct detections of the images are considered. Further increasing the threshold level does not enhance the number of correct detections; it actually increases the number of false alarms. The histogram of distances to the ground truth confirms this interpretation (see e.g. the ES histogram on the lower part of Figure 10.8). It reveals that there is a class of individuals with a center of gravity that is very close to the center of gravity of the ground truth. Classes of distance immediately higher are empty. We have to significantly increase the distance to find other individuals, which are false alarms.

With the edge-oriented term based on the distance map only PSO produces a straight line ROC (see Figure 10.10). The number of correctly identified road signs increases in steps whenever the other two algorithms are used. ES, just like PSO, reaches the maximum rate of correctly identified road signs, whereas CS has a success rate of less than 1.

Figure 10.9. *Examples of detections with an edge-oriented term based on the distance map. The original image is on the left and the best individual can be seen on the right*

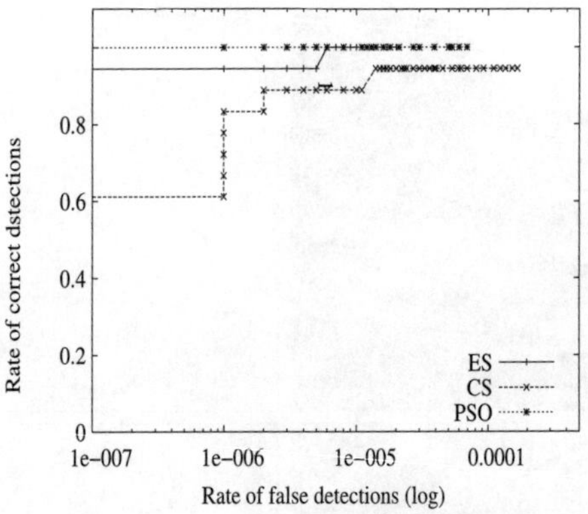

Figure 10.10. *The detector's ROC curve which is created from the edge oriented term that is based on the distance map*

A close examination of the distance histograms confirms that PSO is the best algorithm to use for detecting road signs and this can be seen in Figure 10.11. The correctly identified road signs can be found at a distance which is at most 2 pixels from the center of the ground truth, while for CS and ESs the fitting is less precise.

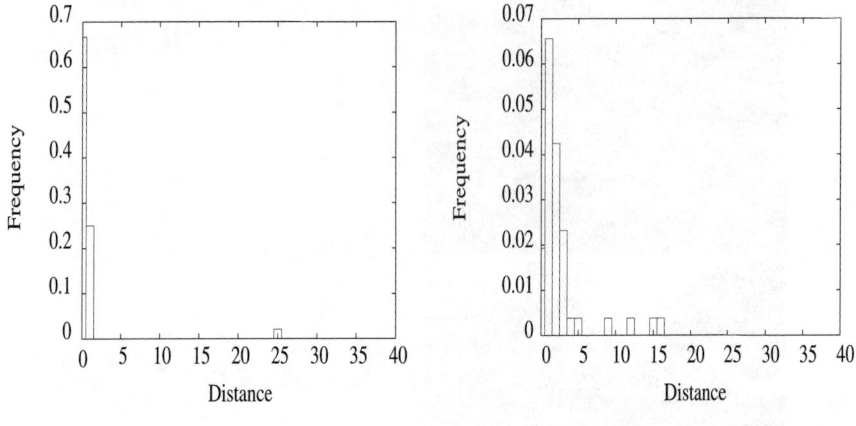

Figure 10.11. *Distance histograms showing the distance of the detector from the road signs. The PSO is on the left, and the CS is on the right. The edge-oriented element is based on the distance map*

The almost binary nature of the ROC curves suggests that the detector is well configured since an increase in the number of correct identifications of road signs does not necessarily mean that the number of false detections will also increase. We can, therefore, rely on the likelihood value to judge the success or failure of the road sign detection process. As a result of this it is quite easy to automate the detector.

Some examples of road signs which have been detected by PSO can be seen in Figure 10.9. The left column shows the original image. The right column shows the best individual after the optimization process. These optimized individuals are shown superimposed either on the gradient's magnitude map (lines one, three and four in Figure 10.9), or on the red color map (line two in Figure 10.9). The top image on the right hand side of Figure 10.9 shows a partly occulted road sign. The image on the second row is an example of noisy red color map. This image also shows the presence of a different number of potential lures (roof tiles and cars in particular). The combination of the *a priori* information and the template make it possible to avoid them. The other two images highlight the detector's ability to locate several objects within the same road scene, on two successive images. It is necessary to develop a tool which can be used to detect several objects in one image if such a detector is to be used on a large scale.

10.7. Conclusion

After introducing the state of the art in road sign detection, we described an off-line detector dedicated to automatic road sign surveys. The associated optimization problem has been dealt with using three different algorithms: evolution strategy (ES), clonal selection and particle swarm optimization (PSO).

Judging from a test on a sequence of 48 typical road scene images, it can be said that PSO, combined with an edge-oriented energy that is based on the distance map, is the best algorithm. It achieves a true-positive rate of 100% with respect to a ground truth made by several human operators. The 18 road signs in the sequence were all correctly located, within a maximum distance of 2 pixels. The use of PSO combined with an edge-oriented energy that is based on the gradient's magnitude as well as the use of ES with the same energy also led to a 100% true detection rate. However, the distances to the ground truth are somewhat more significant with the latter detectors. The use of ES combined with an edge-oriented energy that is based on the distance map is also capable of detecting all of the road signs of the sequence, but at the price of a higher false alarm rate.

Between the two terms proposed to take into account edge-oriented information, the distance map is the one which leads to the most efficient detector. The ROC curves and the associated histograms show that the problem is relatively well defined, i.e a low likelihood level corresponds to the correct identification of a road sign. This therefore makes it possible to define a stopping criterion for the optimization algorithm based on the likelihood level. It is therefore possible to automate the detection process.

If we want to use the detector on a more operational level, it must be possible to process sequences of several thousand images in a reasonable time lapse. The percentage of road images which contain road signs is quite low. For example, CD3 contains 3,000 images for a total of 48 different views of 18 danger road signs. The analysis of the connected components which are calculated from red color maps makes it possible to decide whether a road sign is present in an image or not. It is then quite easy to isolate the images that need to be processed. A good detector must be able to locate several road signs within an image. This can be carried out in a sequential fashion. In this case the areas that have been successfully explored must be avoided. One solution involves deleting, from the images, road signs which have been located. Another solution involves modifying the likelihood energy each time a road sign is identified. These two approaches have been tested by the authors with quite positive results. Finally, and this is probably an interesting optimization problem, it is possible to carry out a parallel search on the existing road signs in an image, provided that there are algorithms that can localize several high quality minima. This is the case of the local form of clonal selection.

10.8. Bibliography

[ALB 02] ALBERTS B., "The adaptive immune system", chapter in *Molecular Biology of the Cell*, pp. 1363–1421, Garland Science, 2002.

[AMI 91] AMIT Y., GRENANDER U. and PICCIONI M., "Structural image restoration through deformable templates", *Journal of the American Statistical Association*, vol. 86, no. 414, pp. 367–387, 1991.

[AOY 96] AOYAGI Y. and ASAKURA T., "A study on traffic sign recognition in scene image using genetic algorithms and neural networks", *Proceedings of the 1996 IEEE 22nd International Conference on Industrial Electronics, Control, and Instrumentation (IECON)*, vol. 3, pp. 1838–1843, 1996.

[BÄC 96] BÄCK T., *Evolutionary Algorithms in Theory and Practice*, Oxford University Press, 1996.

[BET 95] BETKE M. and MAKRIS N.C., "Fast object recognition in noisy images using simulated annealing", *Proceedings of the Fifth International Conference on Computer Vision*, Washington, DC, USA, IEEE Computer Society, pp. 523–530, 1995.

[BOR 88] BORGEFORS G., "Hierarchical chamfer matching: a parametric edge matching algorithm", *IEEE Trans. Pattern Anal. Mach. Intell.*, vol. 10, no. 6, pp. 849–865, IEEE Computer Society, 1988.

[BUC 80] BUCHSBAUM G., "A spatial processor model for object colour perception", *Journal of the Franklin Institute*, no. 310, pp. 1–26, 1980.

[CAS 02] DE CASTRO L. and VON ZUBEN F., "Learning and optimization using the clonal selection principle", *IEEE Transactions on Evolutionary Computation*, vol. 6, no. 3, pp. 239–251, 2002.

[CLE 02] CLERC M., KENNEDY J., "The particle swarm – explosion, stability, and convergence in a multidimensional complex space", *IEEE Transactions on Evolutionary Computation*, vol. 6, no. 1, pp. 386–396, IEEE Press, 2002.

[DAH 04] DAHYOT R., CHARBONNIER P. and HEITZ F., "A Bayesian approach to object detection using probabilistic appearance-based models", *Pattern Analysis and Applications*, vol. 7, no. 3, pp. 317–332, 2004.

[DAS 98] DASGUPTA D., *Artificial Immune Systems and their Applications*, Springer Verlag, 1998.

[ESC 03] DE LA ESCALERA A., ARMINGOL J.M. and MATA M., "Traffic sign recognition and analysis for intelligent vehicles", *Image Vision Comput.*, vol. 21, no. 3, pp. 247–258, 2003.

[ESC 04] DE LA ESCALERA A., ARMINGOL J., PASTOR J. and RODRIGUEZ F., "Visual sign information extraction and identification by deformable models for intelligent vehicles", *IEEE Trans. on Intelligent Transportation Systems*, vol. 5, no. 2, pp. 57–68, 2004.

[GAV 99] GAVRILA D.M., "Traffic sign recognition revisited", *Proc. of the 21st DAGM Symposium für Mustererkennung*, Bonn, Germany, Springer Verlag, pp. 86–93, 1999.

[GRE 89] GRENANDER U. and KEENAN D.M., "Towards automated image understanding", *Journal of Applied Statistics*, vol. 16, no. 2, pp. 207–221, 1989.

[JAN 04] JANEWAY C., TRAVERS P., WALPORT M. and SHLOMCHIK M., *Immunobiology – the Immune System in Health and Disease*, Garland Science, 2004.

[JOL 96] JOLLY M.D., LAKSHMANAN S. and JAIN A., "Vehicle segmentation and classification using deformable templates", *IEEE Trans. on Pattern Analysis and Machine Intelligence*, vol. 18, no. 3, pp. 293–308, 1996.

[KEN 95] KENNEDY J. and EBERHART R.C., "Particle swarm optimization", *IEEE International Conference on Neural Networks*, IEEE, pp. 1942–1948, 1995.

[KEN 01] KENNEDY J., C.EBERHART R. and SHI Y., *Swarm Intelligence*, Morgan Kaufmann Publishers, 2001.

[LAL 95] LALONDE M. and LI Y., Road Sign Recognition, Survey of the State of the Art for Sub-Project 2.4, Report no. CRIM-IIT-95/09-35, Information Technology Research Center in Montreal, 1995.

[LOY 02] LOY G. and ZELINSKY A., "A fast radial symmetry transform for detecting points of interest", *ECCV'02: Proceedings of the 7th European Conference on Computer Vision-Part I*, London, UK, Springer-Verlag, pp. 358–368, 2002.

[LOY 04] LOY G. and BARNES N., "Fast shape-based road sign detection for a driver assistance system", *Proc. IEEE/RSJ International Conference on Intelligent Robots and Systems (IROS2004)*, Sendai, Japan, pp. 86–93, 2004.

[MIG 00] MIGNOTTE M., COLLET C., PEREZ P. and BOUTHEMY P., "Hybrid genetic optimization and statistical model-based approach for the classification of shadow shapes in sonar imagery", *IEEE Trans. On Pattern Analysis and Machine Intelligence*, vol. 22, no. 2, pp. 129–141, 2000.

[MOG 97] MOGHADDAM B. and PENTLAND A., "Probabilistic visual learning for object representation", *IEEE Transactions on Pattern Analysis and Machine Intelligence*, vol. 19, no. 7, pp. 696–710, 1997.

[REC 94] RECHENBERG I., *Evolutionsstrategie'94*, Frommann-Holzboog, 1994.

[SET 96] SETHIAN J., "A fast marching level set method for monotonically advancing fronts", *Proceedings of the National Academy of Sciences*, vol. 93, no. 4, pp. 1591–1595, 1996.

Chapter 11

Metaheuristics for Continuous Variables. The Registration of Retinal Angiogram Images

11.1. Introduction

Image registration is an important tool that is used to resolve any problems that may arise during the analysis of medical images. In the past, scientists have tried to apply traditional minimization strategies to the field of image registration. Such traditional minimization strategies include exhaustive search, gradient descent algorithm, simplex optimization method, simulated annealing algorithm, genetic algorithms and Powell's search [RIT 99, JEN 01].

In most cases image registration is carried out in two phases: image processing and then optimization of a similarity criterion. The aim of the image processing phase is to improve the quality of the image and to extract the relevant information which can be used to improve the optimization phase. The aim of the optimization phase is to find optimal modifications of the image, in accordance with a particular objective function which describes the quality of the image registration. The optimization phase is often carried out by using specific optimization methods. The methods used tend to be local searches, that are unable to find any global optimum.

Since the image processing phase and the calculation of the objective function requires a lot of time, global optimization methods such as metaheuristics (which require a high number of evaluations) are avoided and local optimization methods are used instead since they are quicker and more effective at finding local optima.

Chapter written by Johann DRÉO, Jean-Claude NUNES and Patrick SIARRY.

However, if the case arises that local optima exists, then it may be interesting to use global optimization methods [JEN 01].

The combination of images taken over a certain period of time or of different modalities is frequently used to help doctors diagnose certain diseases. When a set of retinal angiogram images is being taken, it is inevitable that the eyes of the patient will move. Before any quantitative analysis can be carried out on the images, the problems that are associated with any involuntary eye movement must be resolved. The methods used in image registration today means that it is possible to calculate a distortion field which reflects the transformation of structures that are present in any given image.

Metaheuristics are part of a family of algorithms whose main objective is to resolve difficult optimization problems. Metaheuristics are generally used as generic optimization methods which are able to optimize a wide range of different problems, without having to make any significant changes to the algorithm being used.

This chapter introduces the registration of retinal angiograms with the help of metaheuristics. The chapter is divided into six parts. It introduces the framework that is used to solve difficult optimization problems. In this chapter we also look at the problems of image registration and image processing as well as introducing the optimization tools to be used in order to resolve such issues. The results of the experiments carried out in this chapter can be seen in section 11.5, and an analysis of the results is given in section 11.6. The chapter ends in section 11.7 with a conclusion on our results.

11.2. Metaheuristics for difficult optimization problems

11.2.1. *Difficult optimization*

11.2.1.1. *Optimization problems*

The general meaning of the term "optimization problem" can be defined by a set of possible solutions S, whose quality can be described by an objective function f. The aim is therefore to find the best s^* solution with the best $f(s^*)$ quality. After this has been achieved, the aim is then to minimize or maximize $f(s)$. Such optimization problems can also lead to other problems if $f(s)$ changes with time or if it is multi-objective, in other words if there are several objective functions which need to optimized.

Deterministic methods, also known as exact methods, are used to resolve such problems as those mentioned in the previous section in a finite time period. These methods generally need to possess a certain number of characteristics of the objective function, such as the objective function's strict convexity, continuity or its derivability. Examples of such deterministic methods include linear, quadratic or dynamic programming, as well as the gradient method or Newton's optimization method, etc.

11.2.1.2. *Difficult optimization*

It is not possible, however, to solve all optimization problems by using deterministic methods which have been mentioned in the previous section. In some cases it can be problematic for such methods to acquire the characteristics of the objective function, for example there may be a lack of strict convexity. Other problems such as the existence of discontinuities, multi-modality, a non-derivable function and the presence of noise, etc. may exist.

In such cases the optimization problem is said to be difficult due to the fact that none of the exact methods is able to solve the problem within a reasonable time scale in one session. In such cases, it is therefore necessary to rely on the use of approach-based optimization heuristics.

The problems associated with difficult optimization include two types of problems: discrete problems and continuous problems. Discrete problems include NP-complete type issues such as the issue of the traveling salesman. An NP problem is said to be complete if it is possible to describe the problem with the aid of a polynomial algorithm which is present in the form of a sub-set of instances. It is relatively easy to describe a solution to such a problem, but the number of solutions that are necessary to solve the issue increases exponentially with the size of the instance. Up until now, the theory of not being able to solve NP-complete problems by using a polynomial algorithm has neither been proven, nor has it been dismissed. No polynomial algorithm has been developed which can resolve such issues. The use of optimization algorithms makes it possible to find an approach-based solution which can be used within a reasonable time scale.

As far as the continuous problems are concerned, the variables associated with each optimization problem are said to be continuous. This is true for problems associated with identification. In this case the aim is to minimize the errors that exist between a system's model and the experimental observations stemming from the model. There is not a lot of research that has been carried out into this type of problem, but it seems that a certain number of well-known difficulties do exist. Such difficulties include: the existence of numerous variables which lead to the creation of unidentified correlations; the presence of noise or more generally, the presence of an objective function which can only be accessed through a simulation process. More realistically, however, certain problems associated with difficult optimization are made up of both discrete and continuous variables.

11.2.2. *Optimization algorithms*

11.2.2.1. *Heuristics*

An optimization heuristic is an approach-based method which is simple, quick, and one which can be adapted to any given problem. Its ability to optimize a problem with a minimum amount of information is counter-balanced by the fact that such heuristics do not provide any guarantee about the quality of optimality of the best solution which is found.

From an operational research point of view, this slight flaw is not always a problem. This is particularly true in cases when only an approximation of the optimal solution is being researched.

11.2.2.2. *Metaheuristics*

Amongst the heuristics that exist, some of them can be adapted to a large number of different problems without having to make any significant changes to the algorithm itself. When this occurs we use the term metaheuristics. The majority of heuristics and metaheuristics use random processes as ways to gather information and to deal with problems such as combinatorial explosion. In addition to this, metaheuristics are, for the most part, iterative methods. In other words, the same research algorithm is applied several times during the optimization process. This means that the metaheuristics do not use the additional information such as the gradient of the objective function. As far as the metaheuristics are concerned, we are interested in their ability to avoid local optima, and this is made possible by either accepting a degradation of the objective function or by using a population of candidate solutions as a research method. If such a research method is used, it further distances the metaheuristics from the local descent heuristic.

Metaheuristics are generally designed to resolve discrete problems, however, they can also be adapted to resolve continuous problems.

Due to the fact that they can be used to deal with a large number of problems, metaheuristics can also be extended to the following areas:

– Multi-objective optimization (also known as multi-criteria optimization) [COL 02]. In multi-objective optimization it is necessary to optimize several contradictory objectives. The aim of this research is not only to find a global optimum, but also a set of optima according to the rules of Pareto optimality.

– Multi-modal optimization. In multi-modal optimization the aim of the research is to find the best global and/or local optima possible.

– Optimization of sounded problems. Here there is a certain level of uncertainty as far as the objective function is concerned. This uncertainty needs to be taken into consideration when it comes to carrying out research in order to find the global and/or local optimum.

– Dynamic optimization. Here the objective function varies according to time. It is therefore necessary to have an optimum value which is as close as possible to the best optimum value for each step of the phase.

– Parallelization. With parallelization, the aim is to increase the rate of the optimization process by dividing the information that needs to be processed into different units which work together. The problem which arises here is that of trying to adapt the metaheuristics so that they can be distributed evenly amongst the individual units.

– Hybridization. The aim of hybridization is to combine all of the advantages of all the different metaheuristics together.

It should not be forgotten that one of the major advantages of metaheuristics is the way in which they can be easily used to deal with concrete problems. Users of such metaheuristics often want to use effective optimization methods which means that it is possible for them to find a global and/or local optimum with relative precision and within a reasonable time period. One of the problems with metaheuristics is that because there are so many of them, their potential users are unable to make a quick decision in relation to which one should be adopted. Once a particular metaheuristic has been chosen, how is it then possible to simplify its image registration so that it can be easily adapted to a given problem?

11.2.2.3. *Ant colony optimization algorithms*

In operational research, and in the field of difficult optimization, the majority of methods that are used have been inspired by systems that exist in real life, and from the world of biology in particular. The fact that the field of biology often studies the composition and behavior of systems known as intelligent systems means that it is possible to model such systems and then use these systems in real-life computational problems. Such approaches are also referred to as bio-inspired artificial intelligence.

Within biology it has been the field of ethology (the study of the behavior of animals) which has recently led to several significant advances as far as optimization is concerned. Such advances include the development of artificial ant systems. These systems are often studied in robotics, in the classification of living objects and in optimization. The term swarm intelligence is often used to describe these systems, where the intelligence of the entire system is greater than the intelligence of all of the individual parts of the system combined [DOR 03].

As far as optimization processes are concerned, using artificial intelligence has led to the creation of new metaheuristics. The algorithms that are used in optimization form a particular class of metaheuristic which has recently been developed to try and resolve the issues associated with discrete difficult optimization. These algorithms have been inspired by the collective behavior of ants in their colonies. A simple colony of agents (ants) communicate indirectly with one another through changes that are made to their environment (pheromones). By relying on their collective experience, the ants (who communicate with one another) are able to find a solution to a problem that they may have to deal with.

The metaheuristic which has been inspired by the ant colony is currently being formalized. The information that is used to describe this metaheuristic is well known and is as follows: a candidate solution is created (by adding different components to the ant colony algorithm), an indirect memory and a structure (both of which can be compared to the memory and structure of a self-organizing system) are used. Different ideas resulting from research have shown that there is a lot of support for the use of the ant colony algorithm. This metaheuristic could also be described as a distributed system where the interactions between the basic components are carried out by stigmergic processes. Using such processes makes it possible for the algorithm to be used globally which, in turn, makes the algorithm capable of solving difficult optimization problems [DRE 03].

Ant colony algorithms have successfully been applied to several different combinatorial problems and are starting to be adapted to continuous optimization problems [DRE 04]. We would also like to highlight the importance of choosing the correct local research method when it comes to developing algorithms which can be used as an alternative to older, often more specialized, metaheuristics.

11.2.2.4. *Estimation of distribution algorithms*

Estimation of distribution algorithms (EDAs) were originally developed as a variant to evolutionary algorithms [MUH 96]. However, in EDAs there are no breeding or mutation operators. In fact, the population of new individuals is chosen at random according to an estimated probability density function. The information relating to the new population comes from the preceding one. In evolutionary algorithms the relationship that exists between the different variables is implicit, whereas with EDAs such relationships are explicitly estimated. These relationships are worked out by estimating the probability distribution for a selected organism that makes up the population of the best individuals.

The main difficulty encountered when using EDAs is linked to the estimation of the probability distribution, which can be difficult, depending on the model used. Several different models have been developed to solve both continuous and combinatorial optimization problems. It should be pointed out that the distribution model is often based on the normal distribution [LAR 02].

General frameworks, into which these algorithms can be integrated, also exist. An example of such a general framework is the IDEA approach [BOS 99, BOS 00a, BOS 00b]. Ant colony optimization algorithms are also generally seen as being distribution and sampling optimization methods [DRE 04].

11.3. Image registration of retinal angiograms

11.3.1. *Existing methods*

Experiments have shown that one of the methods, based on a local research method, and which is mentioned in [NUN 04a], is not powerful enough to deal with strong inter-image variations. In the majority of scientific publications that deal with the automatic registration of retinal images, such as [BER 99, CAN 99, HAM 00, MUK 01, PIN 98, RIT 99, SIM 01, ZAN 99], it is often said that it is necessary to detect the vascular structures of the eyes before the retinal images can be registered. A retinal image is characterized by a low-level local contrast, however, during the late phase of the angiogram (the phase in which the radio contrast disappears). The particular optimization method that is mentioned in [NUN 04a] is able to overcome such problems that are associated with the detection of vascular structures.

11.3.1.1. *Images of retinal angiograms*

A series of images taken of the back of the eye is obtained by using one of two methods: either by using a fluorescein angiogram or an indocyanine green (ICG) algorithm. These are both useful techniques which are used to evaluate retinal and choroidal circulation. They are also used to diagnose and treat several retinal diseases, such as macular degeneration which is linked to ageing, cytomegalovirus retinitis and diabetic retinopathy [RIC 98].

The two techniques mentioned in the previous section involve injecting a radio contrast into a cubital vein in the arm. The radio contrast which is injected into the cubital vein in the arm is either fluorescein or ICG, depending on the technique that is being used for the angiogram. Once the radio contrast has been added, a close observation of how it spreads over the retinal vessels (for a given amount of time) is made. A retinal angiogram (which is a sequence of 36 images taken as soon as the radio contrast is injected into the arm up until five minutes after) is generally divided into three phases: early, average and late. With the aid of a retinal angiogram it is possible to see the vascular tree or choroidal structures of the eye. It is also possible to detect any existing diseases that may be present within the eye.

The process of image registration is necessary if doctors want to be able to detect and quantify any retinal diseases that may be present (diseases which may affect the fovea, the optic nerve and the vascular structures of the eye). Such areas of the eye are analyzed by ophthalmologists.

The retinal vessels are the only significant visible structure [RIC 98] which is present in all of the images that are used in retinal angiograms. However, local variations of intensity can lead to the creation of several problems such as: the creation of a non-uniform image background; a low-level contrast; eye movements; several types of noise; as well as the presence of blood vessels which possess different variations of light in comparison to the background of the image.

Figure 11.1 shows two examples of three successive images which were taken by a fluorescein and by an ICG angiogram. As the images are heavily sounded it means that the actual process of image registration requires a further phase of filtering that must be applied to the vascular tree.

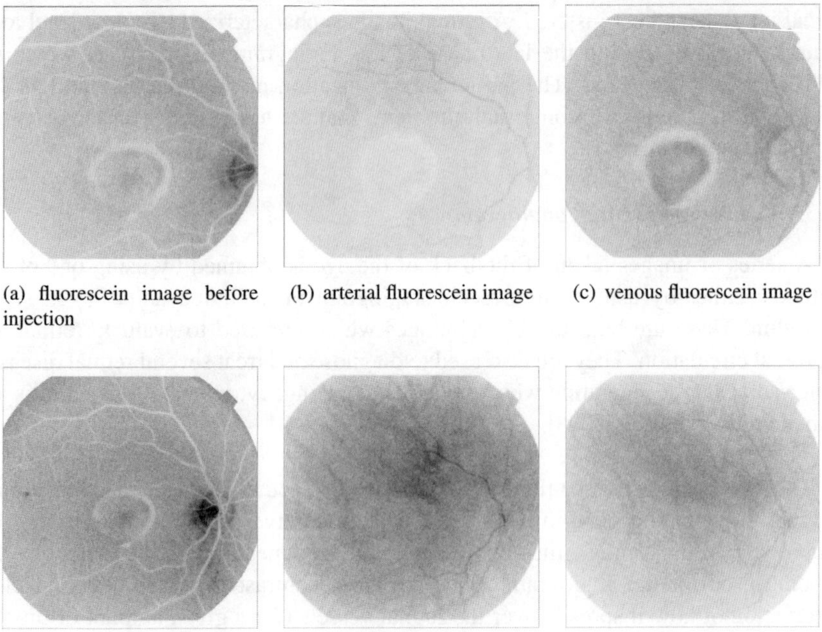

(a) fluorescein image before injection (b) arterial fluorescein image (c) venous fluorescein image

(d) ICG image before injection (e) arterial ICG image (f) venous ICG image

Figure 11.1. *Successive images of angiograms taken from the back of the eye. Top row: images from a fluorescein angiogram. Bottom row: images from an ICG angiogram*

11.3.1.2. *Pre-processing*

The original images of the eye possess a certain number of characteristics which can disrupt the image registration process. It is therefore necessary to carry out some type of pre-processing on the images with the aim of isolating and maintaining the relevant information that is contained in the images.

The filtering phase (or reduction of noise) makes it possible to register the images more effectively. This filter is based on local statistics which are estimated from a neighborhood of 5×5 pixels for each pixel. The main advantage of the filtering phase is that the edges of the vascular structure of the eye are not lost.

During the angiogram, there are generally different variations in intensity, and this is particularly true for the background of an image. Calculating the morphological gradient of the images becomes a necessary process in order to overcome any problems that are associated with the different variations in intensity that arise. The morphological gradient of an image is defined as being the difference between the dilated image and the eroded image.

It is possible to use an additional filtering phase in order to smoothen the gradient image. This phase, known as median filtering, replaces the value of each pixel with the median value of the adjacent pixels in a neighborhood of 3×3 pixels.

More information about the filters which are used can be found in [NUN 03, NUN 04b]. Figure 11.2 shows the pre-processing process that is used.

11.3.1.3. *Traditional image registration methods*

The majority of traditional methods have been used in analyzing eye movement and image registration problems. Such traditional methods can be found in [BAN 96, HAR 94, IRA 94, KIM 01, ODO 94, ZHA 01]. One method which tends to be the most commonly used is the application of multi-resolution methods [ODO 94, ZHA 01]. Such methods are based on optical flow, which calculates the displacement field that exists between two images [NUN 04a]. In the case of multi-resolution methods, all of the research is carried out at increasing resolutions of the images. As far as registering retinal images is concerned, only relevant characteristics from an image are taken [BER 99]. The presence of noise and low-level local contrasts make it difficult to detect the vascular structure of the eye. Images which have been generated by an ICG angiogram have only recently been analyzed by a technique that is based on optical flow.

Several state-of-the-art image registration algorithms can be found in [BRO 92, MAI 98, HIL 01].

11.3.2. *A possible optimization method for image registration*

As part of our study we decided to only work on the translation of the images, a phase that is involved in the image registration process. We decided to focus only on this phase since it is the most important phase of the image registration process and is also the most difficult phase to carry out as far as images of retinal angiograms are concerned. Image registration involves obtaining the best displacement that exists between two images from the sequence of angiogram images.

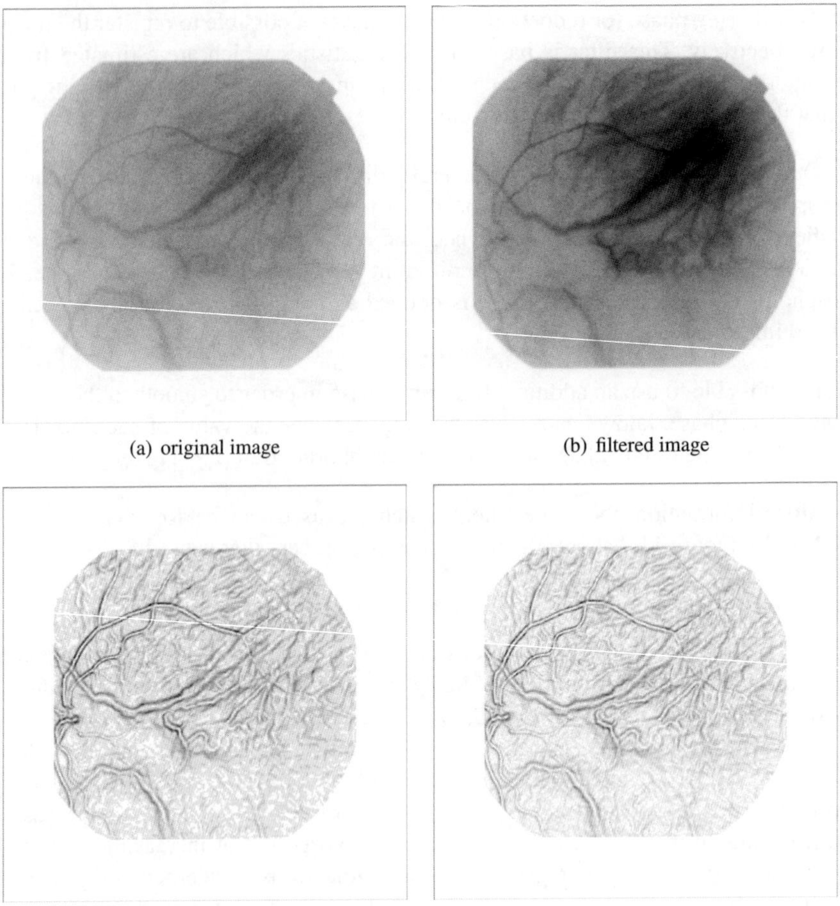

(a) original image (b) filtered image

(c) gradient image (d) filtered gradient image

Figure 11.2. *Pre-processing phase before image registration*

11.3.2.1. *More about the suggested algorithm*

The algorithm is made up of the following phases:

– Wiener filtering of the original images;

– calculating the morphological gradient of the images;

– median filtering (optional phase);

– optimization, this is made up of two parts. First of all the similarity criterion (the sum of the differences of intensity) between two images for the current transformation is calculated. A search is then carried out in order to find the best solution.

11.3.2.2. *The similarity criterion*

Over the past few years, optimization methods based on the intensity of images, and which do not possess any preliminary detection phase, have been used to resolve a large number of image registration problems. Such optimization methods are technically known as iconic methods.

The similarity criterion is the calculation that is used to judge the quality of image registration of two images. In [ROC 99] Roche *et al.* have shown which theories in the field of image registration correspond to a large number of similarity criteria. The aim of this is to be able to understand the uses of such similarity criteria and to be able to choose which particular method can be used to deal with a certain problem. The sum of squared differences [BRO 92], the sum of absolute differences [YU 89], inter-correlation [CID 92], and entropy, etc., are all easy to calculate and do not lead to the creation of complex minimization problems. A lot of algorithms use the sum of squared differences as a measurement of similarity for couplets of images. This measurement is calculated as follows:

$$\text{Precision} = \frac{\sum_{(i,j)=(1,1)}^{(\text{length},\text{width})} \left|I_1(i,j) - I_2(i,j)\right|^2}{(\text{length} \cdot \text{width})} \qquad [11.1]$$

where I_1 and I_2 are two images that need to be registered.

These traditional measurements have different levels of efficiency and precision. However, none of them is really capable of dealing with the relative changes in intensity that occur when moving from one image to another. One of the problems with these measurements is that images (which might be considered as being registered correctly) can still possess errors that are linked to their alignment [ROC 99]. Such alignment errors occur due to the changes in intensity mentioned in the previous sentence. For the purposes of our study, the sum of squared differences is a valid metric since it is calculated from filtered gradient images.

11.4. Optimizing the image registration process

For the purposes of our investigation, the objective function must describe any given registered image in the best way possible. With this in mind, we decided to choose a similarity function as our objective function. The search space to be investigated is the entire range of possible movements associated with image registration.

It is possible to use several optimization techniques in order to carry out the research. As far as our study is concerned, we tested two different methods: a local research method, i.e. the Nelder-Mead Search (NMS) method [NEL 65]; and our own algorithms, the HCIAC and CHEDA algorithms.

11.4.1. *The objective function*

The objective function (defined in section 11.3.2.2) is defined as $f : \mathbb{N} \rightarrow \mathbb{R}$ and must be minimized. Figure 11.3 shows the globally convex aspect of a simple image registration problem.

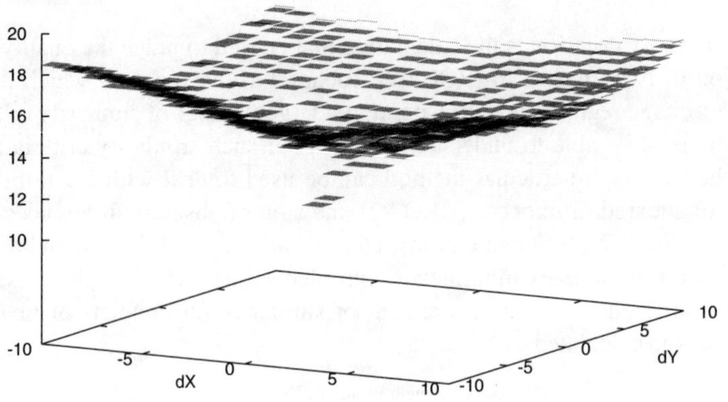

(a) low resolution

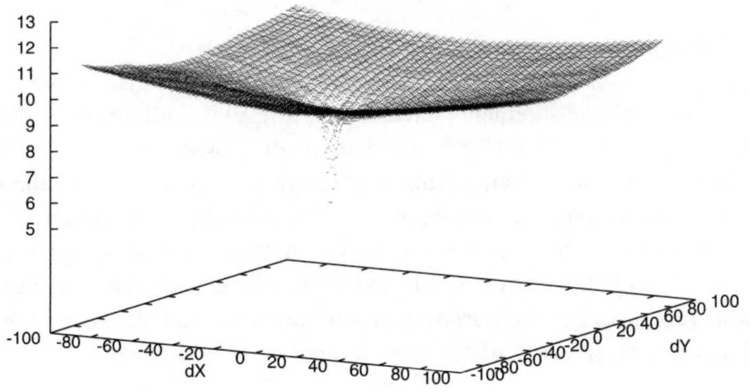

(b) high resolution

Figure 11.3. *Sample from the objective function for a simple image registration problem (described in section 11.5.1)*

If the image registration problem is defined on a set of whole variables then the methods used to solve registration problems need to be adapted so that they can deal with this particular case. In our investigation we simply used a vector curve of the required solution for a particular moment in time. Our tests tended to show that the way in which an algorithm works does not change if the structure of the objective function does not possess any additional information about the structure of the search method that is used. In Figure 11.3 we can see that a simple image registration problem leads to the creation of plateaus for low resolution images. Such plateaus, however, do not change the structure of the objective function which remains globally convex.

11.4.2. *The Nelder-Mead algorithm*

Many algorithms which use a population are not very good at finding local optima in a short time period. Such algorithms are, however, able to localize the regions which contain local optima. Ant colony optimization algorithms are effective when they are used as an additional local research tool. Such a technique is used in the area of discrete optimization with the aim of making all of the algorithms which can be used in this field more competitive with one another [BON 99].

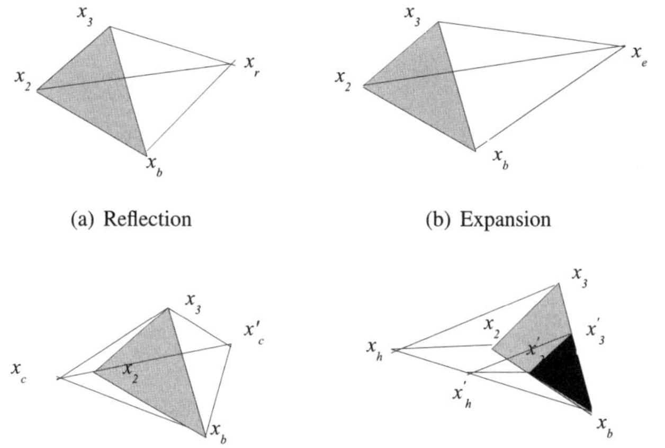

(a) Reflection

(b) Expansion

(c) Contraction (x_c internal, x'_c external)

(d) Multi-contractions

Figure 11.4. *Examples of modifications made to the Nelder-Mead simplex method for a two dimensional problem: reflection, expansion, internal and external contraction and multi-contractions. x_b is the highest point of the minimum value and x_h is the highest point of the maximum value*

The Nelder-Mead algorithm [NEL 65] is an effective and simple local research method which has the advantage of not having to rely on derivative values. The algorithm uses a small population of candidate solutions which are present in the form of a non-degenerated simplex. A simplex is a geometric figure with a non-zero volume of n dimensions, which is a convex envelope or convex hull of $n + 1$ dimensions. The Nelder-Mead method consists of modifying the simplex in four different ways: reflection (co-efficient ρ), expansion (γ), contraction (χ) and shortening (δ) (see Figure 11.4 above). These four types of alterations have been developed so that the Nelder-Mead algorithm follows the gradient of the objective function (see Algorithm 11.1).

Initialize a simplex and calculate the values of the function at the simplex's vertices

Repeat the iterations t, $t \geq 0$ until the stop criterion is reached

Organize the vertices $x_0^t, x_1^t, \ldots, x_n^t$ so that

$$f(x_0^t) \leq f(x_1^t) \leq \cdots \leq f(x_n^t)$$

Calculate the center of gravity

$$\bar{x}^t = \frac{1}{n} \cdot \sum_{i=0}^{n-1} x_i^t$$

Reflection calculate the point of reflection from the equation:

$$x_r^t = \bar{x}^t + \rho(\bar{x}^t - x_n^t)$$

If $f(x_0^t) \leq f(x_r^t) < f(x_{n-1}^t)$ **then** $x_n^t \leftarrow x_r^t$, next iteration

Expansion if $f(x_r^t) < f(x_0^t)$, calculate the point of expansion:

$$x_e^t = \bar{x}^t + \chi(x_r^t - \bar{x}^t)$$

If $f(x_e^t) < f(x_r^t)$ **then** $x_n^t \leftarrow x_e^t$, next iteration

If not $x_n^t \leftarrow x_r^t$, next iteration

Contraction

Exterior If $f(x_{n-1}^t) \leq f(x_r^t) < f(x_n^t)$, carry out process of exterior contraction:

$$x_{oc}^t = \bar{x}^t + \gamma(x_r^t - \bar{x}^t)$$

If $f(x_{oc}^t) \leq f(x_r^t)$ **then** $x_n^t \leftarrow x_{oc}^t$, next iteration

If not go to the shortening phase

Interior If $f(x_{n-1}^t) \leq f(x_r^t) \geq f(x_n^t)$, carry out process of interior contraction:

$$x_{ic}^t = \bar{x}^t + \gamma(x_n^t - \bar{x}^t)$$

If $f(x_{ic}^t) \leq f(x_n^t)$ **then** $x_n^t \leftarrow x_{ic}^t$, next iteration

If not go to the shortening phase

Shorten the simplex method around x_0^t:

$$x_i^t \longleftarrow \hat{x}_i^t = x_i^t + \frac{1}{2}(x_0^t - x_i^t), \quad i = 1, \ldots, n$$

End

Algorithm 11.1. *The Nelder-Mead simplex method*

11.4.3. *The hybrid continuous interacting ant colony (HCIAC)*

The HCIAC algorithm is explained in more detail in Figure 11.5.

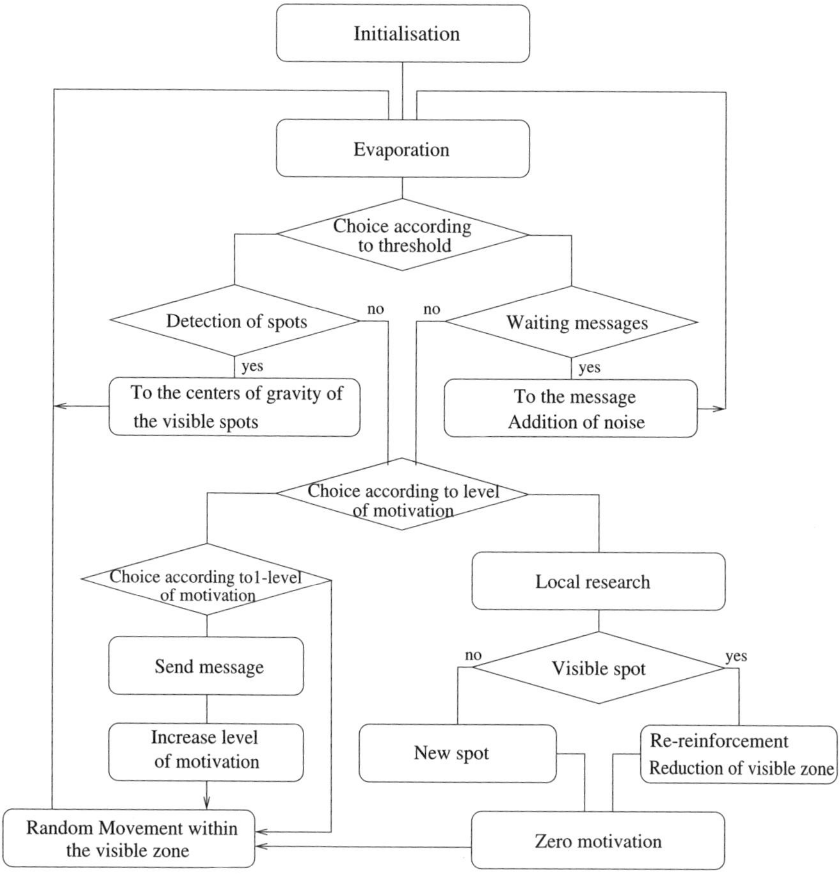

Figure 11.5. *The HCIAC algorithm*

The first stage of the algorithm involves placing η ants in a particular research area following a uniform distribution. The first stage also involves initializing all of the parameters of the ants.

Pheromone spots (marks which can be seen in the research area) are removed when the value $\tau_{j,t+1}$ of each spot j at a time $t + 1$, is calculated according to the following equation:

$$\tau_{j,t+1} = \rho \cdot \tau_{j,t}$$

where ρ represents a persistence parameter.

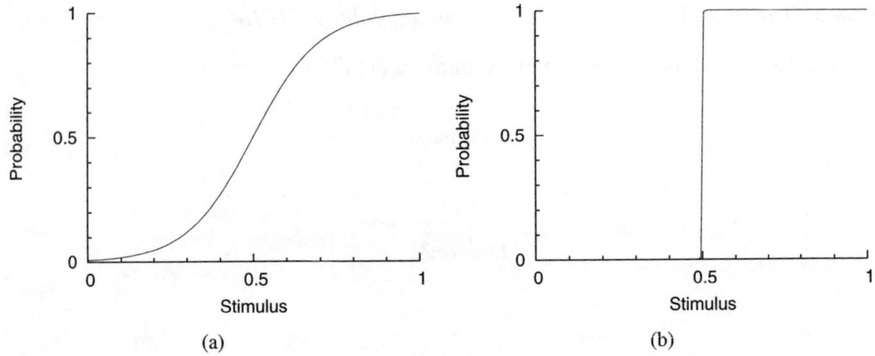

Figure 11.6. *Stimulus/response functions: (a) progressive choice, (b) binary choice*

A decision is then made in accordance with a stimulus/response function (Figure 11.6) and is carried out as follows. An ant chooses a method that it can use to communicate with another ant. Such a communication method is referred to as a channel of communication. Two parameters which are part of the choice function are labeled χ_τ for the threshold level, and χ_ρ for the power of the choice function. These parameters are applied to the entire population of ants. Each individual ant, i, has its own set of individual parameters, χ_i. These individual parameters are initialized during the first phase of the HCIAC algorithm, in accordance with the normal distribution $\mathcal{N}_{\chi_m,\chi_d}$.

If an ant chooses the "trail" channel of communication (which is created by placing pheromone spots in the search space) then the ant will look for a spot within its visible zone π_i. The visible zone is created as part of the first phase of the HCIAC algorithm, in accordance with $\mathcal{N}_{\pi_m,\pi_d}$. If spots do exist, the ant will move towards the center of gravity of the visible spots. If no spots exist then the ant will move on to the decision-making phase.

On the other hand, if an ant chooses the "direct" channel of communication (which consists of sending a message) then the ant will inspect a stack of messages. If messages are available then the ant will move towards the point that is indicated by a message and will then add some noise to the place it has just moved to (in other words, the ant moves randomly within its visible zone). If no messages are available then the ant will move on to the decision-making phase.

If there is no spot nor any message, the ant will make a decision based on its own motivation level ω. The ant makes it decision based on the following aspects which are all initialized for the entire population of ants in the first phase of the HCIAC algorithm: a stimulus/response function, the threshold of motivation ω_ρ and the ant's strength ω_τ. The stimulus ω_i is set to zero during the first phase of the HCIAC algorithm.

The first choice made by the ant leads directly to the use of the "direct" channel of communication and therefore to the management of messages. During this phase another decision is made. This additional decision is taken when a stimulus/response function possesses the same parameters that it had in the previous phase of the algorithm. The exception to this rule is when the stimulus has the value $(1 - \omega_i)$. If an ant decides to disrupt the stack of messages, then a message will be sent to another ant (which is chosen at random) and the overall motivation level will be increased by a small amount ω_δ. If the ant's second choice is to ignore the messages, then the ant will move in any random direction.

If a decision to be made involves a local search, then a Nelder-Mead research method is launched. When such a method is launched the position of the ant is used as the starting point and the radius of the ant's visible zone is taken as the length of the initial phase. The local search is limited to ν evaluations of the objective function. Once the local search has come to an end the ant then looks for visible spots. If there is a spot, it is reinforced in the following way:

$$\tau_{j,t+1} = \tau_{j,t} + \frac{\tau_{j,t}}{2}.$$

The radius of the ant's visible zone is reduced to the distance of the furthest visible spot that can be seen by the ant, from the actual location of the ant itself. If there are no spots then the ant will place new ones in its visible zone. The number of spots that an ant will place in its visible zone is equivalent to the value of the objective function for that specific area. Once the trail channel of communication has come to an end, the motivation level of the ant is set to zero once again.

The final possible phase involves the ant moving randomly within the visible zone. The HCIAC algorithm will stop if it is unable to find the best optimum after θ evaluations of the objective function.

11.4.4. *The continuous hybrid estimation of distribution algorithm*

The IDEA approach determines the general framework of an estimation of distribution algorithm (EDA). Such an algorithm describes three key phases:

1) diversification: this involves randomly choosing a sample. This choice depends on the particular probability distribution that is considered;

2) memory: this involves choosing a probability distribution which best describes the sampled solutions;

3) intensification: selection of the best solutions.

As part of our study and in order to improve the intensification of the IDEA approach, we decided to combine an IDEA type algorithm [BOS 99] with a Nelder-Mead local research method [NEL 65]. This combined algorithm is known as a continuous hybrid estimation of distribution algorithm (CHEDA).

By adopting the IDEA approach, it is possible to use a large number of probability density functions. As far as our study is concerned, we decided to use a multi-variate normal distribution as a basis. Indeed, it is not possible to take any of the dependencies that exist between variables into consideration when a mono-variate distribution is used (see Figure 11.7). Such dependencies are often present in continuous problems which arise in the field of engineering. It is a lot easier to distribute the population of candidate solutions whenever a multi-variate normal distribution is used.

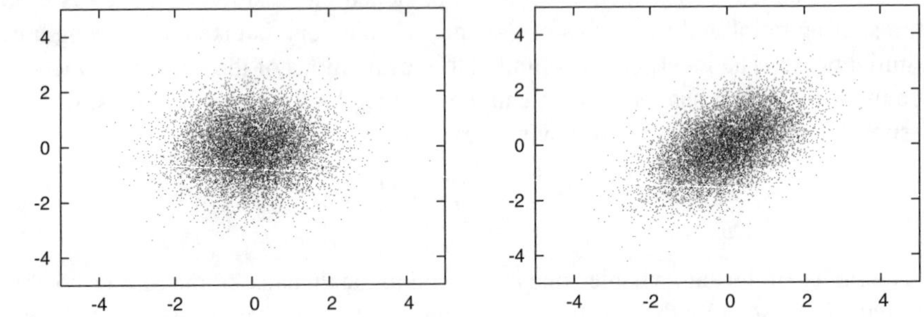

(a) the sum of two monovariate normal distributions (b) a bivariate normal distribution

Figure 11.7. *The difference between the sum of two mono-variate normal distributions and a multi-variate normal distribution. The bi-variate normal distribution makes it possible to consider a correlation that might exist between two variables which are represented in the form of co-variances. The graphs above show the 2D distribution of 10,000 points, with a vector possessing a zero average, a variance of one and a covariance of* $\frac{1}{2}$

The chosen EDA method can be seen in Algorithm 11.2, where $\mathcal{N}_{m,V}$ is the multi-normal distribution, m is the average vector and V is the variance/co-variance matrix. The distribution method that is used becomes truncated, depending on what the boundaries of the search area are.

Three different versions of CHEDA exist, each version having a slightly different intensification phase from the others; the intensification phases are classed as follows: intensification by selection, intensification by local research, as well as intensification by local research and selection. Figure 11.8 shows the differences that exist between the three different intensification phases.

Initialize a population P_0 of π points in $U_{x_{\min},x_{\max}}$

Until the stop criterion

Estimate the m and V parameters of the sample P_{i-1}

Take a population P_i of π points in $\mathcal{N}_{m,V}$

Evaluate P_i

Intensify from P_i

End

Algorithm 11.2. *The EDA algorithm used*

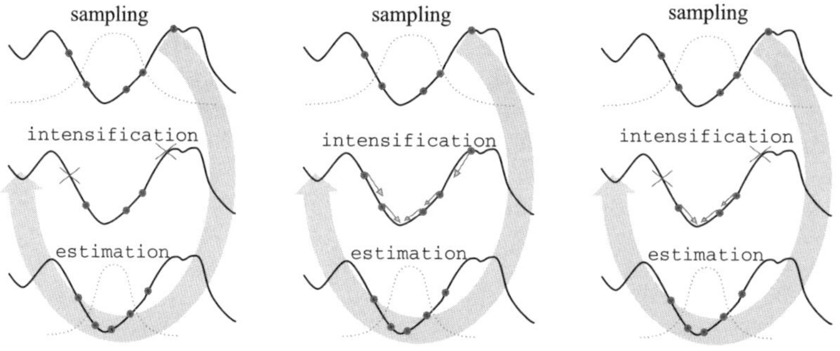

(a) CHEDA$_s$ (b) CHEDA$_n$ (c) CHEDA$_{sn}$

Figure 11.8. *Variates used for the CHEDA methods. The combination of the three phases forms an iteration of the algorithm. This iteration of the algorithm is repeated several times during the optimization process*

Due to the fact that three different intensification phases exist, it means that each version has its own parameters that need to be regulated:

π the number of points of the sample (CHEDA$_s$, CHEDA$_n$ and CHEDA$_{sn}$);

β the proportion of candidate solutions to be chosen from P_i (CHEDA$_s$, CHEDA$_n$ and CHEDA$_s$ n);

ν the maximum number of modifications that can be made to the Nelder-Mead algorithm, as soon as the selection phase has started (CHEDA$_n$ and CHEDA$_{sn}$).

11.4.5. *Algorithm settings*

The HCIAC algorithm is used with the following default settings: $\rho = 0.5$, $\chi_m = 0.5$, $\chi_d = 0.2$, $\pi_m = 0.8$, $\pi_d = 0.5$, $\omega_\delta = 0.1$, $\chi_\tau = \omega_\tau = 0.5$, $\chi_\rho = \omega_\rho = 10$. The crucial parameters, which include the number of ants and the number of iterations of the algorithm, are initialized when $\nu = 3$, and when $\eta = 10$ respectively. These values correspond to a fast-working algorithm (i.e. there is a low number of evaluations). As far as our study is concerned, the maximum number of evaluations possible is fixed at 200 evaluations of the objective function.

RX / Algorithm	0.1		0.5		1.0	
	min	max	min	max	min	max
Optic Flow	7	43*	124	1090*	704	4531*
HCIAC	13	30	86	195	367	826

Table 11.1. *A comparison of the total processing times (in seconds) for the two different optimization methods (indicated as Algorithm in Table 11.1) according to resolution (indicated as RX in the same table). The values are marked with a * when the algorithm has been unable to find any optimum. For the HCIAC algorithm the times which have been noted include the time required for evaluating both the optimization algorithm and for the calculation of the objective function (the evaluation time of the objective function is proportional to the resolution)*

The CHEDA algorithm is used with the following settings: $\pi = 20$, $\beta = 30\%$ and $\nu = 10$. The algorithm will stop after a maximum of 200 evaluations have been carried out.

The NMS algorithm uses its own default settings. The algorithm will stop if the size of the simplex method is less than 10^{-8}, or if it has reached its maximum limit of 200 evaluations.

11.5. Results

11.5.1. *Preliminary tests*

All of the angiograms mentioned in this study have been digitized by a video signal with a resolution of 1,024 $\times$ 1,024 pixels and a grayscale of eight bits per pixel. The different algorithms have been tested using different resolutions, ranging from 10% up to 100% of the size of the original image. The algorithms have also been tested with and without the additional phase of median filtering which is applied to the gradient image.

In order to describe how the optimization algorithm functions and to test its effectiveness, we decided to use a typical registration problem where the global optimum is known (see Figure 11.9).

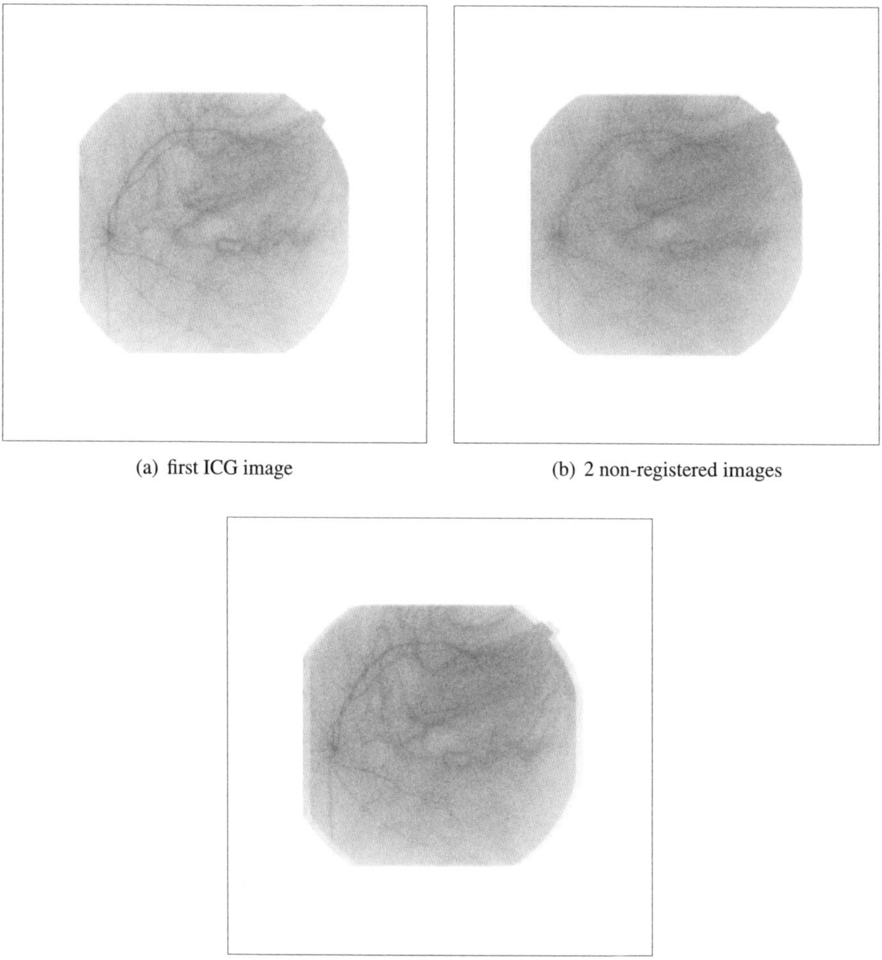

(a) first ICG image

(b) 2 non-registered images

(c) ICG registered images

Figure 11.9. *A typical registration problem which is used to test the accuracy of the algorithms*

In this section we have only mentioned the results of the tests that were carried out on the HCIAC algorithm, in an attempt to avoid overloading the reader with information. We drew the same conclusions for the results that were obtained for the CHEDA algorithm, and these results are explained in more detail in section 11.5.3.

Traditional optimization techniques require much less processing time for low resolution image registration problems. However, for high resolution image registration problems, the processing time can increase dramatically (as can be seen in Table 11.1 and Figure 11.10). The processing time, as far as metaheuristics are concerned, stays practically the same (regardless of the resolution level; see Figure 11.11).

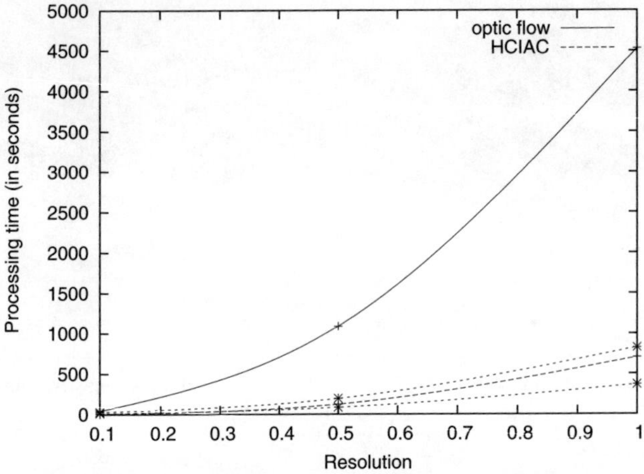

Figure 11.10. *A comparison of the processing times (in seconds) for the two optimization methods: optic flow (indicated by a +) and the HCIAC algorithm (indicated by a ∗). The minimum and maximum processing times for the HCIAC algorithm are illustrated by the dotted lines*

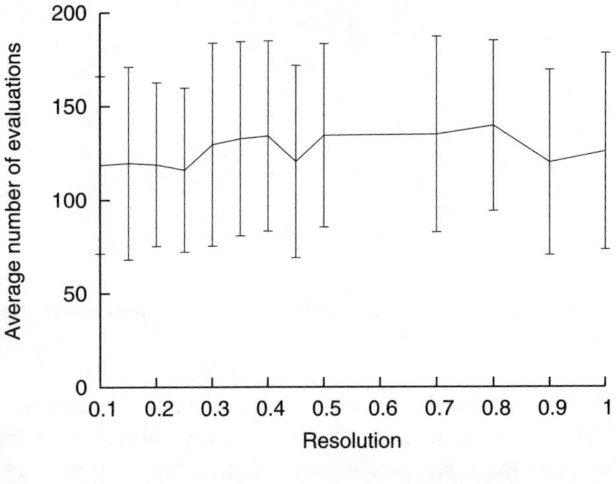

Figure 11.11. *The number of evaluations of the objective function which are necessary for the HCIAC algorithm*

11.5.2. *Accuracy*

In order to test the accuracy of the optimization methods we decided to carry out 50 different tests on the same image registration problem. For the traditional local optimization methods that were used with a low-level resolution, we tested how the level of resolution affected the accuracy of the metaheuristics. The aim of this particular test was to estimate the importance of the additional information which is supplied by a higher resolution level. Furthermore, additional median filtering is often used to improve the optimization phase. However, since processing time is an extremely important factor in resolving image registration problems, we tested how the phase of additional filtering affected the optimization process. In order to carry out this test we used the HCIAC algorithm.

Figure 11.12 shows that the phase of additional filtering is important for high-level resolutions. Standard deviation is an accurate measurement of the accuracy of the algorithm, since the standard deviation indicates the number of images which have been properly registered. The larger the standard deviation, the more often the algorithm will fail to determine how an image should be properly registered.

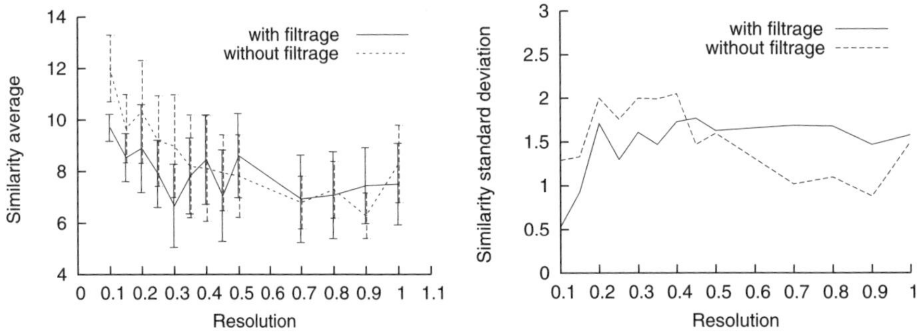

(a) average and standard deviation

(b) standard deviation only

Figure 11.12. *A comparison of the averages and standard deviations of the similarities depending on the level of resolution*

Figure 11.13 shows that when the NMS algorithm is used on its own it is able to find an optimal registered image. This, however, is at the expense of the standard deviation which is proportional to the resolution level of the image. The resolution level of the image has little impact on the accuracy of the HCIAC algorithm.

11.5.3. *Typical cases*

As far as the accuracy of the results is concerned, we tested the metaheuristics on a high-level resolution image (100% the size of the original image) without any

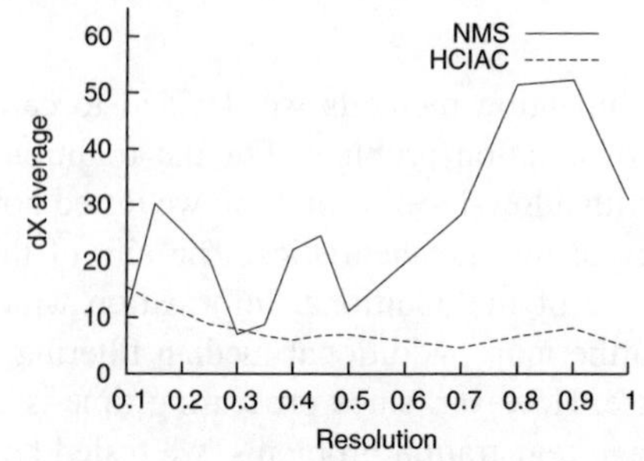

(a) horizontal movement (b) vertical movement

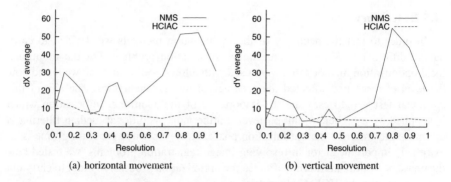

Figure 11.13. *Standard deviations according to optimal image registration, without median filtering*

additional median filtering. We used eight couples of angiogram images and carried out 50 optimizations on each image.

As can be seen in Figure 11.14, the HCIAC is unable to accurately determine the global optimum for each try. In fact, if the HCIAC algorithm always finds a value that is close to the value of the perfectly registered image then the algorithm will produce an inaccurate finding. The two variants of the CHEDA algorithm, however, produce fewer errors and are more accurate in their findings. The variant CHEDA$_{sn}$ is either equivalent to, or better than CHEDA$_s$.

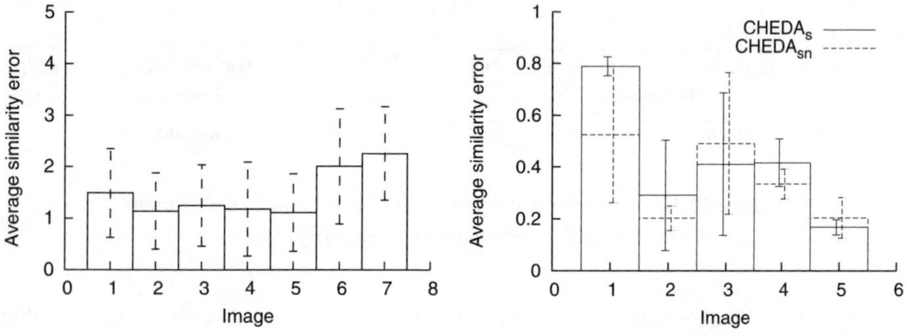

(a) HCIAC (b) CHEDA

Figure 11.14. *Averages and standard deviations of similarities for the different test images*

Figure 11.15 shows that the HCIAC algorithm is more accurate than the NMS algorithm when it is used alone. The CHEDA variants produce a lower standard deviation.

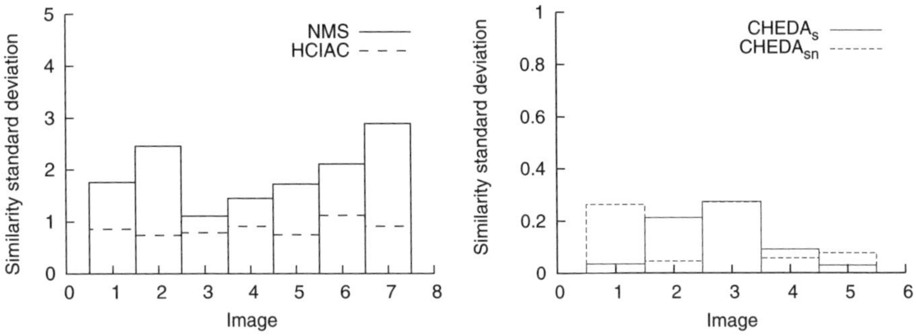

(a) HCIAC and NMS

(b) CHEDA

Figure 11.15. *Standard deviation of similarities for different images using the HCIAC, NMS and CHEDA algorithms*

11.5.4. *Additional problems*

11.5.4.1. *Peripheral image registration*

Some problems that are associated with exact image registration may remain, and to solve such problems a simple translation of an image is not enough to represent a real transformation of the image. Figure 11.16 shows that even if a perfect registration of the image has been obtained, certain errors in the periphery of the registered angiogram image may still exist. It is sometimes better to proceed to the final registration of the image by using an elastic method [NUN 04a]. The fact that the most difficult problem (rigid image registration) is dealt with by metaheuristics means that it is possible to use the optic flow optimization method on the registered image. Our tests have shown that only one transition from the optic flow onto the residual image is required in order to overcome the problems linked with errors that exist in the periphery of the image.

11.5.4.2. *Affine image registration*

In some cases, a rotation transformation and a zoom transformation can be added to the translation transformation. Even though these transformations are generally of a poor quality as far as retinal angiograms are concerned, they can be improved upon by adopting one of the following two approaches:

1) introduce the rotation and zoom transformations as additional parameters of the optimized problem (the problem which has been optimized by one of the metaheuristics);

2) let the final optic flow resolve any additional transformations.

Our tests have shown that if the first approach is adopted, it is possible to carry out large-scale transformations. However, if this approach is used it will dramatically

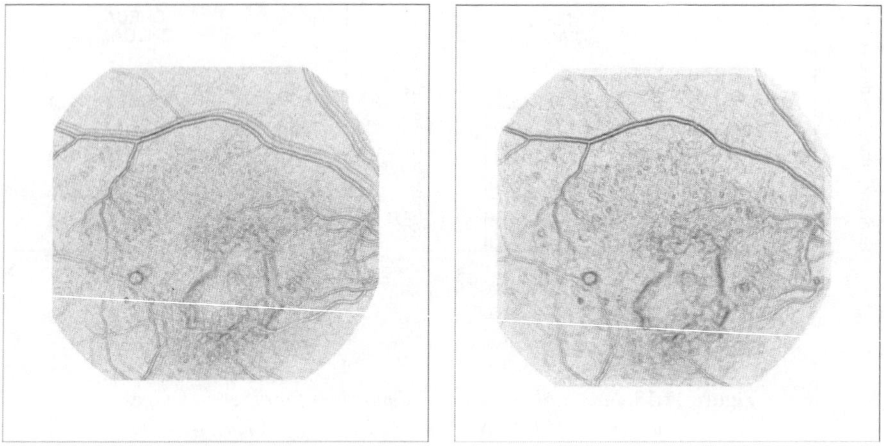

(a) non registered (b) registered

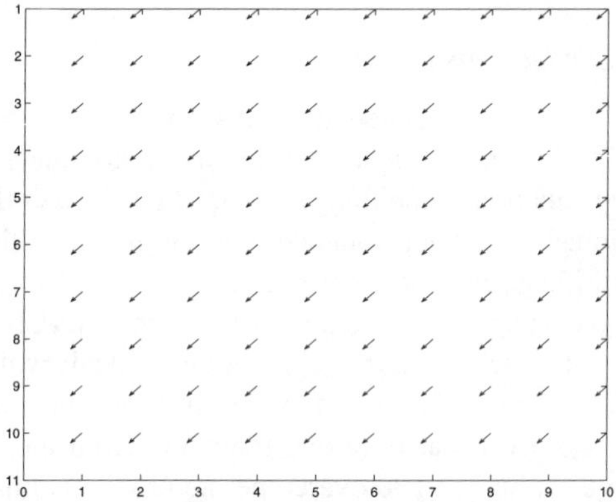

(c) translation

Figure 11.16. *An example of poor peripheral image registration*

increase the complexity of the problem. It is therefore necessary to increase the number of points to be tested, which in turn increases the processing time that is required. On the other hand, if the second approach is adopted, it is possible to carry out transformations without having to increase an algorithm's processing time. However, this method can only be used to resolve small-scale transformation issues.

In the case of retinal angiograms, the rotation and zoom transformations are generally small-scale, in comparison to the translations which are caused by the movement of the eye. It is therefore best to use the optic flow algorithm.

11.6. Analysis of the results

Our results have shown that the use of metaheuristics during the optimization phase of the registration of an angiogram image can be particularly useful as far as high resolution images are concerned. The main advantage of the optimization methods which have been introduced in this chapter is their processing time which is constant, regardless of the resolution of the images. With traditional algorithms, such as the optic flow algorithm, processing time increases as the resolution of the images increases, whereas with the methods introduced in this chapter the processing time remains almost constant. This advantage comes from the fact that metaheuristics use samples from the objective function. Such samples do not rely on the resolution of the image that is to be registered. The processing time for each algorithm is therefore a function of the processing time that is required to find a solution to a problem.

However, processing time is one of the major constraints faced by the algorithms. Our results have shown that, for high resolutions, using additional filtering is not advantageous because it decreases the accuracy of the algorithm and increases its processing time by adding an additional phase to the optimization process. It is therefore necessary to avoid adding additional filtering phases since they are capable of removing any relevant information that can be used by the metaheuristics and thus reduce their ability to find a global optimum.

The Nelder-Mead local research method has the same advantages as the metaheuristics in terms of processing time. However, our results have shown that the accuracy of this local research method decreases when it is used to resolve high-resolution problems. This is due to the fact that local optima are present, and as a result the research method is tricked into believing that these optima are in fact global, not local. This problem occurs quite often when optimization methods are used to resolve image registration issues and one solution that can be used involves adopting a multi-resolution approach. Such approaches require large amounts of processing time. The use of metaheuristics, however, solves the problem linked to the presence of local optima without dramatically increasing the algorithm's processing time.

The CHEDA_{sn} method seems to be the most effective optimization method, closely followed by the CHEDA_s method. The HCIAC algorithm has a relatively poor level of accuracy. A large part of the CHEDA's efficiency is undoubtedly due to the use of a normal distribution, which has a form that is very similar to that of the objective function.

The problems linked to peripheral errors or affine transformations are much less serious than the problems linked to searching for the perfect translation transformation of a retinal angiogram. This problem can only be resolved by using a rigid approach. An elastic image registration method can be used to carry out a residual transformation of a retinal angiogram.

11.7. Conclusion

Image registration is an important concept for several applications which deal with the analysis of medical images, such as the correction of eye movements or multi-modal forms.

Due to the movements that are made by the eyes when a retinal angiogram is being taken, an image registration phase is required in order to improve the quantitative analysis of retinal diseases. The image registration process makes it possible to compare an image with the one that was taken directly before. Image registration also makes it possible to analyze how much a particular retinal disease has progressed over time.

However, the data that comes from a retinal angiogram varies considerably in terms of intensity. Is a pre-processing really necessary to overcome this problem and the problem that is linked to the presence of noise? Wiener filtering, as well as the calculation of the image's gradient, are essential, whereas median filtering is not necessary.

The optimization techniques which have been presented in this chapter have been adapted so that they can be used to work with high resolution images. The HCIAC algorithm is more accurate than the Nelder-Mead research method. The CHEDA algorithm is the most accurate, as it leads to the creation of the highest quality registered images.

One possible way to improve the image registration process as it stands at the moment would be to remove the co-variances from the CHEDA algorithm. This would mean that the vertical and horizontal movements of the eye would not be in correlation with one another, and a sum of the uni-variant distributions would lead to similar results, with a shorter processing time. Another possible improvement would be to use other objective functions that provide a better representation of the problems associated with image registration. An example of such an objective function would be the use of mutual information.

11.8. Acknowledgements

We would like to thank the Intercommunal Hospital in Créteil, France, for supplying us with the images of the retinal angiograms.

11.9. Bibliography

[BAN 96] BANGHAM J.A., HARVEY R. and LING P.D., "Morphological scale-space preserving transforms in many dimensions", *J. Electronic Imaging*, vol. 5, pp. 283–299, 1996.

[BER 99] BERGER J.W., LEVENTON M.E., HATA N., WELLS W. and KINIKIS R., "Design considerations for a computer-vision-enabled ophtalmic augmented reality environment", *Lectures Notes in Computer Science*, vol. 1205, pp. 399–410, 1999.

[BON 99] BONABEAU E., DORIGO M. and THERAULAZ G., *Swarm Intelligence, From Natural to Artificial Systems*, Oxford University Press, 1999.

[BOS 99] BOSMAN P.A.N. and THIERENS D., An algorithmic framework for density estimation based evolutionary algorithm, Report no. UU-CS-1999-46, Utrecht University, 1999.

[BOS 00a] BOSMAN P. and THIERENS D., "Continuous iterated density estimation evolutionary algorithms within the IDEA framework", in MUEHLENBEIN M. and RODRIGUEZ A. (Eds.), *Proceedings of the Optimization by Building and Using Probabilistic Models OBUPM Workshop at the Genetic and Evolutionary Computation Conference GECCO-2000*, San Francisco, California, Morgan Kauffmann, pp. 197–200, 2000.

[BOS 00b] BOSMAN P. and THIERENS D., IDEAs based on the normal kernels probability density function, Report no. UU-CS-2000-11, Utrecht University, 2000.

[BRO 92] BROWN L.G., "A survey of image registration techniques", *ACM Comput. Surveys*, vol. 24, pp. 325–376, 1992.

[CAN 99] CAN A. and STEWART C.V., "Robust hierarchical algorithm for constructing a mosaic from images of the curved human retina", *IEEE Conf. on Computer Vision and Pattern Recognition*, vol. 22, 1999.

[CID 92] CIDECIYAN A.V., JACOBSON S.G., KEMP C.M., KNIGHTON R.W. and NAGEL J.H., "Registration of high resolution images of the retina", *SPIE: Medical Imaging VI: Image Processing*, vol. 1652, pp. 310–322, 1992.

[COL 02] COLLETTE Y. and SIARRY P., *Optimisation Multiobjectif*, Eyrolles, 2002.

[DOR 03] DORIGO M. and STÜTZLE T., "The ant colony optimization metaheuristics: algorithms, applications, and advances", *Handbook of Metaheuristics*, vol. 57 of *International Series in Operations Research and Management Science*, Kluwer Academic Publishers, Boston Hardbound, January 2003.

[DRE 03] DREO J. and SIARRY P., "Colonies de fourmis et optimisation continue. De l'utilité de l'optimisation en général et des colonies de fourmis en particulier, quand l'éthologie et l'informatique se croisent", *Research seminar by the ISBSP*, Créteil, University of Paris 12, 2003.

[DRE 04] DREO J., Adaptation de la méthode des colonies de fourmis pour l'optimisation en variables continues. Application en génie biomédical, PhD Thesis, University of Paris 12, 2004.

[HAM 00] HAMPSON F.J. and PESQUET J.C., "Motion estimation in the presence of illumination variations", *Signal Processing: Image Communication*, vol. 16, pp. 373–381, 2000.

[HAR 94] HART W.E. and GOLDBAUM M.H., "Registering retinal images using automatically selected control point pairs", *IEEE International Conference on Image Processing*, 1994.

[HIL 01] HILL D.L.G., BATCHELOR P.G., HOLDEN M. and HAWKES D.J., "Medical Image Registration", *Physics in Medicine and Biology*, vol. 46, pp. 1–45, 2001.

[IRA 94] IRANI M., ROUSSO B. and PELEG S., "Computing occluding and transparent motion", *IJCV*, vol. 12, pp. 5–16, 1994.

[JEN 01] JENKINSON M. and SMITH S., "A global optimisation method for robust affine registration of brain images", *Medical Image Analysis*, vol. 5, pp. 143–156, 2001.

[KIM 01] KIM M., JEON J.C., KWAK J.S., LEE M.H. and AHN C., "Moving object segmentation in video sequences by user interaction and automatic object tracking", *Image and Vision Computing*, vol. 19, pp. 245-260, 2001.

[LAR 02] LARRAÑAGA P. and LOZANO J., *Estimation of Distribution Algorithms, A New Tool for Evolutionary Computation*, Genetic Algorithms and Evolutionary Computation, Kluwer Academic Publishers, 2002.

[MAI 98] MAINTZ J.B.A. and VIERGERVER M.A., "A survey of medical image registration", *Med. Image Anal.*, vol. 2, pp. 1–36, 1998.

[MUH 96] MUHLENBEIN H. and PAASS G., "From recombination of genes to the estimation of distributions I. Binary parameters", *Lecture Notes in Computer Science 1411: Parallel Problem Solving from Nature*, vol. PPSN IV, pp. 178–187, 1996.

[MUK 01] MUKHOPADHYAY S. and CHANDA B., "Fusion of 2D grayscale images using multiscale morphology", *Pattern Recognition*, vol. 34, pp. 606–619, 2001.

[NEL 65] NELDER J.A. and MEAD R., "A simplex method for function minimization", *Computer Journal*, vol. 7, pp. 308–313, 1965.

[NUN 03] NUNES J.-C., Analyse multiéchelle d'image. Application à l'angiographie rétinienne et à la DMLA, PhD Thesis, University of Paris 12, December 2003.

[NUN 04a] NUNES J.C., BOUAOUNE Y., DELÉCHELLE E. and BUNEL P., "A multiscale elastic registration scheme for retinal angiograms", *Computer Vision and Images Understanding*, vol. 95, no. 2, pp. 129–149, 2004.

[NUN 04b] NUNES J.-C., DREO J. and SIARRY P., "Rigid registration of retinal angiograms through Nelder-Mead optimization", *International Workshop on Electronics and System Analysis, IWESA'04*, 2004.

[ODO 94] ODOBEZ J.M. and BOUTHEMY P., Robust multi-resolution estimation of parametric motion models applied to complex scenes, Report no. 788, IRISA, 1994.

[PIN 98] PINZ A., BERNÖGGER S., DATLINGER P. and KRUGER A., "Mapping the human retina", *IEEE Trans. on Medical Imaging*, vol. 17, pp. 606–619, 1998.

[RIC 98] RICHARD G., SOUBRANE G. and YANNUZZI L.A., "Fluorescein and ICG angiography", *Thieme*, 1998.

[RIT 99] RITTER N., OWENS R., COOPER J., EIKELBOOM R.H. and SAARLOOS P.P.V., "Registration of stereo and temporal images of the retina", *IEEE Trans. On Medical Imaging*, vol. 18, pp. 404–418, 1999.

[ROC 99] ROCHE A., MALANDAIN G., AYACHE N. and PRIMA S., "Towards a better comprehension of similarity measures used in medical image registration", *MICCAI*, 1999.

[SIM 01] SIMO A. and DE VES E., "Segmentation of macular fluorescein angiographies, a statistical approach", *Pattern Recognition*, vol. 34, pp. 795–809, 2001.

[TAL 02] TALBI E.-G., "A Taxonomy of Hybrid Metaheuristics", *Journal of Heuristics*, vol. 8, no. 5, pp. 541–564, 2002.

[YU 89] YU J.J.-H., HUNG B.-N. and LIOU C.-L., "Fast algorithm for digital retinal image alignment", *IEEE Ann. Int. Conf. Engineering Medicine Biology Society, Images Twenty-First Century*, vol. 2, pp. 374–375, 1989.

[ZAN 99] ZANA F. and KLEIN J.C., "A multimodal registration algorithm of eye fundus images using vessels detection and Hough transform", *IEEE Trans. on Medical Imaging*, vol. 18, pp. 419–458, 1999.

[ZHA 01] ZHANG Z. and BLUM R.S., "A hybrid image registration technique for a digital camera image fusion application", *Information Fusion*, vol. 2, pp. 135–149, 2001.

Chapter 12

Joint Estimation of the Dynamics and Shape of Physiological Signals through Genetic Algorithms

12.1. Introduction

The aim of this chapter is to introduce an optimization technique which is based on genetic algorithms (GA). This optimization technique will be used in order to estimate brainstem auditory evoked potentials (BAEPs). We must point out that in certain abnormalities these physiological signals are generally highly non-stationary and are also corrupted by heavy noise, basically due to the electroencephalogram activity (EEG).

Estimating the BAEPs relies on several models, relative to both their dynamics and their shape. In this chapter, a definition of BAEPs will be given as well as an explanation on the way they are generated. An insight into the techniques used in estimating the BAEPs will then be introduced in section 12.3. The principle of GAs will be reviewed in section 12.4. The use of such algorithms to deal with the problems related to BAEP non-stationarity is described in sections 12.5 and 12.6.

Chapter written by Amine Naït-Ali and Patrick Siarry.

12.2. Brainstem auditory evoked potentials

BAEPs are low energy electrical signals, generated when stimulating the auditory system by acoustical impulses. They are mainly used to ensure the earliest possible diagnosis of acoustic neuromas.

An acoustic neuroma is in fact a benign tumor which might lead, in some cases, to the death of a patient. One of the first signs of this disease in patients is single-sided deafness. Figure 12.1 shows this type of tumor, which can clearly be seen on a nuclear magnetic resonance (NMR) image. Although this imaging modality provides precise and detailed information on the location and features of the tumor, the study of BAEPs remains a non-invasive and inexpensive test. Therefore, the use of BAEPs is not only limited to the detection of acoustic neuromas, they can also be used to confirm the integrity of the auditory pathways or, if need be, to diagnose certain disorders (such as multiple sclerosis).

In some other clinical applications, the BAEPs are used for the purpose of clinical monitoring; for instance, to study the effects of certain substances on the auditory system, and also to help surgeons to preserve the auditory pathways during surgery.

Generally speaking, the aim of the clinician during a BAEP recording consists of analyzing the BAEP by trying to reduce any possible error. An efficient BAEP analysis regarding a clinical sense depends on the quality of the BAEP recognition parameters, usually provided by an expert system. This expert system uses a set of extraction algorithms, recognition algorithms and decision algorithms. In fact, an accurate clinical interpretation of the BEAP is achieved only if its estimation is properly performed. For this purpose, prior useful signal features related to BAEPs and to other noise signals such as EEG, EMG, are required.

This information then allows the scientists to model this type of physiological phenomena.

As can be seen in Figure 12.2, a BAEP is characterized by five major waves which are denoted by I, II, III, IV/V.

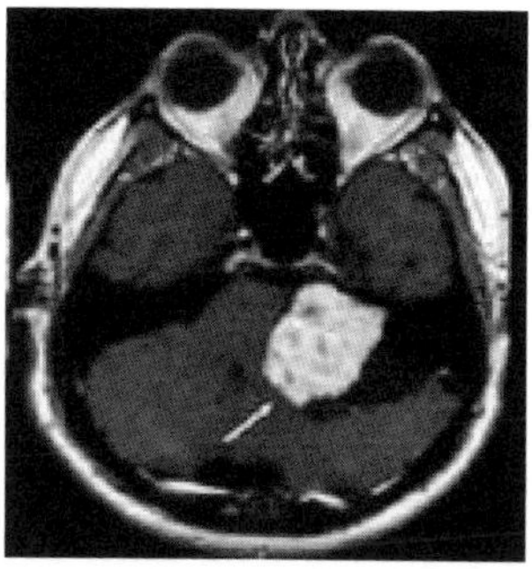

Figure 12.1. *An image of the brain, obtained by ultrasound imaging, showing an acoustic neuroma (Photo: Baylor College of Medicine)*

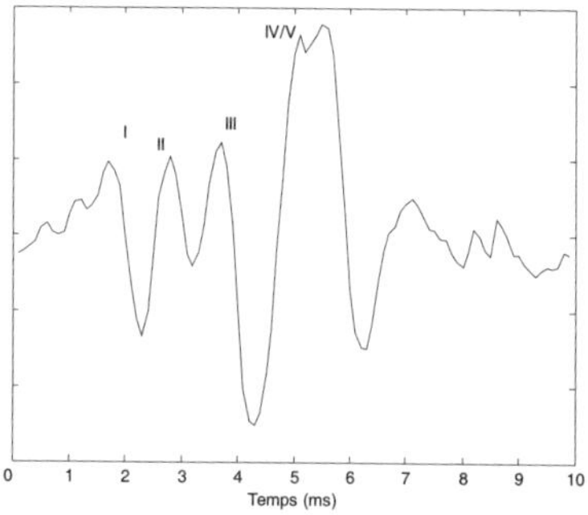

Figure 12.2. *A real BAEP*

12.2.1. *BAEP generation and their acquisition*

The acquisition of a BAEP is carried out as follows: the auditory system is excited by a set of acoustic impulses at a frequency which is generally less than 30 clicks per second. For higher frequencies, the BAEPs might be distorted due to the superposition of late evoked potentials. Therefore, some useful techniques using a kind of aperiodic stimulations have been proposed in the literature.

The responses to the stimulations, which can be sampled at a frequency of 10 kHz, are naturally corrupted by the EEG. Generally, each response is recorded over less than 10 ms by using some electrodes which are properly located on the vertex and on the mastoid (see Figure 12.3). After the denoising process, BAEP wave recognition is then performed. As is well known, the different noises corrupting the BAEPs are essentially physiological (e.g. EEG or EMG).

12.3. Processing BAEPs

The traditional methods used in estimating the BAEPs consist of averaging the recorded responses. The signal-to-noise ratio can reach record values from -20 to -30 dB. In practice, it is generally impossible to observe a BAEP directly from only a single response, even after the filtering process. As is known, the traditional averaging technique is based on the stationarity hypothesis. In other words, it is assumed that the BAEP (i.e. useful signal) in each response is time-invariant and that the noise (i.e. the EEG) is a zero-mean stationary signal.

Therefore, even if this technique seems to be simple, it can achieve excellent results in some cases, namely, when the subject is healthy and the recordings are carried out under appropriate conditions. In this case, among 800 and 1,000 responses are generally required to extract an averaged BAEP. This procedure assumes that during the acquisition phase, the patient is somehow relaxed. However, if the subject is pathological and if the recordings are not carried out under appropriate conditions then a larger number of responses is required in order to reduce the energy related to the noise in comparison to the energy related to the BAEP. Moreover, it is important to mention that the EEG may change its spectral and statistical characteristics over recording time, and that the EMG can be added to recordings during patient's muscular contractions. In some pathological cases, even if it is possible to reduce the effect of the noise during the averaging process, the BAEP non-stationarity regarding both its dynamics and its shape leads to an unrecognizable smoothed average signal. In this case, an objective analysis becomes a critical task, in particular for latency measurement (i.e. the time taken between the moment of stimulation and reaching the maximum value of a BAEP wave) or the measurement of conduction times (i.e. duration I-III, duration I-IV).

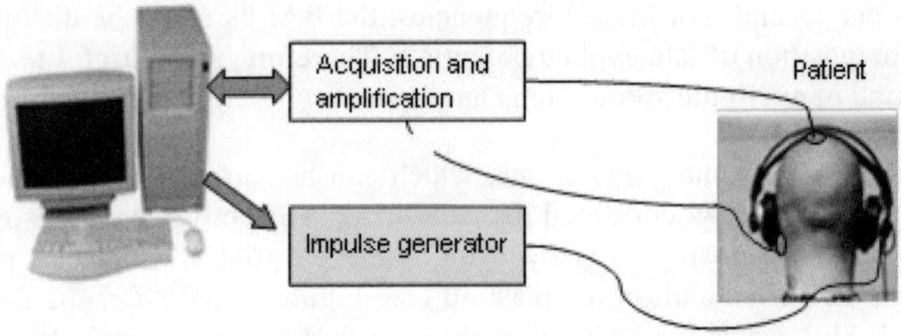

Figure 12.3. *The acquisition of BAEPs*

Several signal processing techniques dedicated to extracting BAEPs have been suggested in the relevant scientific literature. For example, some techniques are based on the weighted averaging principle [DAV 97] or on adaptive filtering [YU 94a] and [YU 94b]. These techniques generally use information based on the characteristics of noise and do not focus enough on the dynamics of the BAEP. The main reason for this is due to the fact that a direct BAEP observation from a given recorded response is almost impossible, as has been explained previously.

The approach that has been developed in this chapter is mainly based on modeling the non-stationarity (i.e. the dynamics) of the BAEPs, by taking into account the fact that the energy of the noise may be reduced during the averaging process. The parameters of the models that have been studied will be determined by using GAs. The reader can also refer to [NAI 02], [CHE 05] and [NAI 06] in which thesimulated annealing (SA) approach has been used as an optimization technique.

The use of this metaheuristic can be justified by the fact that models may be dynamic and complex (due to the non-convexity of the criteria to be optimized). In terms of computing, only a single code is required for the whole set of models. The advantage here consists of avoiding an increasing number of algorithms and also means that it is possible to considerably reduce the memory space required for storing code. It should also be pointed out that GAs can be easily adapted for use on multiprocessor platforms.

In the following section, the principle of a GA is introduced. It will be used throughout this chapter to deal with the problem of estimating BAEPs.

12.4. Genetic algorithms

GAs were developed in 1979 by J. Holland and his colleagues at the University of Michigan. The basic idea of GAs was inspired by the theories of natural selection and genetics. In order to understand how GAs work, it seems important to be aware of some of their basic principles which will then be applied to the problem that we are facing in relation to estimating the BAEPs (see Table 12.1).

Let us assume that a population is made up of N individuals. In terms of genetics, each individual is characterized by a chromosome. At a lower level, the chromosome is made up of genes. The first generation is considered as a first generation able to evolve in a given environment in order to produce other generations by following the rules of reproduction, crossover and mutation. In other words, certain individuals will disappear from one generation to the next, (i.e. the weakest will disappear). The strongest will be able to reproduce with no modifications: the child is the clone of his parent. On the other hand, certain individuals from the same generation will be able to crossover so that the future generation can have similar characteristics. Certain genes of certain individuals can be changed and replaced by genes from the search space.

Population	A set of potential solutions in a generation (m). $\mathbf{d}_1^{(m)}, \mathbf{d}_2^{(m)}, \ldots \mathbf{d}_K^{(m)}$ $\mathbf{d}_i^{(m)}$ is a vector with M parameters to be determined.
Chromosome or individual	A potential solution $\mathbf{d}_i^{(m)}$: $\mathbf{d}_i^{(m)} = \left[d_{i,0}^{(m)} \quad d_{i,1}^{(m)} \quad \ldots \quad d_{i,M-1}^{(m)} \right]^t$
Gene	An element of a potential solution. For example: $d_{i,n}^{(m)}$, n=0,…M-1.
Reproduction	A potential solution in a generation (m-1) is maintained in the next generation (m).
Breeding	Two potential solutions of a given generation (m-1) are combined to generate two other solutions for the future generation (m). Example: $\mathbf{d}_i^{(m-1)}$ and $\mathbf{d}_j^{(m-1)}$ can produce $\mathbf{d}_i^{(m)}$ and $\mathbf{d}_j^{(m)}$: $\mathbf{d}_i^{(m)} = \left[\underbrace{d_{i,0}^{(m-1)} \; d_{i,1}^{(m-1)} \ldots}_{Elements\ of\ d_i^{(m-1)}} \; \underbrace{\ldots d_{j,M-2}^{(m-1)} \; d_{j,M-1}^{(m-1)}}_{Elements\ of\ d_j^{(m-1)}} \right]$ $\mathbf{d}_j^{(m)} = \left[\underbrace{d_{j,0}^{(m-1)} \; d_{j,1}^{(m-1)} \ldots}_{Elements\ of\ d_j^{(m-1)}} \; \underbrace{\ldots d_{i,M-2}^{(m-1)} \; d_{i,M-1}^{(m-1)}}_{Elements\ of\ d_i^{(m-1)}} \right]$
Mutation	If $\mathbf{d}_i^{(m-1)}$ is a potential solution in a generation, the mutation can occur in the following generation in order to generate $\mathbf{d}_i^{(m)}$ by modifying one of its elements: Example: $\mathbf{d}_i^{(m-1)} = \left[d_{i,0}^{(m-1)} \quad d_{i,1}^{(m-1)} \quad \ldots \quad d_{i,M-1}^{(m-1)} \right]^t$ $\mathbf{d}_i^{(m)} = \left[d_{i,0}^{(m)} \quad s_{i,1}^{(m)} \quad \ldots \quad d_{i,M-1}^{(m)} \right]^t$ The element $d_{i,1}^{(m)}$ has been replaced by $s_{i,1}^{(m)}$.

Table 12.1. *The principle of GAs*

12.5. BAEP dynamics

In this study we consider that the BAEPs vary over the time from one response to another according to random delays. This problem, known as "jitter" (i.e. desynchronization of the signals), which can be due to a physical or physiological origin, is not new in terms of signal processing. There are many different methods of dealing with similar problems, such as the techniques that are used to solve some problems related to radar detection, or even in some specific biomedical engineering applications. Unfortunately, these methods cannot be adapted to our problem, especially because the recording conditions are particularly poor (i.e. very low SNR). We must point out that BAEP desynchronization in each single response is only an assumption. However, if such an assumption turns out to be true, a phenomenon known as smoothing, which occurs during the averaging process, will be unavoidable. The distortion which is caused by smoothing can lead to quite serious consequences depending on the nature of desynchronization (distribution, variance). In such situations, as mentioned above, two distinct waves from a BAEP may be transformed into one single wave.

In order to describe this phenomenon according to a mathematical formula, we assume that at each ith stimulation, a signal $x_i(n)$, is recorded. This signal can be explained as follows:

$$x_i(n) = s(n + d_i) + b_i(n) \qquad [12.1]$$

where:

$s(n)$ is the useful signal (BAEP) that we want to estimate;

$b_i(n)$ is the noise, corrupted by the EEG during the ith acquisition;

d_i represents the time delay of each signal $s(n)$ (in relation to the moment of stimulation).

For M stimulations, averaging leads to:

$$\bar{x}(n) = \underbrace{\frac{1}{M}\sum_{i=0}^{M-1} s(n+d_i)}_{A} \quad + \underbrace{\frac{1}{M}\sum_{i=0}^{M-1} b_i(n)}_{B} \qquad [12.2]$$

From equation [12.2], it is clear that term B (which is related to the noise) can be neglected if the statistical average of the noise is zero or close to zero ($E[b(n)] \approx 0$). If the noise is not a zero-mean process, several situations can arise:

1. The noise is a 1st order stationary process: in this case, the averaging provides an offset and a pre-filtering is necessary. Therefore, whatever the energy of this component is, its influence will be completely eliminated.

2. The noise is stationary but its statistical average moves towards a low energy signal.

3. The residual noise signal provided by the averaging process is not correlated to the useful signal (i.e. the BAEP). In such cases, a simple filtering over the BAEP bandwidth is necessary.

4. The components of the averaged noise overlap with the BAEP components. Pre-filtering can only reduce the noise's influence.

What happens in real situations? The EEG signal is a cortical activity and is considered as a major noise source in BAEP clinical routines. It should also be pointed out that this signal has been largely studied in specialized scientific literature. It has been studied from different points of view according to pre-defined objectives and under certain conditions. The different characterizations and models that have been suggested in the literature cannot be directly applied to the context in which we are working.

In order to show how the EEG influences the estimation of BAEPS (i.e. in terms of averaging) we take the example of a small experiment which consists of averaging 800 acquisitions from an actual EEG. In this experiment we consider that BAEPs are totally absent in each acquisition. In Figure 12.4 we have superimposed several BAEP realizations in order to provide an idea of the energy produced by each of these realizations in comparison with the energy of the signal that is obtained through the averaging process (see Figure 12.5). The first comment that can be made when comparing the energy produced is that, after averaging, the energy has been reduced by 400% (which tends to greatly reduce the influence of noise on the averaged signal). This example was chosen randomly; other similar examples could also be illustrated to show that the noise average decreases according to a non-linear manner. Of course, this decrease depends on the nature of the EEG, which depends on the recording conditions that are relative to a given patient.

It is widely known that the properties of the EEG tend to vary whenever the patient passes from one state to another (i.e. tense patient, relaxed patient, patient who is asleep). Furthermore, movements such as eyes blinking and muscular contractions also contribute to the slowing down of the rate at which the EEG is attenuated by the averaging process.

Now that we are sure that the influence of the noise during the averaging process is almost negligible for a given number of responses, a possible distortion of the BAEP signal can only occur due to its unstable feature. Consequently, in these conditions the characterization of the EEG is not important when it comes to estimating the BAEPs.

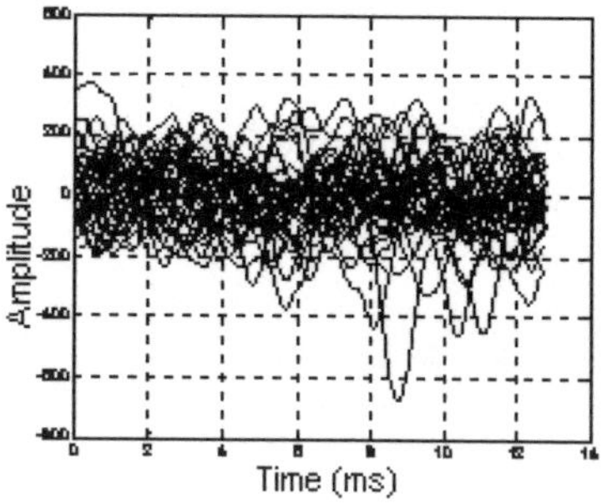

Figure 12.4. *The superposition of several creations*

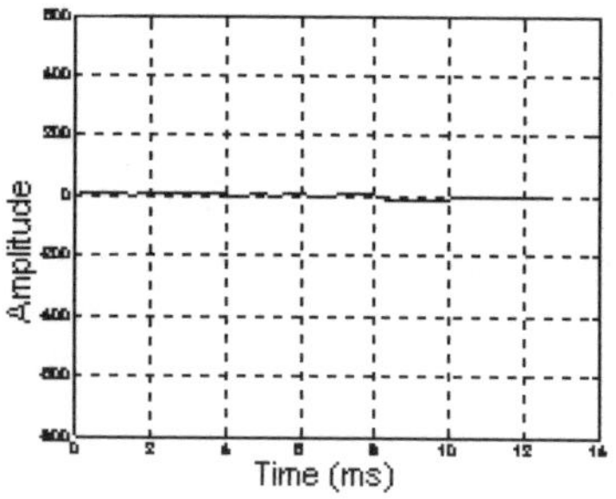

Figure 12.5. *Averaging of 800 creations*

It should also be pointed out that if it is possible to identify the way in which the BAEPs vary throughout time, it would probably make it possible:

1. to estimate the BAEP that corresponds to each stimulation;

2. to determine the dynamics of the BAEPs.

The second point mentioned above could also be used as an indicator in some techniques of functional exploration.

Under these conditions the averaged signal can be explained as follows:

$$\overline{x}(n) = \frac{1}{M}\sum_{i=0}^{M-1} s(n+d_i) \qquad [12.3]$$

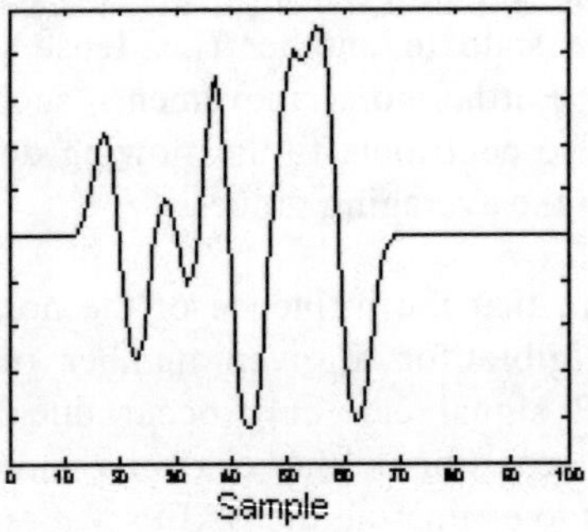

Figure 12.6. *A model of a BAEP used in simulations*

The issue that now arises is the following: how is it possible to estimate the delay parameters d_i? If we want to find the best solution for a given criterion, then the use of an optimization algorithm is necessary.

If the recorded M signals and their direct averages are already known, then the issue of optimization can be dealt with as follows: find the best set of parameters d_i, which maximizes the energy of the averaged signal.

This condition is only validated if the BAEPs are aligned. In mathematical terms this can lead to maximizing the following energy equation:

$$f_{\mathbf{d}} = \sum_{n=0}^{N-1}\left(\frac{1}{M}\sum_{i=0}^{M-1} x_i(n-d_i)\right)^2 \qquad [12.4]$$

where:

$\mathbf{d}$=[$d_0, d_1, d_2, \ldots\ldots, d_{M-1}$] represents the vector of the delay parameters,

N is the number of samples in each response.

The optimization problem can thus lead to the minimization of the following equation:

$$J_{\mathbf{d}} = -\sum_{n=0}^{N-1}\left(\frac{1}{M}\sum_{i=0}^{M-1} x_i\left(n-d_i\right)\right)^2 \qquad [12.5]$$

It is clear that equation [12.5] is neither quadratic nor convex. If we adapt the problem to two dimensions (i.e. estimation of two delays d_0 and d_1), the objective function will introduce an aspect which is similar to that which can be seen in Figure 12.7 (simulations obtained from the BAEP which can be seen in Figure 12.6).

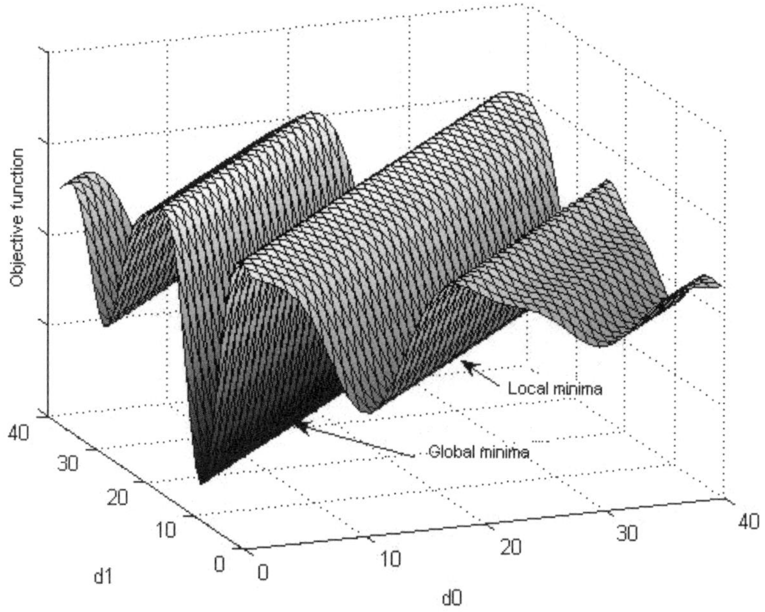

Figure 12.7. *The estimation of two delay parameters*

However, we can see that the global optimum of this objective function is not unique. The reason for this is that the BAEPs can be synchronized at distinct moments.

In order to guarantee the uniqueness of the solution it is important to keep the signals still. An example is illustrated in Figure 12.8 which shows a two-dimensional problem (*M=2*). The first BAEP is fixed while the second is moving. At each point the sum of the two signals is recalculated.

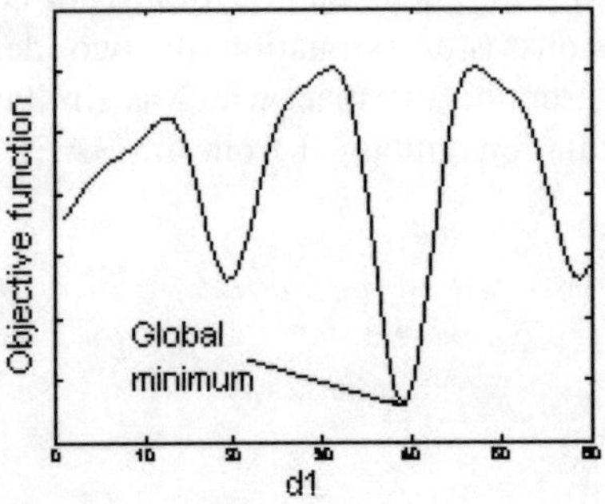

Figure 12.8. *The estimation of two delay parameters by fixing one in relation to the other*

In this case, the objective function is only one minimum point (the alignment of the two signals). This problem can be extended to the case *M=3* when determining the first signal. The resulting objective function from this analysis shows several local minima and only one global maximum, the latter corresponding to the moment when the set of signals is synchronized. This can be seen in Figure 12.9.

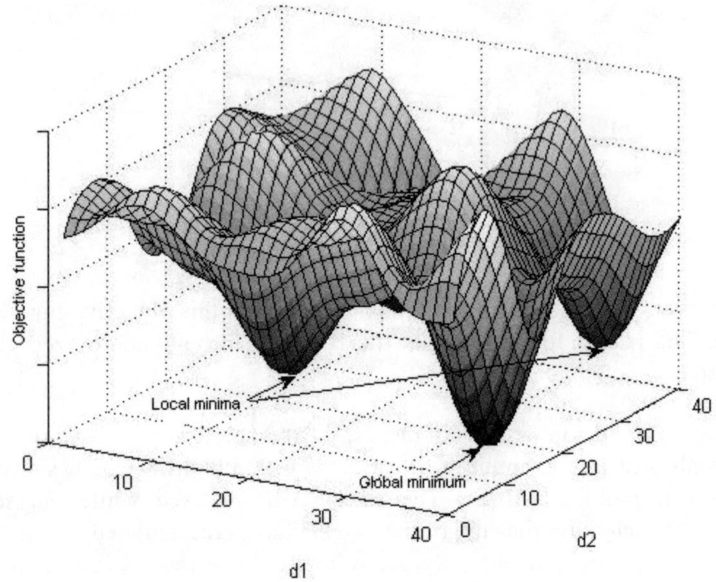

Figure 12.9. *The estimation of two delay parameters which are calculated in relation to the first signal, used as a reference*

The generalization of the problem with M dimensions also requires that one of the responses be considered as a reference signal, for example the first response. After the convergence of the optimization algorithm, the set of BAEPs will be systematically aligned with the first signal.

12.5.1. *Validation of the simulated signal approach*

This phase of simulation is fundamental. It enables us to validate the proposed algorithm (summarized in Table 12.2) when the acquisition conditions have been verified, and it also makes it possible to check the convergence of a signal towards a global minimum. All of the simulations are created from a model of the BAEP, which is corrupted by an 8th order autoregressive (AR) noise. The AR model is often used to model the EEG. In terms of energy, the effect of the AR model becomes negligible after the averaging process, in relation to the energy of the useful signal (BAEP). First of all, Gaussian random perturbations are considered; then some non-Gaussian cases are analyzed.

1. *Choose, at random, K vectors which represent potential solutions. Each solution is made up of M delays which correspond to M responses,*
2. ***Reproduction stage:** generate K other potential solutions with the help of crossover and mutation operators,*
3. *Evaluate the objective function in each of the potential solutions,*
4. ***Selection stage:** take the best K solutions amongst the K+k solutions so that the following generation can be produced,*
5. *Save the best solution,*
6. *If the number of maximal generations is not reached, go back to 2,*
7. *Solution = best point found, stop the program.*

Table 12.2. *The genetic algorithm applied to the estimation of BAEP delays*

12.5.1.1. *Random Gaussian distribution desynchronization*

Simulations are carried out on 200, 400 and 800 responses. Each BAEP is randomly moved according to Gaussian distribution and is then corrupted with an AR noise. As has been previously mentioned, the phenomenon of Gaussian desynchronization systematically leads to a smoothing of the waves during averaging. Figure 12.10 shows that the two waves II and III, as well as the complex IV/V, have been overlapped (test carried out on 200 responses). The problem which is then raised by the clinician is the difficulty in extracting the clinical parameters (latencies and conduction time) from such a signal, which makes any analysis and interpretation of such parameters extremely difficult to carry out. Figure 12.11 shows that it is possible to restore the shape of the BAEP after the synchronization of the responses using the solutions which result from the optimization of equation [12.1]. The waves go back to their original shape, as well as to their original amplitudes and original latencies.

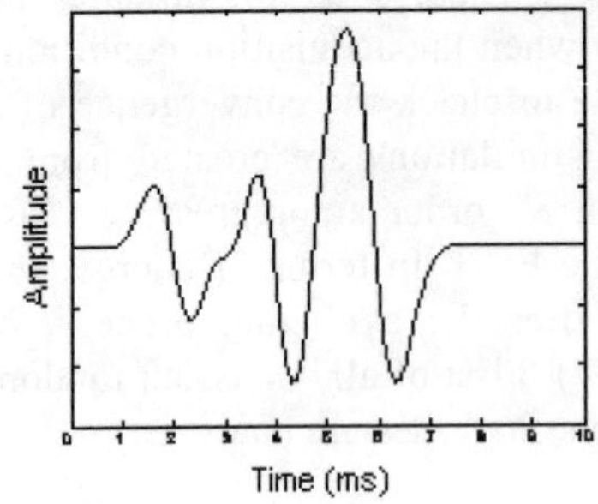

Figure 12.10. *Averaged signal based on 200 acquisitions*

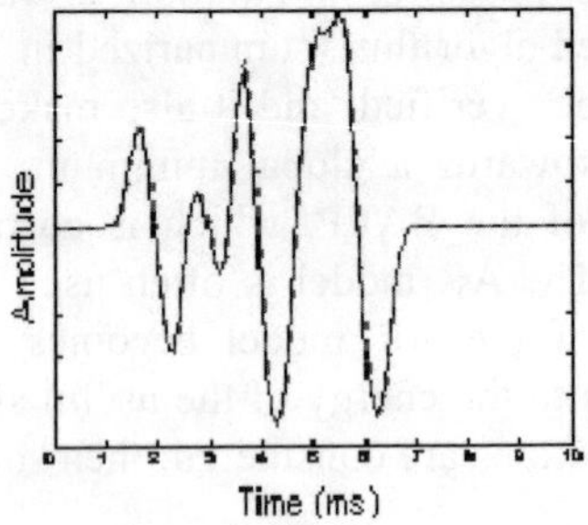

Figure 12.11. *Signal obtained after synchronization*

Such a result would not have been possible with basic techniques such as adapted filtering. In fact, adapted filtering can only be used if the signal-to-noise ratio is favorable.

Figure 12.12 shows the set of BAEPs to be synchronized in the shape of an evolutionary surface. For illustrative reasons, we will not show the noise that corrupts each response. The objective is to highlight the alignment that is obtained after synchronization (see Figure 12.13).

We are now going to analyze the delay vector which disturbs the BAEPs and then compare it with the optimal solution provided by the GA which minimized in equation [12.3].

To analyze and use the results it must be pointed out that the solution required for the synchronization of the BAEPs is the opposite value of the delay vector. When analyzing and using the results, the analysis can be compared to a physical system that has been unbalanced by a force in one direction; then the system is rebalanced by applying a second force in the opposite direction. Figure 12.14 compares the unbalancing vector with the (inversed) solution after the convergence of the GA. Figure 12.15 represents the convergence of the GA in relation to 500 generations.

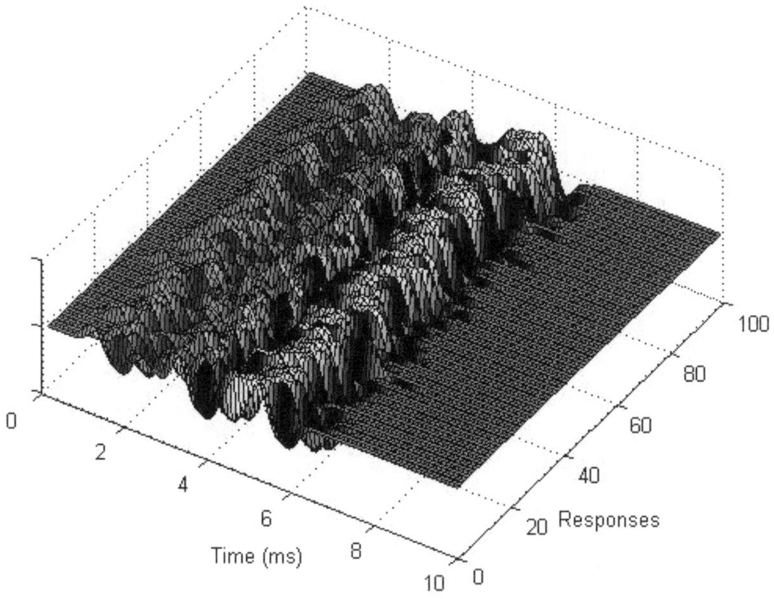

Figure 12.12. *2D representation of BAEPs before synchronization*

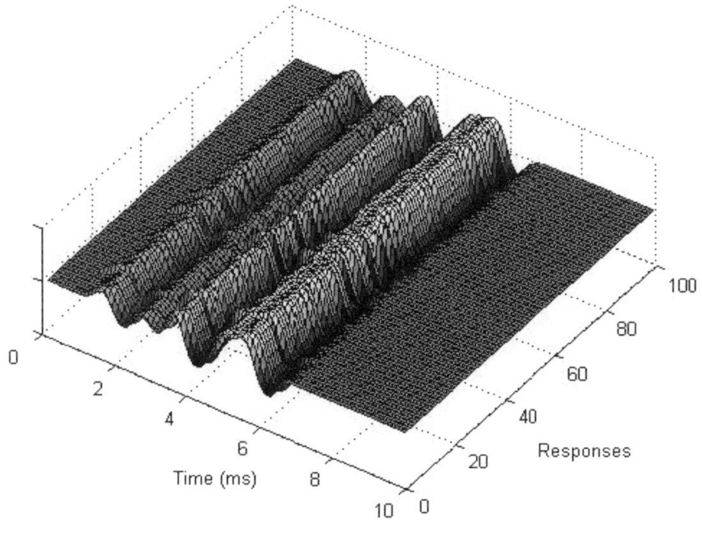

Figure 12.13. *2D representation of the signals after synchronization*

When it comes to experiments with 400 and 800 responses, processing time clearly becomes more significant and is characterized by non-linear growth. This behavior is essentially due to the stochastic nature of the algorithm.

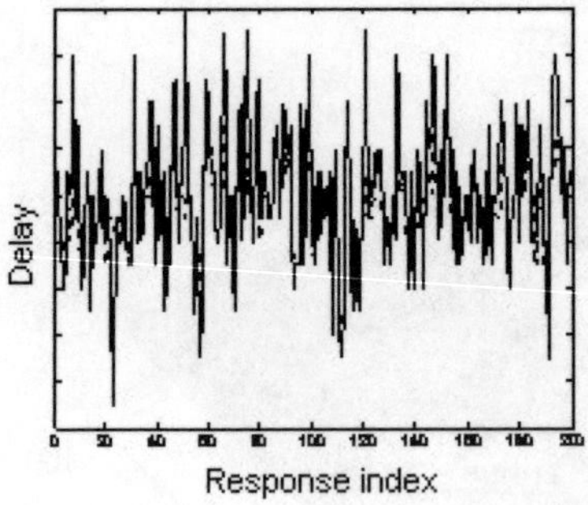

Figure 12.14.
[-] Original delay signal, [...] estimated signal, used for the synchronization of 200 BAEPs

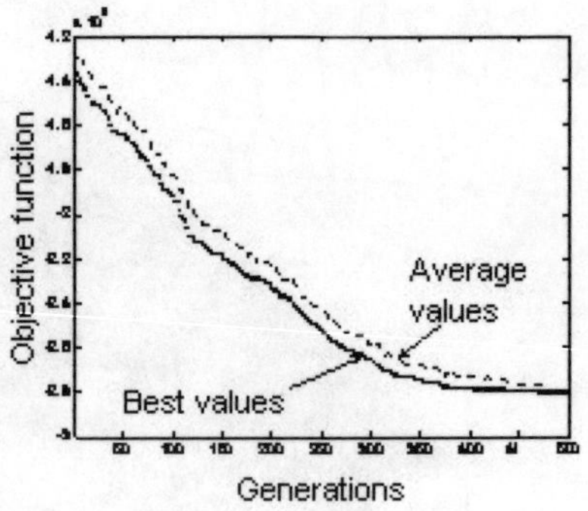

Figure 12.15.
Convergence curve on 500 generations

It is clear that the convergence time is not the only criterion used for evaluating the performances of the optimization algorithm. Other parameters, such as variance as well as the bias of the averaged signal after synchronization, can be taken into consideration. Tests on the same set of signals that were initially used can be carried out by using these other parameters. The fact that the GA is of a stochastic nature means that the results will undoubtedly be different from one test to another.

By superimposing the estimated signals from each test, as shown in Figure 12.16, we can gather some information about the variance of the estimator. This illustration clearly shows that the dispersion of the solutions is minimal. The bias can be evaluated by simply calculating the average of the results from the different tests and by comparing the average with the ideal solution (which is represented by the BAEP model in Figure 12.17). With a very low variance and an almost zero bias, it is possible to consider the estimator as consistent.

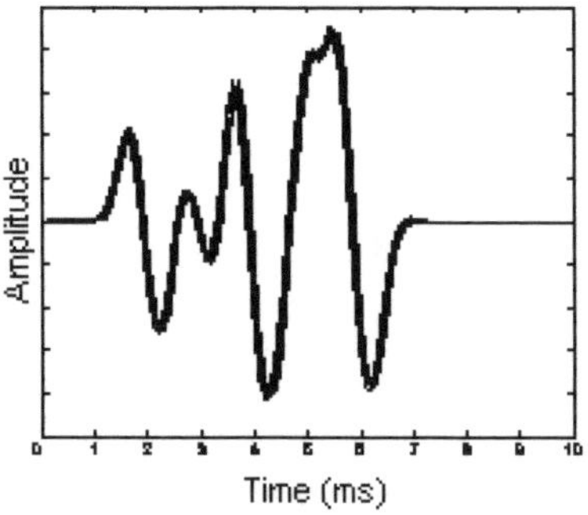

Figure 12.16. *Analysis of the estimator's variance*

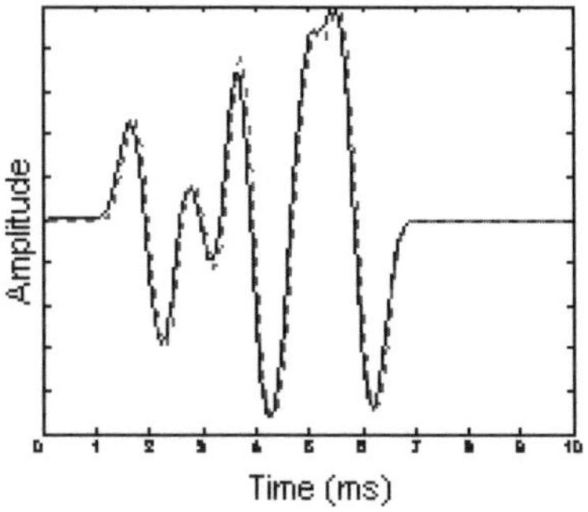

Figure 12.17. *Analysis of the estimator's average*

12.5.1.2. *Non-Gaussian desynchronization*

Now that we have studied the general case of Gaussian randomization of the BAEPs, we are going to have a look at some unusual cases.

Let us consider that the shape described by the delays does not emanate from a Gaussian distribution as was the case previously, but could describe any shape (an example is given in Figure 12.18). In this shape we consider that the set of BAEPs are delayed in only two stages. Everything occurs as if a normal acquisition was taking place without the BAEPs being delayed at the beginning; suddenly, the BAEPs move in one direction (with a continual delay), then move in the other direction before going back to their initial state. This sudden movement might occur in certain surgical procedures, in particular when the surgeon is working on auditory pathways.

Two-dimensional illustrations of the BAEPs are shown in Figure 12.19 (signals before synchronization) and in Figure 12.20 (signals after synchronization).

We have seen that when it comes to Gaussian desynchronization, the averaging process leads to a smooth signal. This phenomenon cannot occur when randomization is commonplace. Averaging in such situations can lead to the generation of false supplementary waves.

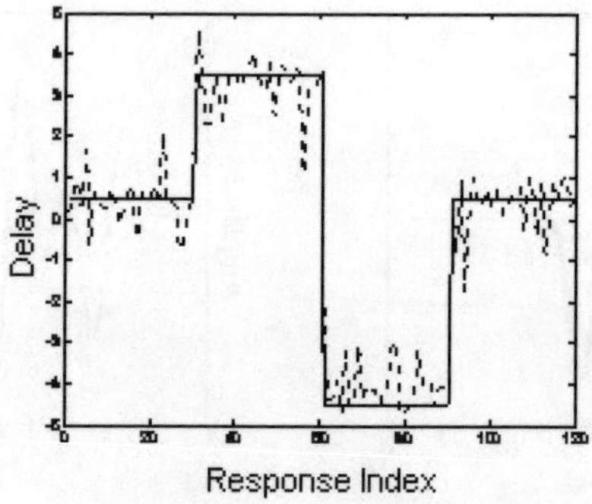

Figure 12.18. *[-] Delay signal, [...] estimated synchronization signal*

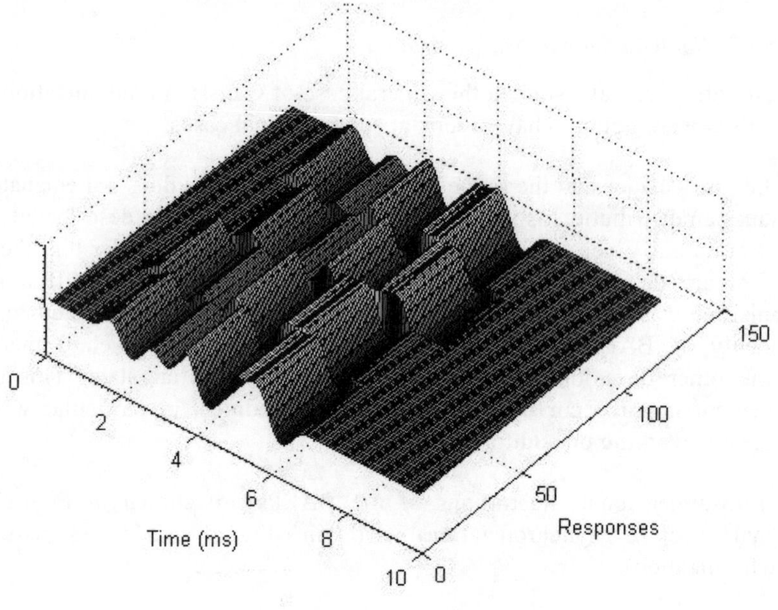

Figure 12.19. *2D representation of the BAEPs before synchronization*

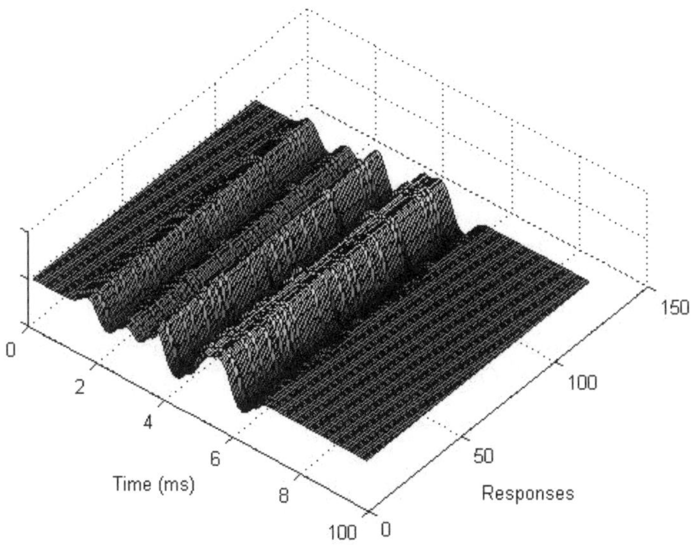

Figure 12.20. *Representation of the BAEPs after synchronization*

As far as tackling the problem of desynchronization is concerned, we have illustrated other atypical cases in Figures 12.21 and 12.22. The objective is to establish that the algorithm that is used ensures the convergence towards the global minimum, whatever the nature of the delay. The only difference is that the number of generations required by the GA is variable. This variability in the number of generations required by the GA leads to problems associated with the stop criterion.

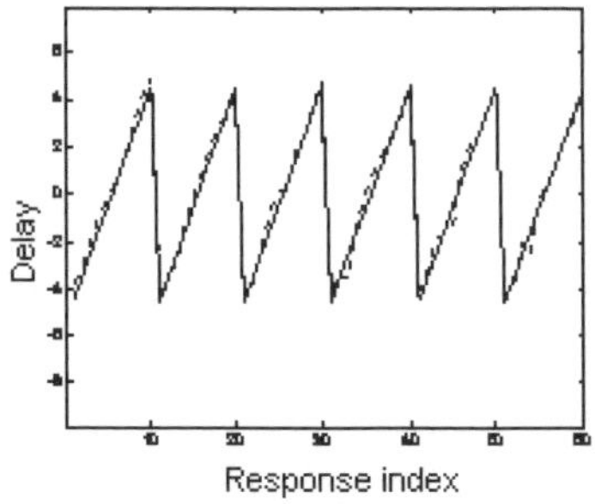

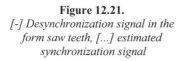

Figure 12.21.
[-] Desynchronization signal in the form saw teeth, [...] estimated synchronization signal

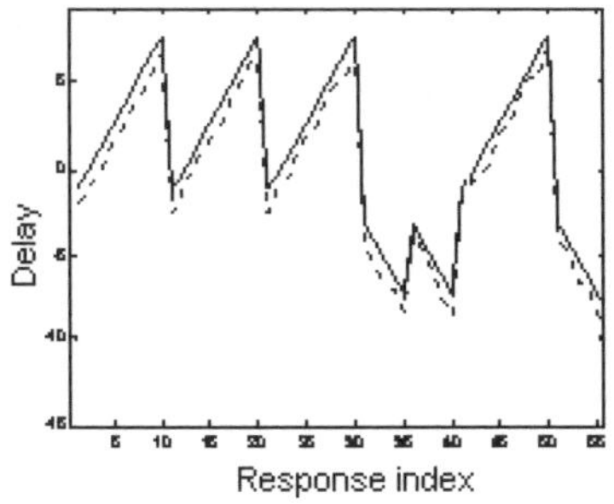

Figure 12.22.
[-] Desynchronization signal in the form of rugged saw teeth, [...] estimated synchronization signal.

12.5.2. *Validating the approach on real signals*

Simulations have shown that the approach that we have just mentioned allows for convergence towards the optimal solution regardless of the nature and the dynamics of the BAEP. The use of this approach on real signals can only be efficient if the hypotheses and recording conditions are verified. The problem that then arises in the case of real signals is that the type of non-stationarity of the BAEPs is unknown. It is impossible to confirm that the BAEP signals are desynchronized. The BAEPs cannot be observed in one given acquisition due to the energy coming from the EEG. As a consequence, when this approach is applied to real signals, it is initially assumed that the signals are delays. The reliability of this hypothesis is then checked afterwards.

Two different cases arise:

1. Traditional (or direct) averaging leads to an BAEP with an unrecognizable shape, while corrected averaging leads to the formation of a normal-shaped BAEP (which can be identified by its five waves). In this case we can confirm that the BAEP dynamics is a "delay". Under no circumstances should the clinicians settle for only the corrected signal because if they do they run the risk of overlooking information that is linked to a particular disease; this information is introduced in the delay vectors which are used to correct the average signal. This vector could therefore be used as an indicator of a potential disease.

2. The BAEP signal that has been averaged by the traditional method is unrecognizable. This case may occur due to one of the following two reasons:

a. the BAEP dynamics is not a delay signal: in this case, we should modify the dynamic model of the BAEP;

b. the number of M acquisitions is not sufficient to reduce the energy of the averaged noise: it is therefore necessary to increase the number of acquisitions, which also means an increase in processing time.

Figures 12.23 and 12.24 illustrate a case that corresponds to a real recording.

The patient is 70 years old and is stimulated at an intensity of 90 dB. The only clinical information that we have shows that this patient suffers from vertigo and from a loss in hearing. At this stage, the clinical information we have is poor and does not allow us to establish an objective correlation between the shape of the signal and the patient's illness. An ideal situation would be to be able to identify the disease by simply looking at the shape of the BAEP. Unfortunately, we are still quite far from reaching this goal and in order to do so, a radical review in the way BAEPs are recorded is needed. Nevertheless, the results that have been obtained from real signals have been encouraging as our objective was to have access to hidden information. This hidden information provides information relating to the movement of the BAEPs. With this objective in mind, it will enable future studies to be carried out in a more determined rather than statistical manner. As we can see in Figure 12.23, wave II seems to have been smoothened with wave III when only traditional averaging is used. The separation of these two waves is clearly visible after the synchronization process (see Figure 12.24).

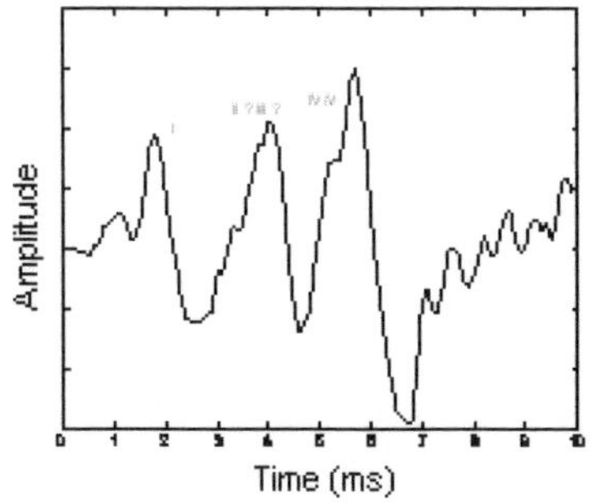

Figure 12.23. *Signal obtained by averaging 800 acquisitions*

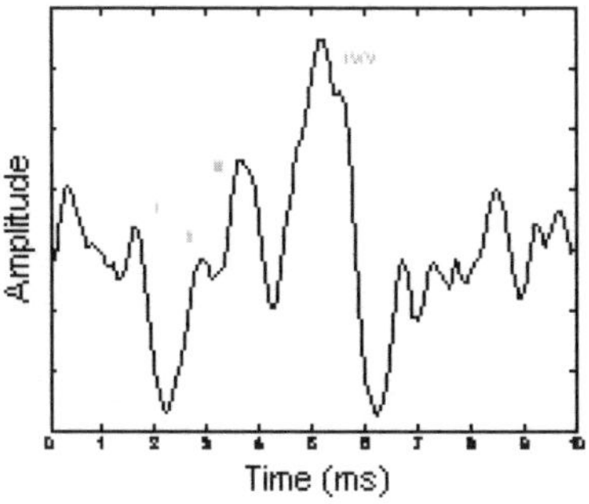

Figure 12.24. *Corrected signal (separation of waves II and III)*

12.5.3. *Acceleration of the GA's convergence time*

In the method that has been described up to this point, the number of parameters (delays) to be estimated corresponds exactly to M responses that are used to extract the BAEP.

The question that can now be asked is the following: is it possible to estimate M delays by using fewer parameters? It is possible, but on the condition that we can parametrically model the curve that is described by the delays.

Let us take the example in which the delays vary according to a sinusoidal curve. In this case, it is possible to use a total of three parameters (instead of 800). It is a question of the frequency of the sinusoid, its amplitude and its phase. If the curve that is described by the delays is slightly more complex than the sinusoid, then we can think of using a larger scale sinusoid, (i.e. two, three, … k sinusoids).

Generally speaking, a delay-curve modeled by k sinusoids needs $3.k$ parameters that have to be identified. This approach is seen by some as following the principle of the Fourier transform approach. It is clear that a compromise in relation to processing time must be taken into consideration. This approach can only be efficient if $3.k<M$.

What has just been mentioned can be described in what follows below.

The delay corresponding to the ith acquisition can be expressed by:

$$d_i = \sum_{k=1}^{K} A_k \sin\left(2\pi f_k i + \phi_k\right) \qquad [12.6]$$

where:

A_k is the amplitude of the kth sinusoid,

f_k is the frequency of the kth sinusoid,

ϕ_k is the phase of the kth sinusoid,

K is the number of sinusoids.

The problem once again comes back to estimating the set of parameters A_k, f_k and ϕ_k in relation to equations [12.5] and [12.6]. In this case, by inserting the model of equation [12.6] within equation [12.5] we obtain the following:

$$J_{\mathbf{d}} = -\sum_{n=0}^{N-1}\left(\frac{1}{M}\sum_{i=0}^{M-1} x_i\left(n - \sum_{k=1}^{K} A_k \sin\left(2\pi f_k i + \phi_k\right)\right)\right)^2 \qquad [12.7]$$

The introduction of the delay model means that the optimization algorithm must be adapted because here the optimization problem is not completely combinatory. This is because the search space depends on the nature of the parameters to be identified:

– The values of the amplitudes A_k are limited by the frequency of the samples, as well as by the duration of the ith acquisition, which leads to the creation of a combinatory case.

– The frequencies f_k are positive and are able to take any value in $\Re^+$. The problem is therefore continuous.

One possible adaptation is to discretize the search space depending on the precision that is desired, even if the desired precision has the negative effect of increasing processing time.

Other optimization techniques could also be used, such as ant colonies [DOR 96], [DOR 97], [DRE 02] or the particle swarm optimization technique which tends to be used the most [KEN 97].

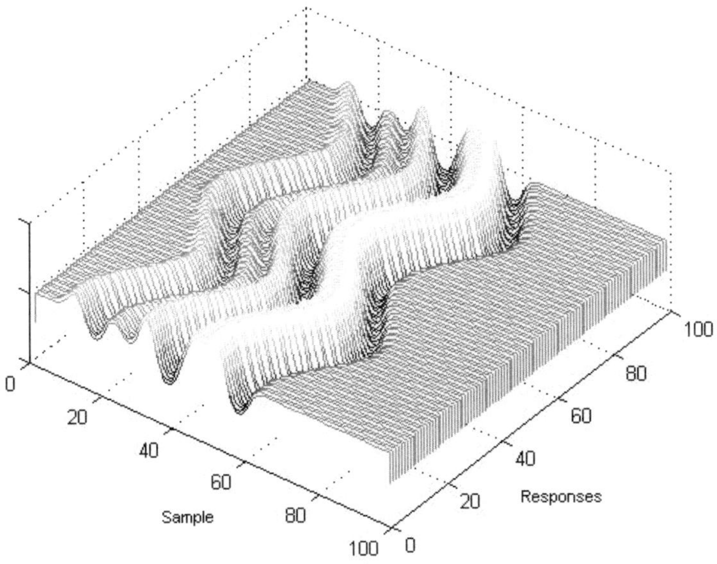

Figure 12.25. *An example of a sinusoidal delay model which leads to a distortion of the BAEP*

Figure 12.25 illustrates an example in which the BAEPs are delayed by following a sinusoidal curve. Does this occur in practice? Perhaps the curve is not exactly sinusoidal but BAEPs with a low frequency are subject to possible movements.

In certain clinical tests when analyzing the movement of wave V over time, it is also possible to study the fatigue of the auditory system following stimulation that has gone on for several hours.

In certain studies it is also possible to analyze the effect that certain substances have on the auditory system, such as sodium chloride or Amikacin, in the shape of a BAEP.

Generally speaking, this approach seems appropriate in such situations and in particular when tens of thousands of responses are used. For a given protocol, this approach, in fact, does not depend on the number of acquisitions.

The results which come from the simulations on delay models of order 2 (i.e. two sinusoids) and order 5 (i.e. five sinusoids) are shown in Figures 12.26 and 12.27 respectively.

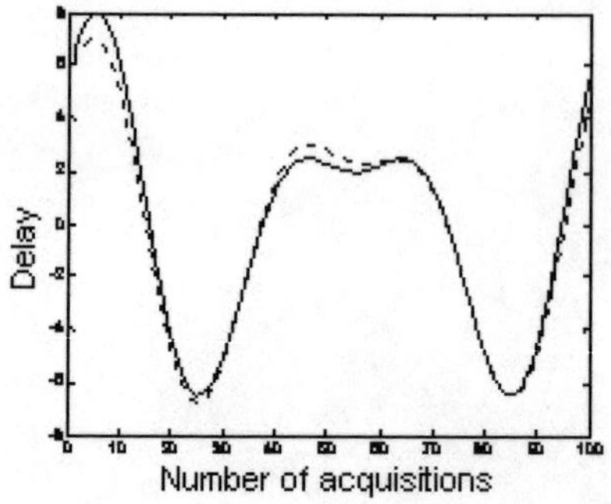

Figure 12.26. *Delays according to two sinusoids, [-] original signal.[...] signal estimated by the model*

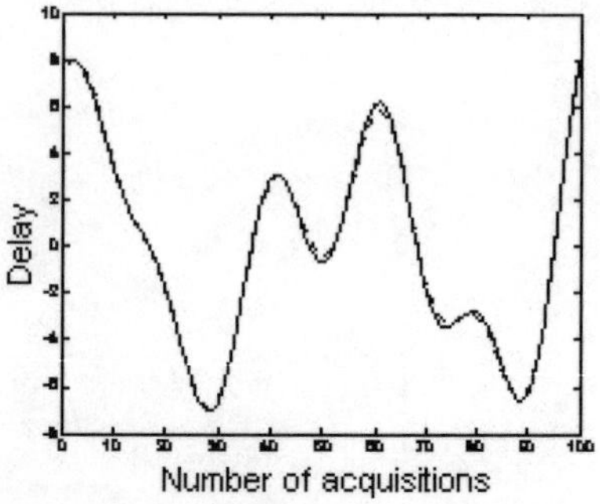

Figure 12.27. *Delays according to five sinusoids, [-] original signal, [...] signal estimated by the model*

12.6. The non-stationarity of the shape of the BAEPs

It has been noticed that in certain BAEP clinical examinations, and in particular when it comes to retro-cochlear diseases, the shape of the BAEPs is unusual and sometimes completely unrecognizable. BAEPs tend to change shape quite randomly over time. During this transformation, the notion of the compression and dilation of the shape of the BAEP over time should be taken into consideration. As a consequence, the traditional averaging of the responses could lead to an overall deformation of the BAEP, thus making it impossible to work with. This deformation is not strictly related to the smoothness of the waves as was mentioned when describing the previous type of non-stationarity. However, deformation could lead to the production of other virtual waves, which in turn may lead to a false clinical interpretation.

In basing our judgment on equation [12.8], we assume that the BAEP is compressed or dilated according to a factor known as a_i, which is a random factor with an unknown distribution. When a_i is greater than 1, the BAEP tends to be compressed and when a_i is less than 1 the BAEP becomes longer. The factor a_i varies over time and plots an uncharacteristic two-dimensional surface as can be seen in Figure 12.28.

The signal that results from the average of these BAEPs leads to a disordered BAEP (Figures 12.29 and 12.30).

$$x_i(n) = s(a_i.n) + b_i(n) \qquad [12.8]$$

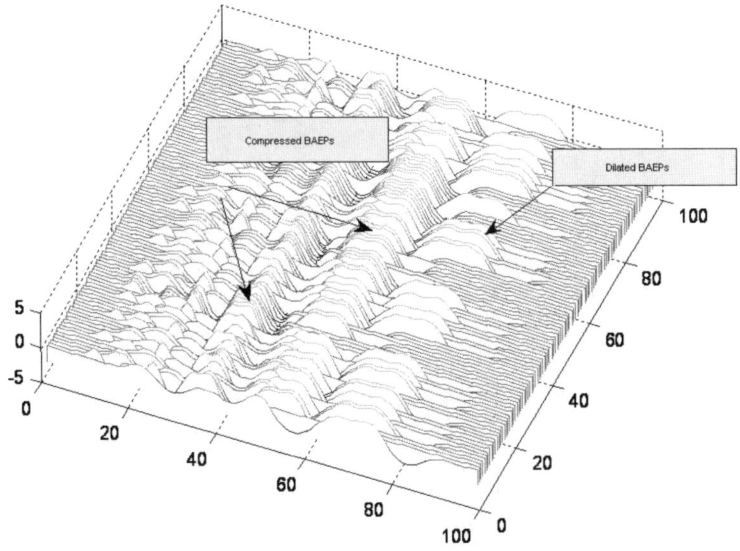

Figure 12.28. *Two-dimensional surface which is obtained from uncharacteristic non-stationarity due to the compression and dilation of the BAEPs*

Our aim is to invert the function that leads to the creation of such a signal. This inversion must ensure the synchronization of the set of BAEPs so that they can converge to become one single shape. The solution involves dilating or compressing the M responses in such a way that the energy of the averaged signal is at its maximum. Once again we experience another optimization problem, a problem that we wanted to resolve through the use of GAs. The objective function requires polyphase-type filters. The dilation of each acquisition must be ensured by the principle of interpolation, whilst compression is ensured by sub-sampling. In each iteration of the GA, several combinations of polyphase filters are applied. The equation to be minimized is as follows:

$$J_{\mathbf{d}} = -\sum_{n=0}^{N-1} \left(\frac{1}{M} \sum_{i=0}^{M-1} x_i \left(a_i n\right) \right)^2 \quad [12.9]$$

It is clear that clinicians are unable to make any objective diagnosis by solely basing their judgment on the averaged and corrected signal. The curve that describes the evolution of the compression/dilation parameters represents an important indicator that the clinician can use to quantify the changes that a BAEP can experience from one stimulation to another.

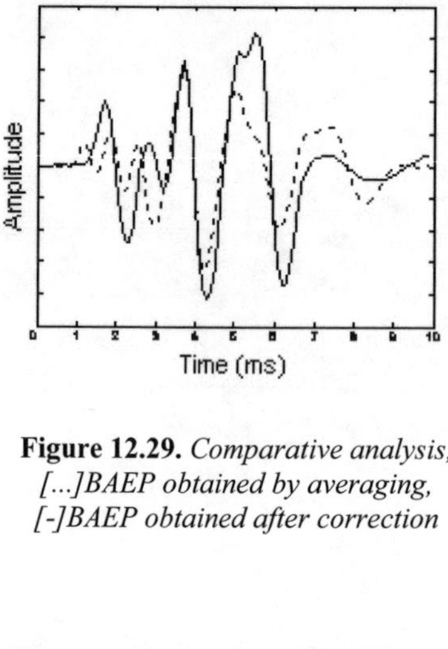

Figure 12.29. *Comparative analysis, [...]BAEP obtained by averaging, [-]BAEP obtained after correction*

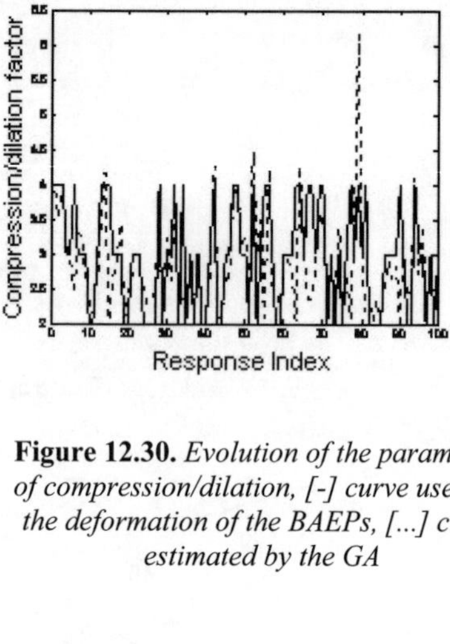

Figure 12.30. *Evolution of the parameters of compression/dilation, [-] curve used for the deformation of the BAEPs, [...] curve estimated by the GA*

The two dynamics of BAEP mentioned before this point have dealt with the process of desynchronization, and the idea of compression and dilation.

These two characteristics can be joined together in the same equation, as can be seen in equation [12.10]. A possible generalization of these two characteristics can be made and this can be seen in equation [12.11]. In equation [12.11] each recorded BAEP is considered to be deformed by a non-linear system which varies over time according to one operator (which is not strictly linear), and which is noted as Ξ.

$$x_i(n) = s(k_i.n + d_i) + b_i(n) \quad [12.10]$$

$$x_i(n) = s(n) \Xi h_i(n) + b_i(n) \quad [12.11]$$

Without being too optimistic, we believe that the increasing processing power of computers will make it easy to explore these types of models.

12.7. Conclusion

Metaheuristics often use a lot of computer processing time; but their efficiency in terms of optimizing non-linear equations provides a considerable advantage. As we have seen in this chapter, only one algorithm is needed and is efficient for solving several types of non-stationarity of the BAEPs. In certain situations, and in particular when the signals are characterized by a slow dynamics, we have shown that the convergence towards a global optimum can be effectively accelerated. The results could be further improved by introducing further degrees of freedom in the criteria that need to be optimized.

12.8. Bibliography

[CHE 05] CHERRID N., NAIT-ALI A., SIARRY P., "Fast simulated annealing algorithm for BAEP time delay estimation using a reduced order dynamic model", *Med. Eng. and Phys.*, vol. 27, Issue 8, pp. 705-711, 2005.

[DAV 92] DAVILA C., MOBIN M., "Weighted averaging of evoked potentials", *IEEE Trans. Biomed. Eng.*, vol. 39, pp. 338-45, 1992.

[DOR 96] DORIGO M., MANIEZZO V., COLORNI A., "The ant system: optimization by a colony of cooperating agents", *IEEE Trans. Syst. Man Cybern*, vol. 26, pp. 29-41, 1996.

[DOR 97] DORIGO M., GAMBARDELLA L., "Ant colony system: a cooperative learning approach to the travelling salesman problem", *IEEE Trans. Evol. Comp.*, vol. 1, pp. 53-66, 1997.

[DRE 02] DRÉO J., SIARRY P., "A new ant colony algorithm using the heterarchical concept aimed at optimization of multiminima Continuous Functions", *Proceedings of the Third International Workshop on Ant Algorithms (ANTS"2002)*, vol. 2463, pp. 216-221, Brussels, 2002.

[KEN 95] KENNEDY J., EBERHART R., "Particle swarm optimization", in *Proc. IEEE Int'l. Conf. on neural networks*, pp. 1942-1948, 1995. Piscataway, NJ.

[NAI 02] NAIT-ALI A., SIARRY P., "Application of simulated annealing for estimating BAEPs in some pathological cases", *Med. Eng. and Phys.*, vol. 24, pp. 385-392, 2002.

[NAI 06] NAIT-ALI A., SIARRY P., "A new vision on the averaging technique for the estimation of non-stationary brainstem auditory evoked potentials: application of a metaheuristic method" , *Comp. in Biol. and Med.,* vol. 36, pp. 574-584, 2006.

[QIU 94] QIU W., CHAN F., LAM F., POON P., "An enhanced approach to adaptive processing of the brainstem auditory evoked potential", *Australas. Phys. Eng. Sci. Med.*, vol. 17, pp. 131-5, 1994.

[YU 94a] YU X., "Time-varying adaptive filters for evoked potential estimation", *IEEE Trans. Biomed. Eng.*, vol. 41, pp. 1062-71, 1994.

[YU 94b] YU X., ZHANG Y., HE Z., "Peak component latency-corrected average method for evoked potential waveform estimation", *IEEE Trans Biomed. Eng.*, pp. 1072-82, vol. 41, 1994.

Chapter 13

Using Interactive Evolutionary Algorithms to Help Fit Cochlear Implants

13.1. Introduction

The surgical technique which enables profoundly deaf people with a fully functional cochlea to hear again was developed some 40 years ago [LOI 98]. During the surgical procedure, the surgeon inserts a very thin silicon filament which bears several electrode inserts, into the cochlea of the patients. The aim of this procedure is to stimulate the auditory nerve. The electrodes are connected to an antenna which is surgically placed under the skin, just behind the patient's ear (see Figure 13.1).

In order to activate the electrodes, the patient wears a small apparatus called a BTE (for behind the ear) which looks like a hearing aid. The BTE is made up of two microphones that are connected to a digital signal processor (DSP) which transforms the received signal into electric pulses which are sent to the electrodes. The BTE is connected to a second exterior inductive antenna which works in collaboration with the antenna that is implanted under the patient's skin, thanks to the use of a powerful magnet. The impulses that are emitted by the DSP are transmitted to the electrodes which have been implanted by the two inductive antennae (see Figure 13.1).

The objective of the interface that is created is to stimulate the auditory nerve with the aim of restoring the patient's hearing to a certain level so that they are able to understand spoken language. The question of how to stimulate the auditory nerve is very important. It is tackled by adjusting the parameters of the DSP.

Chapter written by Pierre COLLET, Pierrick LEGRAND, Claire BOURGEOIS- RÉPUBLIQUE, Vincent PÉAN and Bruno FRACHET.

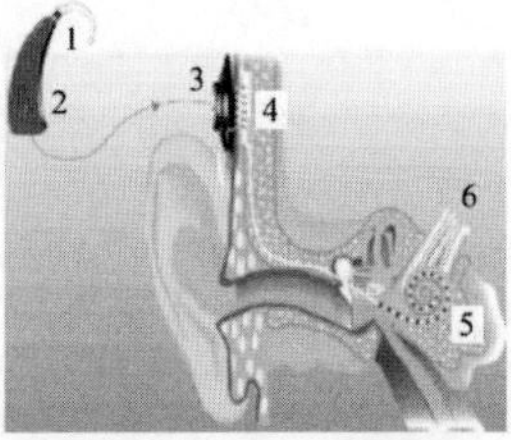

Figure 13.1. *Principle of a cochlear implant: 1) microphone; 2) processor; 3) external antenna (fixed and centered by a magnet); 4) internal antenna inserted under the skin and containing a magnet; 5) electrodes inserted into the cochlea; 6) auditory nerve*

The main problem is linked to the large number of parameters that need to be tuned, given that there are many causes for loss of hearing; for example, congenital deafness, traumatic deafness (due to an accident) or deafness due to illness. Other factors, such as the age of the patient, the number of years between the beginning of deafness and the implantation, the depth of insertion of the electrodes inside the cochlea, also add to this problem, making it very difficult to solve. The experience of the practitioners often leads to excellent results, such as patients who are able to follow a telephone conversation and enjoy listening to music. However, in some cases, no good results can be achieved for some unknown reason.

In this chapter, an interactive evolutionary algorithm is used to optimize the parameters of the DSP for patients for whom implantation is a failure.

This research was the doctoral thesis of Claire Bourgeois-République [BOU 04]. Further research has been carried out on her original research as part of the French Ministry of Health's RNTS project known as HÉVÉA. This work was carried out at the ear, nose and throat department of Bobigny University Hospital in France.

13.1.1. *Finding good parameters for the processor*

The aim of adjusting the DSP's parameters is to get the implant to fit the patient, and eventually enable him/her to distinguish between relevant information that is heard in speech in order to improve understanding. All of this should be possible without causing any discomfort to the patient, in other words it should be at an acceptable auditory level [LOI 00]. Originally, cochlear implants were only made up of one or two electrodes. In some cases this led to a vast and surprising improvement in the hearing of certain patients, despite the small number of electrodes. Thanks to the minimization process that exists nowadays, the designers of cochlear implants have been able to develop cochlear implants with 9, 15, 16, 20, 22 and even 24 electrodes [COC, MED, AB, MXM]. Results have been better than ever, but the number of parameters to be tuned has increased accordingly, along with the increased power of the embedded microprocessor.

The principal parameters which are available include the following:

– for each electrode:

- a range of sound frequencies which will activate the electrode,

- the minimum intensity threshold under which the patient will not be able to feel any sound sensation (referred to as *T* for threshold),

- the maximum intensity threshold which the patient can endure for a long period of time (referred to as *C* for comfort level);

– the number of electrodes which are activated simultaneously;

– the gain in hearing for each sound frequency;

– the sensitivity of the patient for each sound frequency;

– the use of one or two microcomputers;

– the type of stimulation chosen, etc.

In an implant which has 20 electrodes, there are easily hundreds of different parameters that need to be tuned, and that can have an important role when it comes to enabling the patient to understand everyday conversation.

13.1.2. *Interacting with the patient*

One important aspect that needs to be taken into consideration is the subjective nature of the interactions that take place with the patients. Evaluating the different parameters of the DSP and the cochlear implant depends on the patient being able to recognize different feelings, which can sometimes be quite difficult to express. Furthermore, evaluating the success of a cochlear implant can be quite biased. This is known as the *Pygmalion effect*, which means that patients tend to place surgeons and experts, who have years of experience in adjusting the fittings of the cochlear implants and DSP, on a pedestal because it is these people who enabled the patient to hear. As a result, the patients imagine that any adjustments made to the parameters of the DSP will be beneficial.

Then, there are often several sets of parameters which are available on the processor and which can be selected by a micro-switch that is located on the BTE worn behind the ear. It is common practice to refer to the current fitting as *P1* and the previous fitting as *P2* in case the most recently tested parameters are rejected by the patient (due to tiredness or lack of comfort, etc.). The possibility of moving from the old set of parameters to the new set also enables the patient to compare the two sets of parameters.

Currently, adjusting the parameters of the DSP and the cochlear implant takes place in the following way:

1) The fitting expert asks the patient if the last fitting was better or worse than the one carried out before that. The practitioner will then use the best recorded fitting as a basis from which he can start to work.

2) The expert then loads the fitting that needs to be improved (*P1 or P2*) into proprietary software to tune the implant.

3) The expert tries to find the problematic parameters, by carrying out a series of tests with the patient to check if the patient can recognize consonants, vowels and syllables.

4) Thanks to his experience, the expert then changes certain parameters and carries out the same tests to see if the adjustment which has been made has led to any improvement. This is where the Pygmalion effect usually comes in; if the adjustment has not worked or has had a negative effect on the patient's hearing, the patient will often find it difficult to say so, mainly because he does not want to disappoint the expert. Another factor should also be taken into consideration: patients may have difficulties in communicating, since, after all, it should not be forgotten that these patients are profoundly deaf. The age of the patients also varies.

5) The previous point is repeated until the expert and the patient are both satisfied with the adjustments that have been made.

6) The previous best fitting is loaded into the *P2* memory and the new fitting to be tested is loaded into the *P1* memory.

It takes between 45 mins and 1 hour to find new satisfactory parameters. Then, the patient has to evaluate the new fittings over a period of several weeks in order to take neuro-plasticity into consideration. If the patient is not satisfied with the new fitting after this period of several weeks, another appointment will be made so that the fitting can be adjusted again. In some cases, evaluating the patient's ability to understand spoken language is carried out by speech therapists in hospital. If this is the case, evaluation normally takes longer than 1 hour.

Computer scientists who are specialized in optimization will see a problem in the protocol which has been described above: the optimization which is carried out is a type of local research. The expert who adjusts the fittings of the cochlear implants tries to improve on the best fitting that has been found to date. The expert will use this best fitting as his basis and will only change some of the parameters in order to improve them. This means that the new fitting will be very similar to the previous one. If the problem is multimodal, then the expert runs the risk of becoming trapped by a local optimum and the patient's hearing will never get better.

13.2. Choosing an optimization algorithm

The process that was described in the previous section is long and does not look like it can be optimized with an algorithm.

However, an actual discussion with a patient led to the possibility of such an optimization taking place. The patient said that he was capable of immediately detecting if the new fitting was good or not. However, we still cannot count on thousands of evaluations to find the best possible fitting, so a deterministic algorithm could not be used. Among stochastic algorithms, we must find one that will not be much affected by local optima, but that is still able to converge quite rapidly in order to improve obtained results within a limited number of evaluations.

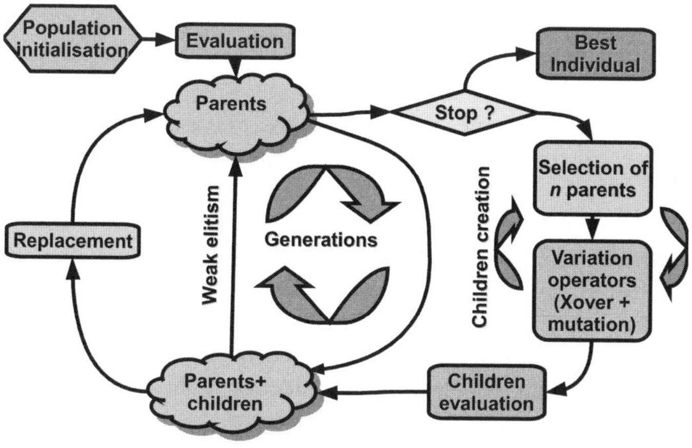

Figure 13.2. *The evolutionary loop*

Evolutionary algorithms (see Figure 13.2) have experienced a significant development and growth since their revival at the beginning of the 1990s:

1) An initial population is created and then evaluated (by a user in the case of an interactive algorithm) with the aim of creating a population of parents.

2) If we are not satisfied (non-verified stopping criterion), then n parents are selected (from the best parents) to create m children with the help of genetic operators.

3) The new individuals who are created are then evaluated.

4) Finally, the weakest individuals are eliminated thanks to the use of a replacement operator. This phase makes it possible to reduce the size of the population to its initial size. The loop is restarted at point two with a new generation of parents.

The use of evolutionary algorithms has become increasingly widespread in many different domains and their increased use towards the end of the 1990s means that they are much better understood now. As a result, it has become possible to use them to solve different interactive problems [COL 04], as can be seen in the work published by Takagi [TAK 98, TAK 99, TAK 01] and by [HER 97, BIL 94 DUR 02]. Rwo approaches can be used to reduce the number of evaluations: the Parisian approach [COL 00] (described in Chapter 2 and which dramatically reduces the number of evaluations) or Krishnakumar's micro-GAs [KRI 89] that use populations of a few individuals only. In his tutorials on evolution strategies, Thomas Bäck shows that it is possible for evolution strategies to produce better results than human experts for a number of evaluations of the same order than the number of real parameters that need to be optimized [BAE 05].

These recent results have shown that the theories which suggested that evolutionary algorithms were to be used when all else failed, and required tens of thousands of evaluations in order to produce an average result, are no longer valid. A well designed evolutionary algorithm can now find interesting solutions with a reduced number of evaluations. Evolutionary algorithms seem to be more suited to solving multimodal problems than other stochastic methods such as simulated annealing, even if there are only a few individuals in the population. In the next sections of this chapter it is assumed that the reader possesses some knowledge of standard evolutionary algorithms (for an introduction to this topic it is recommended that the reader looks at [DEJ 05]). The descriptions that are given for this type of algorithm will also highlight any changes that have been made to them when they are to be used interactively.

13.3. Adapting an evolutionary algorithm to the interactive fitting of cochlear implants

In an interactive evolutionary algorithm it is the human user who evaluates each individual that is suggested by the algorithm. As far as the optimization of the parameters of a cochlear implant is concerned, it is the patient who evaluates the fitting of the implant based on a particular series of words or syllables. It is therefore not possible to rely on a large number of evaluations, because in a domain like this fatigue, as well as the physiological and physical behavior of the patient, can also influence the results.

According to Thomas Bäck's results [BAE 05], an evolutionary algorithm should perform as well as, if not better, than a human expert on a number of evaluations equivalent in size to the number of real parameters that need to be optimized. Therefore, 100 evaluations would allow an EA to perform better than a human on a problem with 100 real parameters. If it is possible to reduce the evaluation time for each parameter from 45 minutes down to only 5 minutes, then it would take 8 hours to complete the evaluation of the 100 parameters, which is not unreasonable over two days. Knowing that the presented work only optimizes 30 parameters, we can hope to do better than a human expert during a two day fitting session.

It is also necessary to take certain psychological aspects into consideration. For example, if the convergence speed of the algorithm is correctly tuned for the 100 evaluations, (i.e. 8 hours to carry out the fitting), then the patient could be disheartened by this because over 8 hours progression would be very slow. The idea is then to break the experiment into several partial optimizations and have some "restarts" [JAN 02].

Doing this also brings another advantage: all optimization algorithms (including Eas) have a strong tendency to converge prematurely on local optimae, meaning that optimizers usually implement many techniques to delay the convergence of the algorithm. If several restarts are used, convergence is, on the contrary, no longer a problem. In interactive evolution, we want an algorithm to converge very quickly, so that the patient sees an improvement over the very few evaluations of one run, and keeps up his spirit. After several restarts, we can use the best individuals of the previous runs (which will hopefully have converged on different solutions) as the initial population of afinal fitting session.

So, rather than fighting against premature convergence, this work will encourage it (by using a micro-GA, for instance).

13.3.1.*Population size and the number of children per generation*

There are two possibilities which arise on a fixed number of evaluations: either we choose to create many children per generation over a few generations, or we create a few children per generation over many generations.

Of these two possibilities, the latter will most favor convergence. This means that a steady-state type of replacement will be used (or a $(\mu + \lambda)$ replacement strategy, with a heavily reduced λ (number of children) [BAE 95]). Then, in order to avoid using too many evaluations in the initial population, the population size can be reduced as in micro-GAs [KRI 89].

Now, as far as the optimization of cochlear implants is concerned, there are too many parameters to hope to optimize them all. Fitting experts have recommended starting with the optimization of the minimal T and maximum C thresholds for each electrode, meaning two variables per electrode. For MXM cochlear implants which have 15 electrodes, this means that 30 real variables need to be optimized over 100 evaluations, giving good chances of performing better than a human expert.

13.3.2. *Initialization*

There is one major constraint that needs to be respected: the maximum value for the stimulation of each electrode (C value) must never be exceeded, so as not to damage the patient's auditory neurons. For each new patient, the first appointment with the fitting expert consists of a "psychophysical" test in order to determine this maximum level for each electrode. A minimum intensity level (T value) is also determined because if an electrode is stimulated under this intensity level it means that a patient would be unable to hear anything.

Then, the initialization of each individual is carried out by taking two random values within the predetermined $[T, C]$ interval for each electrode.

13.3.3. *Parent selection*

Parent selection is different from parent replacement (see section 13.3.5) in that a parent can be selected several times.

A 0.9 stochastic tournament [BLI 95] is used, that chooses 2 parents at random, and returns the better of the two with a 90% probability. This selection is preferred to a proportional selection that would rely on the fitness landscape of the problem (which is unknown here).

13.3.4. *Crossover*

The genome is made up of real values, which means that it could have been possible to use some type of barycentric crossover. However, since the aim is to make the intervals evolve, adopting this style of crossover would actually have progressively reduced the width of the intervals.

Therefore, the method used for crossovers comes from binary genetic algorithms. Genes are exchanged between parents after a particular crossover point (the locus, chosen at random) has been reached. A single-point crossover is used because according to experts, we can expect a high level of epistasis between the different electrodes. Using multi-point crossover points would therefore have been disruptive and would have turned the crossover into some kind of macro-mutation.

Determining the locus is carried out electrode by electrode in the hope of not breaking any good combinations of genes that exist, meaning that *T* and *C* values are not separated. Since the evolutionary algorithm is a $(\mu + \lambda)$ type of algorithm, with a number of children that is smaller than the population size (see Figure 13.2., crossover is therefore used for the creation of each child (100% probability).

13.3.5. *Mutation*

Mutation is also used with a 100% probability level on each child that has been produced by the crossover process. In the evolutionary algorithm, each gene has a 10% probability of being mutated. Since there are 30 genes, each child will undergo three mutations on average. These figures may seem high but due to the high level of epistasis that exists, modifying one threshold on a genome would have a limited effect on the general evaluation process. This high rate of mutation lets the algorithm keep some kind of exploratory character, despite the low number of evaluations that take place.

13.3.6. *Replacement*

A steady-state like replacement is used, in order to promote fast convergence. However, where a strict steady-state would create only one child that would replace the worst of the parents, several children will be created, turning this replacement operator into a kind of $(\mu + \lambda)$ replacement, even though the number of children is very small.

13.4. Evaluation

Until now, evaluating a patient's ability to understand speech has been the test for a new fitting. The first method involved sending the patient back home with the new fitting stored in the *P1* memory and the previous fittings stored in the *P2* memory. This meant that the patient was able to compare the two different fittings in his own environment. The other method involved a speech therapist who carried out an evaluation on the patient using intensive tests taking over one hour to complete.

However, it was impossible to apply an interactive evolutionary algorithm with any of these two methods since the evaluation time was much too long. A new evaluation protocol was developed using calibrated sentences taken from Professor Lafon's cochlear lists [LAF 64] that contain syllables that are both representative of the French language and supposed to be discriminant cochlear-wise. Ten sentences (a total of 78 words) were chosen to evaluate a patient's understanding of the French spoken language. Here are the sentences (with their translation, even though their meaning is clearly secondary):

Se réveiller chaque matin peut être un plaisir.
(Waking up each day can be a pleasure)

La cravate garde encore du prestige pour certains.
(The tie still retains prestige for some people)

Il ne restait que de l'eau à boire.
(There was only water left to drink)

Il s'est fait aider pour porter ses bagages.
(He got someone to help him carry his luggage)

Les chiens gardent les villas contre les voleurs.
(Dogs protect villas against thieves)

Il existe des perles fines et des perles de culture.
(There exist both natural and cultured pearls)

L'enfant appelait sa mère parce qu'il avait peur.
(The child called his mother because he was frightened)

On aspire et expire par la bouche.
(We breathe in and out through our mouth)

L'intelligence permet à l'homme de comprendre.
(Intelligence enables man to understand)

Les parfums doivent avoir une odeur agréable.
(Perfumes should have a nice smell)

Maximizing the understanding of a patient is, of course, one of the main objectives of the evaluation process but it must also be comfortable enough for patients to use in their everyday life.

The overall evaluation is therefore the weighted mean of the comfort level of the cochlear implant (marked over 10) and of the patient's understanding. For the tests which are described in the next section, the comfort mark is multiplied by 2.2 which means the overall comfort mark is out of 22. The total number of words that were recognized (out of 78) is added to this mark in order to get a total overall mark of

100. This total mark out of 100 will be used as the patient's evaluation for the evolutionary algorithm. Understanding has the predominant mark, while comfort is not totally ignored.

The evaluation procedure typically lasts about four minutes. Four minutes is clearly not long enough to obtain a precise evaluation of a patient's ability to hear/understand speech, but it enables us to carry out 100 tests in 6h40', or 1h20' per run, if the 100 evaluations are divided into 5 runs. The aim of this reduced evaluation protocol is different from the complete evaluation protocol which is carried out by experts. Since experts cannot test as many configurations as evolutionary algorithms (patients may only have as many as ten appointments in one year) they need to work on as precise as possible evaluation of their patient's hearing.

Evolutionary algorithms work in a different way. They are stochastic processes that test many different fittings. They do not need very precise evaluations of the results, but need to be roughly guided towards a good solution. In fact, getting only a rough estimation of the patient's hearing may improve the efficiency of the algorithm, as it may help to skip over local optima.

13.5. Experiments

A certain number of experiments were carried out, which led to surprising results in many respects. The first set of experiments described in the next part of this section was carried out by Claire Bourgeois-République as part of her doctoral thesis at the University of Bourgogne and was published in several articles [BOU 04, BOU 05a, BOU 05b]. Other, more recent, tests were carried out as part of the RNTS HÉVÉA project which was supported by the French Ministry of Health.

13.5.1. *The first experiment with patient A*

The first experiment using the algorithm mentioned in the previous section was carried out with a patient whose hearing became progressively worse until the patient became totally deaf in 1983. In 1991 the patient was first implanted with a one-electrode implant in the right ear. This implant was later replaced with a 15-electrode MXM implant, which was activated by an external processor the size of a walkman that clipped to the belt. The MXM implant with 15 electrodes led to average results. In 2003, the same patient then bought a behind the ear (BTE) miniaturized processor. As its name suggests, this miniaturized processor is worn behind the ear. However, the results were not as good as the results of the walkman-style processor, so the patient decided to stop using the BTE processor.

13.5.1.1. *Psychophysical test*

The psychophysical test (which is used to determine the CT and C thresholds for each electrode) was performed by an expert practitioner (see Table 13.1). Electrodes 10, 11 and 12 were not functional (the patient was unable to hear anything regardless of the power of the stimulation) and therefore not activated.

Electrode	1	2	3	4	5	6	7	8	9	10	11	12	13	14	15
T	6	6.5	6.5	9	9	9	8	8	8	0	0	0	7	6	5
C	9.5	13	13	18	20	21.5	21.5	18	16.5	0	0	0	12	10	9

Table 13.1. *The set of T and C parameters*

13.5.1.2. *Evaluation of the expert's fitting with the BTE and the hearing aid*

Before the experiments were started, the best fittings previously obtained by the expert after 10 years on the walkman processor and BTE processor were tested against the 4-minute evaluation function that will be used for the evolutionary algorithm.

As a reference point, the weighted mark for the walkman processor is 53/100, whereas the weighted mark for the BTE is 48.5/100. These marks confirm the patient's observation: the BTE performs worse than the walkman-size processor.

All the experiments below are done with the BTE processor.

13.5.1.3. *Experiment 1 and results*

The aim of this first set of experiments was not to actually obtain good results, but to tune the parameters of the algorithm (determining the optimal size of the population, the number of children per generation, the selection pressure, etc.. Successive tests were then carried out on the algorithm with the aim of varying these parameters.

The first test uses a population of three individuals, with three children produced per generation. This test also uses a stochastic tournament selection with a probability of 0.8. The mutation rate of each parameter is fixed at 0.1 and the crossover rate of each parameter is fixed at 1.

During this first experiment, 12 different fittings were evaluated by the patient. Evaluating one fitting (preparation and evaluation) lasts a little less than four minutes. The results which were obtained for this experiment can be seen in Table 13.2.

Fitting	1	2	3	4	5	6	7	8	9	10	11	12
Mark	44.2	21.2	9.2	31.4	55.6	46.4	74.8	74.8	58.4	81	81	79.8

Table 13.2. *Results of experiment 1*

The first line refers to the number of particular fittings to be evaluated and the second line refers to the fitting's weighted mark. There are three fittings per generation, so fittings 1 to 3 refer to the first generation, fittings 4 to 6 to the second generation, and so on.

The first three evaluations correspond to the individuals which were part of the initial population and which were created randomly. The process of artificial evolution begins at fitting 4. Fitting 5 has a mark that is higher than the expert's best fitting. Fittings 7 and 8 have almost identical parameters and have identical marks. The algorithm seems to converge from the fitting 10 onwards with a very good mark of 81. The patient is happy and appreciates the speed of the evaluation procedure.

13.5.1.4. *Experiment 2 and results*

For this second experiment, the number of individuals of the population and the number of children per generation are doubled with the aim of delaying the premature convergence that was observed in experiment 1. The initial population is therefore increased to six individuals and four children are produced per generation.

Fitting	1	2	3	4	5	6	7	8	9	10
Mark	24	17	30	19	53.2	37.4	22.6	24	33.4	32

Fitting	11	12	13	14	15	16	17	-	-	-
Mark	9	27.4	34	34.5	12	27	32	-	-	-

Table 13.3. *Results of experiment 2*

The marks of the first six individuals are generally low. However, what is rather surprising is that one of the six random individuals (i.e. one of the first six fittings) has a mark which is higher than the mark obtained by the expert (53 vs 48.5). As can be seen in Table 13.3, the genetic operators are unable to find any suitable individuals. This is why the algorithm is stopped voluntarily after fitting 17, for the well-being of the patient. Results are not as good and the patient becomes increasingly tired.

13.5.1.5. *Experiment 3 and results*

During this experiment the population is reduced again to three individuals and only two children are produced per generation. In order to prevent any premature convergence (as was the case during experiment 1), mutation rate is increased to 0.6. A roulette-wheel type of selection is then tested. This type of selection is tested as it may have a stronger selection pressure on a problem where fitness varies a lot between individuals.

Fitting	1	2	3	4	5	6	7	8	9	10	11
Mark	54	33	26.5	48	52	51.6	54.6	62.8	59.6	65.6	60.1

Fitting	12	13	14	15	16	17	18	19	20	21	22
Mark	60	72	69.4	53.4	73	67	50.1	62	68.3	67.3	65

Table 13.4. *Results of experiment 3*

From the first three individuals that were created randomly (fittings 1 to 3), the first shows once again a mark that is slightly higher than the mark obtained by the expert. Table 13.4 shows that the marks of the individuals steadily increase. From the fitting 5 onwards, all of the marks are higher than 50. The best individual is the fitting 16 which has a mark of 73.

13.5.1.6. *Experiment 4 and results*

For the fourth experiment, the size of the population is four individuals and four children are produced per generation. The mutation rate is 0.1 and the selection mode used for choosing the parents is once again tournament selection.

Fitting	1	2	3	4	5	6	7	8	9	10	11	12
Mark	59.4	62.2	57.3	58.9	57	62.3	65	73	75.3	65.2	83.1	68

Fitting	13	14	15									
Mark	75.4	91	91.5									

Table 13.5. *Results of experiment 4*

Table 13.5 shows that the individuals that were randomly chosen from the population (fittings 1 to 4) achieve an average mark of 59.5 (this mark is much higher than the mark of 48.5 that was achieved for the BTE by the expert, even after many years). All the other values are above 56.5. The best marks from this experiment and from all of the experiments are 91 and 91.5, and were achieved by the 14th and 15th individuals respectively (in other words more than 90% of the words from Professor Lafon's list were understood). The patient is extremely happy and is astonished by such good results.

13.5.1.7 *Experiment 5 and its results*

The size of the population is fixed at five individuals and two children are produced per generation. The selection mode which is used to choose the parents is, once again, a tournament selection and the mutation rate is 0.1.

Table 13.6 shows that two out of the five individuals which make up the initial population (fittings 1 to 5) achieve marks above 70. However, because of the evolution of the algorithm, it is impossible to find one particular individual which is better than the rest. The algorithm is stopped during the evaluation of the 23rd fitting.

Fitting	1	2	3	4	5	6	7	8	9	10	11	
Mark	18.6	53	70.1	9	71.9	58.4	60.3	58	51	57.3	48.2	

Fitting	12	13	14	15	16	17	18	19	20	21	22	23
Mark	36	36.2	50	29	33.5	50.3	40.2	44.5	48.3	49.3	45.2	50

Table 13.6. *Results of experiment 5*

13.5.2. *Analyzing the results*

Figure 13.3 shows how the value of the best individual evolves during the evaluation process, which takes place in each experiment. As far as evolution is concerned, experiments 1, 3 and 4 led to the best results (there was an increase in the marks obtained by each individual). The good number of individuals to have in a population seems to be 3 or 4. The good number of children to have per population is quite low (2 or 3). These values validate the hypotheses that were made earlier in this section.

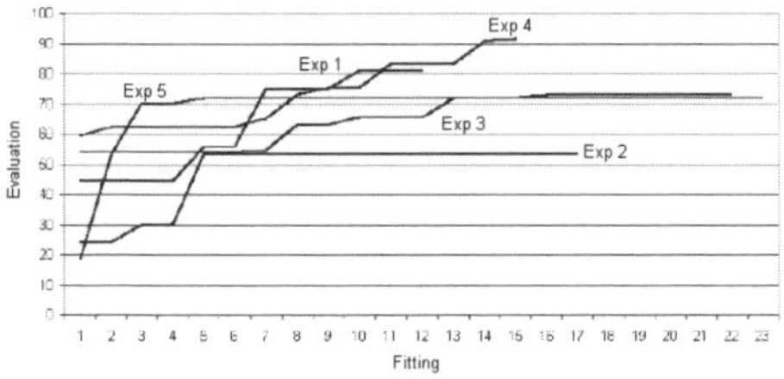

Figure 13.3. *Evolution of the best individual during the evaluation processes which were carried out for each experiment*

Medically speaking, it seems rather strange that the random fittings (the initial population) would yield better results than the results obtained by the expert practitioner. Some explanation is given below on this particular point.

The T and C values of the best and worst individuals have been plotted for each electrode in Figure 13.4. The dark lines refer to the minimum and maximum values given by the psychophysical test (see section 13.5.1.1). The dotted curve refers to the T and C values of the worst fitting, and the thin lines refer to the best fitting (with a mark of 91.5).

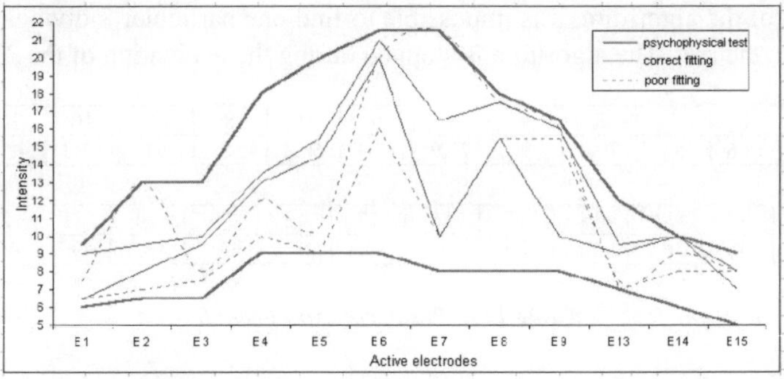

Figure 13.4. *T and C values of the best and worst fittings*

The medical team, and in particular the experts who specialize in cochlear implants, were surprised that this particular fitting was the best because of the small $[T,C]$ intervals observed for electrodes 3, 4, 5, 6, 12, 14 and 15 (0; 0.5 and 1). Usually, the expert generally tries to feed the auditory nerve with maximum information, and therefore tries to maximize the $[T,C]$ interval for each electrode. This means that the expert usually sets the T and C values to the same values as those that are determined during the psychophysical test. The expert might sometimes reduce these values if he has the feeling that too much information may saturate the auditory nerve, but this happens rarely. In this case, the interactive evolutionary algorithm minimized the values of C-T for practically all the electrodes except 1, 7, (8) and 9, which does not make much sense. Nevertheless, doing this enabled us to obtain better results than if all the C-T values had been maximized (the expert's fittings).

Several issues therefore arise:

– is minimizing the $[T,C]$ interval equivalent to de-selecting an electrode?

– would there be a problem of interference between the electrodes (diaphony)?

– would the problem be combinatorial (i.e. would certain combinations of electrode work better than others)?

13.5.3. *Second set of experiments: verifying the hypotheses*

A second set of experiments was then carried out, with fittings that were not created by the interactive evolutionary algorithm, but were created in order to validate certain hypotheses. The tests were carried out with the same patient and with the same evaluation protocol.

It should be pointed out that a period of one month elapsed between the two tests. During this one month break, the patient went back to using his old walkman-style processor, meaning that he was therefore unable to physiologically adapt to any new fittings (neuroplasticity). This means that the evaluations of the two experiments can therefore be compared with one another. In the next part of this section the first set of experiments will be referred to as *C1* and the second set of experiments will be referred to as *C2*.

13.5.3.1. *Experiment 7: is the minimization of C-T equivalent to de-selecting an electrode?*

For this experiment all of the electrodes apart from electrodes 1, 7 and 9 are reduced to intensities which are much lower than the liminary intensities (*T*). This means that the patient will not hear anything from these electrodes.

However, the differences between *T* and *C* have been maximized for electrodes 1, 7 and 9 (see Figure 13.5).

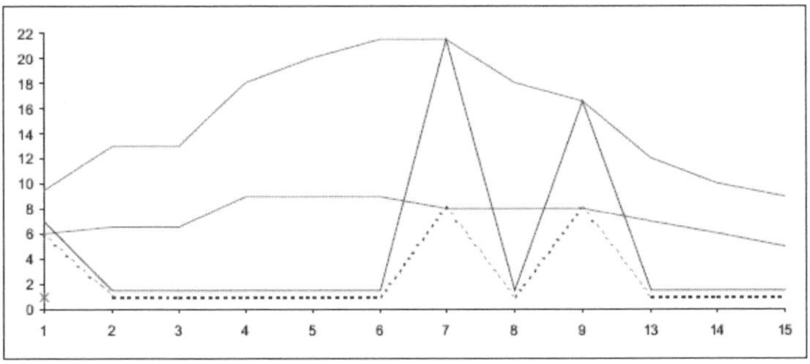

Figure 13.5. *Experiment 7*

The resulting weighted mark of this fitting is 82, with a level of understanding of 90%. In other words, the patient was able to understand 90% of the words from Professor Lafon's sentences. This fact alone seems to confirm three points:

1) Minimizing the difference of *C-T* is equivalent to de-selecting electrodes, since similar results were obtained as for the best individual.

2) This fitting (which can be compared to the best fitting that was evaluated a month earlier) still produces a good result.

3) All of the electrodes, apart from electrodes 1, 7 and 9 (which were maximized) have been minimized in order to produce a good result. The problem thus seems to be quite binary (even if the values of each interval were refined, it would only lead to a slight improvement as far as the quality of the result is concerned).

13.5.3.2. *Experiment 8: a study of electrode 8 and its influence on the fitting*

In the *C1* set of experiments, the evolutionary algorithm set a medium interval on electrode 8. In this test, electrode 8 was maximized along with electrodes 1, 7 and 9, using the values from the psychophysical test.

The resulting weighted mark of this fitting is 81. The patient finds that the fitting is slightly less comfortable than the previous fitting, although the level of understanding is exactly the same. Electrode 8 seems to play a rather neutral role as far as understanding spoken language is concerned.

13.5.3.3. *Experiment 9: is there any diaphony between the electrodes?*

In order to investigate this hypothesis, the even-numbered electrodes are de-activated (both *T* and *C* are set to a value below the *T* liminary intensity) and the odd-numbered electrodes are maximized (by using the values from the psychophysical test). The aim of this action is to increase the spacing between active electrodes (see Figure 13.6).

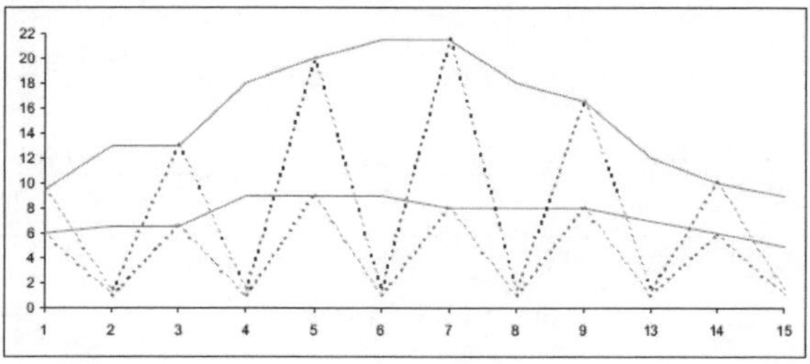

Figure 13.6. *Experiment 9*

The resulted weighted mark of this fitting is 78.8. The patient stated that this fitting was less comfortable than the others. This result is obviously not as good as the results that were obtained for experiments 7 and 8. Adding other electrodes to electrodes 1, 7 and 9 does not seem to improve the fitting. The result, however, is still better than the expert's result of 48.5 that was obtained for the BTE.

13.5.3.4. *Experiment 10: a wider spacing of the electrodes*

In order to further decrease the risk of interference, in this experiment we decided to activate one in every three electrodes instead of every two electrodes. Electrodes 7 and 9 remained activated and the maximized electrodes are now electrodes 1, 4, 7, 9, and 15 (electrodes 10 to 12 are non-functional); see Figure 13.7.

The result is astonishing. The fitting is not very comfortable and the weighted mark for this fitting is only 58.5. First of all, these facts refute the idea that interference between the different electrodes could exist, but above all, if maximized electrodes are compared with those of experiment 7 (where only electrodes 1, 7 and 9 were activated), the difference being the addition of electrodes 4 and 15.

This information suggests that this binary problem is combinatorial. In other words, there are certain combinations of electrodes that work better than others. Another conclusion which can be drawn from these facts is that the activation of certain electrodes has a negative effect on a patient's understanding of spoken language (adding electrodes 4 and 15 degraded speech understanding).

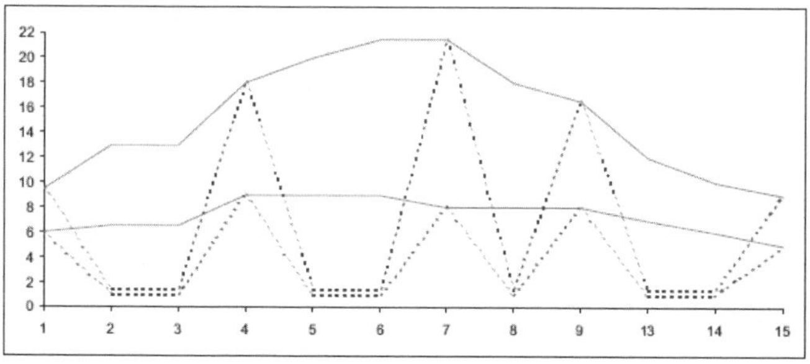

Figure 13.7. *Experiment 10*

13.5.3.5. *Experiment 11: evaluation of the best individual from C1*

Experiment 11 investigates the reliability and accuracy of the evaluation protocol.

The parameters of the best fitting from the *C1* set of experiments (with a mark of 91.5), which were carried out a month earlier, are tested once again.

The result of the vocal test is once again very good (94% of the words are understood), and even slightly better than one month earlier, but the fitting is evaluated as slightly less comfortable, which accounts for a weighted mark of only 86.2. This mark is slightly lower than the previous weighted mark of 91.5, but it is still the best fitting that was found during the *C2* set of experiments.

13.5.3.6. *Experiment 12: an evaluation of the expert's fitting*

For this experiment the original expert's best fitting (in which more or less all of the electrodes were maximized, and which obtained a mark of 48.5 during the C1 set of experiments) is uploaded into the BTE.

The result of the vocal test is poor (only 33% of the words were understood), and the comfort mark, which was given by the patient, was only 4/10. This means that the weighted mark this time was only 41.8, a mark which is much worse than that obtained during the C1 set of experiments.

In the period of one month, the best fitting generated by the interactive evolutionary algorithm went from a mark of 91.5 to 86.2, and the expert's fittings went from a mark of 48.5 to 41.8. The absolute values have decreased between C1 and C2, but the difference remains about the same (43 against 44.4). Repeating these two fittings showed that the quick evaluation procedure is pretty reliable, because results are reproducible after one month, during which the patient used his old processor (and therefore could not adapt to other settings).

13.5.3.7. *Other experiments*

In order to check whether the evolutionary algorithm actually added some value, other tests were performed with randomly chosen values for T and C. Results were average to poor (but often higher than 41.8, the expert's best fitting) and comfort was often criticized by the patients. The fact that many random fittings obtained better results than the expert's fitting may be explained by the fact that by maximizing the $[T,C]$ interval of all electrodes, the expert also maximizes the influence of detrimental electrodes such as electrode 4 for the tested patient. By doing so, if there is only one detrimental electrode for a patient, the expert is sure to obtain a relatively bad speech understanding result, while a random fitting that would skip detrimental electrodes would very likely perform better.

13.5.4. *Third set of experiments with other patients*

One thing stands out from the previous results: in several cases the random values for the T and C parameters gave a result which was equal to or better than the expert's best fitting (who understandably was trying to maximize the $[T,C]$ interval for all of the electrodes).

A new series of tests was then started with four other patients in order to confirm or refute this observation. Unlike the previous experiments (where the evaluation process was a very short process with only ten sentences taken from the Lafon corpus), auditory speech evaluation (ASE) tests and vowel consonant vowel (VCV) recognition tests were used. The test is longer, but it makes it possible to determine a patient's hearing ability in a much more detailed manner.

It would take too long to provide a complete summary of the tests in this chapter. The results of the tests are shown as percentages of recognized VCVs in Table 3.7.

Patient	Mark for expert's fitting	Best random fitting	Number of tests
A	31%	**33%**	3
B	43%	**50%**	3
C	16%	**25%**	3
C	20%	**27%**	9
A	**33%**	31%	3
A	**33%**	27%	3
D	20.5/22	**21/22**	3

Table 13.7. *Results of tests shown as percentages of recognized VCVs*

In almost all cases, if three random fittings are tested, it is nearly always possible to find a result that is equal to or better than the expert's fitting.

In the final case (patient D), another marking system was used. This marking system was based on the ASE test. For this patient the best fitting that was obtained by the expert can be seen in Figure 13.8. The mark corresponds to the psychophysical test in which the intervals for each electrode are maximized.

In this experiment the evolutionary algorithm was briefly tested (only six evaluations due to the time that was required to complete the evaluations) and the best individual obtained an ASE mark of 22/22 (in comparison to a mark of 20.5/22 for the fitting which was set by the expert). Once again there were some astonishing values that were recorded for the electrodes (see Figure 13.9).

Given the very low number of evaluations, it is not possible to consider the algorithm as having played an important role in achieving such a result. However, analyzing the results is quite interesting because the intervals for all of the electrodes are quite small. Some of the electrodes were deselected and these included electrodes 5, 8, 11, 12 and 13. This goes against the recommendations of the experts and the developers of cochlear implants, who have been working in this field since the development of cochlear implants first took place over 40 years ago. However, once such electrodes have been removed, this supports the observation made by the first patient who had a better understanding of spoken language when only 3 out of the 12 electrodes were activated.

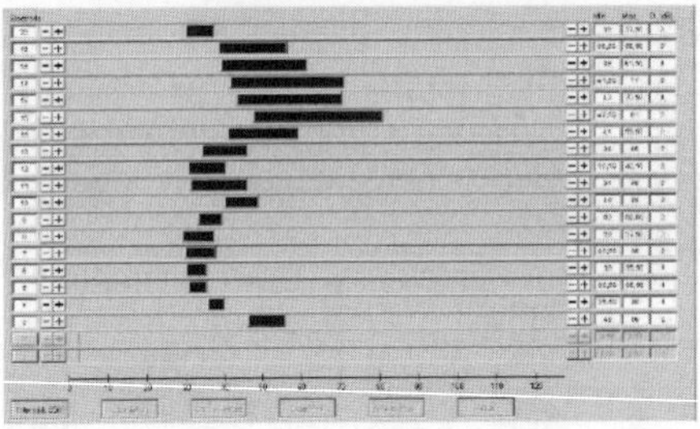

Figure 13.8. *The best fitting obtained by the expert for patient D: each rectangle represents the [T, C] interval for each electrode*

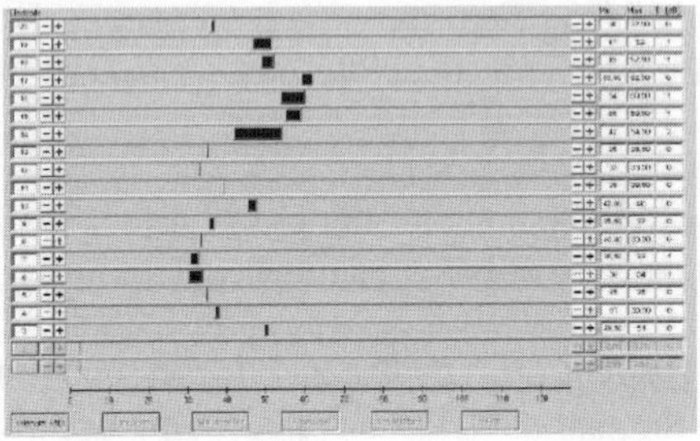

Figure 13.9. *The best individual for patient D: each rectangle represents the [T, C] interval for each electrode*

13.6. Medical issues which were raised during the experiments

It should be pointed out that our study does not focus on a wide range of different cases and because of this the results cannot be used on a more general level. However, throughout our study certain issues have been raised and we think it would be interesting to try and find a solution to such issues. As far as the first set of experiments is concerned, they were all carried out on only one patient.

1) In studying the graphs which show the progression of the different evolutionary tests, and which also show the results produced from the second set of tests, it is possible to evaluate a patient's hearing (with the help of an optimization algorithm) in less than four minutes. This fact alone goes against the recommendations made by speech therapists. Speech therapists believe that it is impossible to evaluate a patient's hearing in less than 20 minutes.

The speech therapists are, of course, correct. They are correct in the sense that a four minute evaluation will not lead to the same quality of conclusions as a one-hour evaluation. However, in pragmatic terms, an evaluation procedure must be used in relation to the quality of the optimization algorithm that is also used. An evolutionary algorithm does not benefit from any of the experience or intelligence that an expert has. The quality of a very refined evaluation process (and therefore long evaluation) would be wasted if such a rough algorithm were to be used.

Therefore, for such algorithms, it is better to have more evaluations than more precise evaluations. A more refined and longer evaluation process would benefit a human expert who works in this field, since the expert is able to provide a better interpretation of the results from the tests and therefore provide a better evaluation of a patient's hearing.

2) Patient D seems to be able to evaluate a new fitting within seconds. However, experts say that a patient can only give a true evaluation of any new fitting at least one or two weeks after the new fitting has been installed (this is probably due to neuroplasticity).

Once again, the experts are more than likely correct if the aim is to carry out a more detailed evaluation of the patient's hearing. As far as a rough evaluation is concerned (which is fine for an evolutionary algorithm) it is not problematic to evaluate the fitting only seconds after it was uploaded in the processor (but one example may not be enough to generalize this conclusion).

3) With patient D it was possible to carry out 89 tests with different parameters, in a period of one and a half days, with results that were good enough to guide an interactive evolutionary algorithm where speech therapists and other experts working in this field believe that after a period of two hours any evaluations that are carried out are no longer relevant, due to patient fatigue. The same comments as above may apply here.

4) With two patients, it has been possible to obtain a similar or higher level of understanding of spoken language when some of the electrodes were either minimized or deselected where experts, as well as cochlear implant manufacturers, believe that it would make more sense to maximize the number of activated electrodes and to maximize their range of stimulation.

This point still remains a mystery and requires further investigation in order to provide a more accurate explanation.

On a less general note, it seems that for patient A the problem is combinatorial, i.e. certain combinations of electrodes work better than others. This can be seen in the example where two electrodes (electrodes 4 and 15) were added to a set of electrodes (1, 7 and 9). With this particular combination of electrodes, the patient's understanding of the spoken language actually got worse. If this hypothesis were proved, it would lead to a big problem, because:

a) the problem would face a combinatorial explosion. With nine electrodes, the number of possible combinations is $2^9 = 512$. With 15 electrodes, the number of possible combinations is $2^{15} = 32{,}768$, and with 22 electrodes, the number of possible combinations is a massive $2^{22} = 4{,}194{,}304$;

b) since the expert has no way to determine which electrode will have a positive or negative effect on speech understanding this means that, in theory, it would be necessary to test all electrode combinations in order to find the best combination. This, however, is impossible in practice.

As far as the third set of tests is concerned, it seems that one in every three random parameters taken from the T and C thresholds leads to a result that is equal to, if not better than the result which is obtained by the expert (who maximizes the [T,C] intervals). This point alone suggests that there are numerous fittings that are satisfactory for the patient.

13.7. Algorithmic conclusions for patient A

To start with, the contribution of the implemented evolutionary algorithm seems to be real, looking at Figure 13.3, even though it might seem surprising to achieve such good results in such a short period of time. However, if the problem is combinatorial (where certain combinations of electrodes enable a patient to understand spoken language better), then the chances of finding a good combination whilst using an interactive evolutionary algorithm for 100 evaluations are far from being zero:

1) in the case of patient A, three electrodes out of 15 were non-functional. This means that out of the remaining 12 activated electrodes, there were a total of 2^{12} = 4.096 different combinations of electrodes;

2) if we assume that there are over 100 random evaluations, then the chances of finding the correct combination is one in 40;

3) as is suggested in the random tests carried out during the third set of experiments, it is possible for different combinations of electrodes to carry out a good evaluation. If there are ten satisfactory combinations, then 100 random evaluations would have one chance in 4 of finding one of these ten good combinations;

4) it should not be forgotten that having one chance in 4 of finding a good combination (using 100 evaluations) is obtained solely by random research.

If it is assumed that an evolutionary algorithm can perform better than a random search, the probability of finding a good fitting suddenly comes closer to 1. However, all of this is only possible if the algorithm is capable of carrying out many evaluations in a short period of time.

Finally, there is one final element that changes the probability of finding a good fitting from "very probable" to "extremely probable": the interactive evaluation given by the patient may be of very high quality, and this is very important because evaluation is what guides the evolutionary algorithm. During the 15th French evolutionary workshop (JET) organized by the French Society for Artificial Evolution, a competition was organized with the aim of solving the famous MasterMind board game. This was a very difficult task to complete because the problem was a 13 x 13 size problem (13 positions, 13 colors), meaning that it was necessary to find a good combination from among 3.10^{14}, i.e. approximately 300 thousand billion. On 1,000 games, the best evolutionary algorithm was able to find the correct combination in an average of only 19.5 evaluations.

This feat was only possible because the information given by the MasterMind game's evaluation function is very rich (number of correct colored pegs in the correct place, and number of correct colored pegs in incorrect places), and that the evolutionary algorithm found a way to exploit it correctly.

As far as cochlear implants are concerned, if the interactive evaluation that is given by the patient is of high quality, then this could significantly simplify the problem. All works on interactive evolutionary algorithms seem to support this idea [TAK 05].

Furthermore, micro-population evolutionary algorithms have also proved to be efficient [KRI 89], and the number of variables to optimize was lower than the number of available evaluations. This set of hypotheses seems to reinforce the idea that the good results were not obtained through chance.

The only way it is possible to confirm this would be to carry out a significant number of complete tests with new patients.

13.8. Conclusion

Fitting a cochlear implant is a problem for which it is very difficult, if not impossible to find a suitable solution deterministically within a limited time period for at least two reasons:

– the objective function cannot be modeled and can vary a lot due to the fact that it is dependent on the patient and linked to the subjective evaluation of his sensations;

– the search space is large enough to eliminate any guarantee of optimality.

The work and the findings that are presented in this chapter describe a way to help the expert practitioner by using a micro-population interactive evolutionary algorithm. The first tests are very promising and the results raised several issues related to the current evaluation protocol that is used by experts in the field. The results also seem to question the seemingly evident idea that more electrodes would allow the patient to understand better, and that maximizing the $[T,C]$ interval of all electrodes may not be a good idea, if for a patient some electrodes are detrimental to speech understanding.

However, it goes without saying that this work is only the beginning. It is now necessary to build on this research by carrying out numerous clinical experiments.

13.9. Bibliography

[AB] Advanced Bionics: http://www.cochlearimplant.com

[BAE 95] BAECK T., *Evolutionary Algorithms in Theory and Practice*, New-York, Oxford University Press, 1995.

[BAE 05] BAECK T., "Tutorial on evolution strategies", *Genetic and Evolutionary Computation Conference Gecco'05*, 2005.

[BIL 94] BILES J., "GenJam: a genetic algorithm for generating jazz solos", *Proceedings of the International Computer Music Conference*, San Francisco, 1994.

[BLI 95] BLICKLE T., THIELE L., "A mathematical analysis of tournament selection", ESHELMAN L. J., (Ed.), *Proceedings of the 6th International Conference on Genetic Algorithms*, Morgan Kauffmann, p.9-16, 1995.

[BOU 04] BOURGEOIS-RÉPUBLIQUE C., Plateforme de réglage automatique et adaptatif d'implant cochléaire par algorithme évolutionnaire interactif, PhD Thesis, University of Bourgogne, France, 2004.

[BOU 05a] BOURGEOIS-RÉPUBLIQUE C., FRACHET B., COLLET P., "Using an interactive evolutionary algorithm to help with the fitting of a cochlear implant", *MEDGEC (GECCO)*, 2005.

[BOU 05b] BOURGEOIS-RÉPUBLIQUE C., VALIGIANI G., COLLET P., "An interactive evolutionary algorithm for cochlear implant fitting: first results", *SAC*, p. 231-235, 2005.

[COC] Cochlear: http://www.cochlear.com

[COL 00] COLLET P., LUTTON E., RAYNAL F., SCHOENAUER M., "Polar IFS + Parisian GP = efficient inverse IFS problem solving", *Genetic Programming and Evolvable Machines*, vol. 1, number 4, p. 339-361, 2000.

[COL 04] COLLET P., Vers une évolution interactive, from his thesis and Habilitation à Diriger des Recherches, June 2004, University of the Littoral Côte d'Opale.

[DEJ 05] DEJONG K., *Evolutionary Computation: a Unified Approach*, MIT Press, 2005.

[DUR 02] DURANT E. A., Hearing aid fitting with genetic algorithms, PhD Thesis, University of Michigan, USA, 2002.

[EA] The Association for Artificial Evolution: http://ea.inria.fr.

[HER 97] HERDY M., "Evolutionary optimization based on subjective selection – evolving blends of coffee", *Proceedings of the 5th European Congress on Intelligent Techniques and soft Computing, EUFIT'97*, 1997.

[JAN 02] JANSEN T., "On the analysis of dynamic restart strategies for evolutionary algorithms", *Parallel Problem Solving from Nature*, p. 33-43, 2002.

[KRI 89] KRISHNAKUMAR K., "Micro-genetic algorithms for stationary and non-stationary function optimization", *SPIE: Intelligent Control and Adaptive Systems*, vol. 1196, Philadelphia, PA, 1989.

[LAF 64] LAFON J., "Le test phonétique et la mesure de l'audition", *Ed. Centrex*, Eindhoven, 1964.

[LOI 98] LOIZOU P., "Introduction to cochlear implants", *IEEE Signal Processing Magazine*, p. 101-130, 1998.

[LOI 00] LOIZOU P., POROY O., DORMAN M., "The effect of parametric variations of cochlear implant processors on speech understanding", *Journal of Acoustical Society of America*, p. 790-802, 2000.

[MED] MEDEL, http://www.medel.com

[MM] Mastermind competition: http://tniyurl.com/gaa7y

[MXM] MXM Labs: http://www.mxmlabs.com

[TAK 98] TAKAGI H., "Interactive evolutionary computation: system optimization based on human subjective evaluation", *Proceedings of the IEEE Intelligent Engineering Systems (INES'98)*, Vienna, Austria, 1998.

[TAK 99] TAKAGI H., OHSAKI M., "IEC-based hearing aid fitting", *Proceedings of the IEEE Conference on Systems volume 3*, 1999.

[TAK 01] TAKAGI H., "Interactive evolutionary computation: fusion of the capabilities of EC optimization and human evaluation", *Proceedings of the IEEE,* Vol.89, number 9, 2001.

[TAK 05] TAKAGI H., "Tutorial on interactive evolutionary algorithms", 2005.

List of Authors

Djedjiga AÏT AOUIT
Ecole Polytechnique of Tours
France

Olivier ALATA
Signal, Image and Communications Laboratory
University of Poitiers
France

Sébastien AUPETIT
Computer Science Laboratory
University François Rabelais
Ecole Polytechnique of Tours
France

Claire BOURGEOIS-RÉPUBLIQUE
LE2I
University of Bourgogne
France

Stéphane CANU
INSA – Rouen
France

Francis CELESTE
DGA
CEP
Arcueil
France

Pierre CHARBONNIER
LRPC
Strasbourg
France

Christophe COLLET
National College of Physics
Strasbourg
France

Pierre COLLET
Computer Science Laboratory
University of Littoral Côte d'Opale
Calais
France

Frédéric DAMBREVILLE
DGA
CEP
Arcueil
France

Johann DRÉO
Laboratory of Imagery, Signals and Intelligent Systems
University of Paris 12
Créteil
France

Guillaume DUTILLEUX
LRPC
Strasbourg
France

Bruno FRACHET
Avicenne Hospital
Bobigny
France

Jean-Pierre LE CADRE
French Research Institute in Computing and Risk Systems
Rennes
France

Pierrick LEGRAND
Computer Science Laboratory
University of Littoral Côte d'Opale
Calais
France

Gaëlle LOOSLI
INSA – Rouen
France

Jean LOUCHET
INRIA
Rocquencourt
France

Nicolas MONMARCHÉ
Computer Science Laboratory
University François Rabelais
Ecole Polytechnique de Tours
France

Amine NAÏT-ALI
Laboratory of Imagery, Signals and Intelligent Systems
University of Paris 12
Créteil
France

Jean-Claude NUNES
Laboratory of Signal and Image Processing
University of Rennes 1
France

Christian OLIVIER
Signal, Image and Communications Laboratory
University of Poitiers
France

Abdeljalil OUAHABI
Ecole Polytechnique de Tours
France

Vincent PÉAN
Center of Technological Resources
Innotech
Bobigny
France

Patrick SIARRY
Laboratory of Imagery, Signals and Intelligent Systems
University of Paris 12
Créteil
France

Cécile SIMONIN
DGA
CEP
Arcueil
France

Mohamed SLIMANE
Computer Science Laboratory
University François Rabelais
Ecole Polytechnique de Tours
France

Index